MITRE

11 Strategies of a World-Class

Cybersecurity Operations Center

Kathryn Knerler, Ingrid Parker, Carson Zimmerman

MITRE | SOLVING PROBLEMS
FOR A SAFER WORLD®

About MITRE

Protecting the digital enterprise against sophisticated cyber adversaries requires strategy, timely information, and 24/7 vigilance. As a not-for-profit company pioneering in the public interest, MITRE works in partnership with an innovation ecosystem of government, private sector, and academia to secure cyber systems. In our 60+ years of catalyzing change through partnership, we never lose sight of the human factor behind every complex system and innovative solution. MITRE draws from a wealth of deep technical expertise to address the ever-evolving challenges in cybersecurity.

Why? We know that working in partnership to protect organizations is crucial to national security, critical infrastructure, economic stability, and personal privacy. The guidance we share with the cybersecurity community continues to advance the field's science and practice. Operating without commercial conflicts of interest, we're working to arm a worldwide community of cyber defenders with vital information to thwart network intruders.

As part of our cybersecurity research in the public interest, MITRE has a long history of developing standards and tools used by the broad cybersecurity community, such as STIX™, TAXII™, and CVE®. Our MITRE ATT&CK® framework, which provides a free online knowledge base of cyber adversary behavior, is used worldwide.

Our expert staff continues to partner and collaborate on many cybersecurity resources and innovations. The 11 Strategies of a World-Class Cybersecurity Operations Center is a practical guide to enhancing digital defense for SOC operators—and an embodiment of MITRE's mission of solving problems for a safer world.

About the Authors

This book was a fully collaborative effort among the three primary authors. The order of names on the front is alphabetical and does not reflect a difference in level of contribution.

Carson Zimmerman was the author of the first edition of this book, *Ten Strategies of a World-Class Cybersecurity Operations Center.* Throughout both versions of the book, many additional colleagues contributed their time, expertise, and advice. Please see the acknowledgements for the full list of those names.

Kathryn Knerler

Kathryn has decades of experience in cybersecurity. Her experience includes cyber analysis, incident response, and network security architecture. She is a Department Manager and Senior Principal Cybersecurity Architect in MITRE Labs' Cyber Solutions Innovation Center. She specializes in cyber threat intelligence and advising executives in operationalizing threat defense strategies. Prior to MITRE, she advanced from incident responder to Program Director of Computer Incident Advisory Capability (CIAC) at Lawrence Livermore National Laboratory (LLNL). Kathryn has a B.S. in Electrical Engineering, a M.S. in Cybersecurity, and an M.B.A.

Ingrid Parker

Ingrid has worked in cybersecurity roles spanning from operational hands-on analysis through engaging with CISO's of large federal departments and agencies. She is currently the Chief Engineer for the Homeland Security Enterprise Division at MITRE where she provides expertise across a range of cybersecurity topics and integrates new ideas, capabilities, solutions into the work programs. Prior to joining MITRE Ingrid worked as a malware, forensic, and cyber threat intelligence analyst for Northrop Grumman and served in the U.S. Army as a systems administrator and network engineer. Ingrid has a B.F.A. in Graphic Design and a M.A. in Information Management.

Carson Zimmerman

Carson has been working in cybersecurity for nearly 20 years. In his current role at Microsoft, Carson leads an investigations team responsible for defending the M365 platform and ecosystem. Previously at The MITRE Corporation, Carson specialized in cybersecurity operations center architecture, consulting, and engineering. In his early days at MITRE, Carson worked in roles ranging from CSOC tier 1 analysis, to secure systems design consulting, to vulnerability assessment. Carson has a B.S. in Computer Engineering and an M.S. in Information Systems.

Second Edition Acknowledgments

This edition of the now *11 Strategies of a World-Class Cybersecurity Operations Center* is the result of collective decades of experience and work, and the authors are grateful for the technical mentors and experts who have helped them along their career paths. No cybersecurity expert can know everything in such a rapidly evolving era of technology and its use, and we all have those we consult with when we have challenges. Thank you to all those we worked with, leaned on, and learned from over the years.

The authors would explicitly like to thank the following for their contributions to the second edition:

- Kristin Esbeck for her significant contributions to the cyber threat intelligence (CTI) strategy. Her perspective and knowledge around CTI qualities and use elevated and focused the strategy beyond the bits and bytes of indicators of compromise.
- Stacey Myers for her initial organizing as this effort was getting started, content ideas and suggestions for the overall outline of the book and contributions and edits on initial chapter drafts.
- Roman Daszczyszak II for his threat hunting topic ideas, supporting logistics for getting the book effort underway, and overall content brainstorming.
- Dan Aiello for his dedication to reviewing every chapter and providing substantial feedback that greatly improved the quality of the entire book.
- Eric Clopper, Daryl Baker, and Lou Montella for their technical edits and improvement suggestions.
- Troy Townsend for his belief in our work and his support in getting us over the finish line.
- Jackie Dufresne for her content on cyber deception.
- Mari Spina for her thoughts on cloud related content.
- Linda McCall, Sidney Gibson, and Phil Owen for their feedback on early chapter drafts.
- Nick Johnson for review of references early in the process.
- Laura Temple and David Cleary for their publishing and editing expertise.
- Tony Catlin for developing both the internal graphics and the cover. You made us look good!

This edition of the book would not have been possible without the support of MITRE leaders who believed in our work. Thank you for the continued steadfast leadership support from Eric Clopper. In addition, thank you to Daryl Baker, Leslie Anderson, and Troy Townsend for their persistence in championing funding, project leadership, and general support. In addition, thank you to Mindy Rudell and Jeff Picciotto for their encouragement and support.

This book is dedicated to our families for the many nights and weekends over the past two years (and more) we spent burning the mid-night oil (often literally).

Kathryn: Thank you to my amazing husband, Dave, for his unwavering belief in me and sense of humor, and for keeping our family fed and so much more; and to Max, Jack, and Sophie. To my one-of-a-kind technical mentor, John Dias, for his encouragement and strategies for dealing with technical people, which are why I am still in cybersecurity.

Ingrid: Thank you to my wonderful husband Steve, who made sure I had the space, time, and mental energy needed to work on this book. I truly appreciate your patience and understanding as I continued to tell you the book would be done "soon."

Carson: Thank you to my amazing wife, Kristin, whose patience, understanding, and support have been essential to making this book a reality; to my son who has wondered when this book will ever finish. To my family members in the stars, every day is precious. To every cybersecurity professional that I've met over the years who has ever pulled 14-hour days or long weekends cleaning up messes in cyberspace, your experiences and wisdom made this book possible. Again.

First Edition Acknowledgments

We retained some content from the first edition, and we are very grateful for those enduring contributions in the second edition. In a dynamic field, it is a real testament to the talent of the individuals who contributed to the first edition for some of the material to remain relevant. Due to the significant way this second edition was restructured, it is difficult to pinpoint all the exact usage from the first edition. For that reason, we believe it important to re-acknowledge all those who contributed to the first edition, authored by Carson Zimmerman. His first edition acknowledgements are here.

There are many individuals whose hard work has contributed to the creation of this book.

First of all, I would like to recognize Eric Lippart, whose many years of work in computer network defense (CND) contributed to every aspect of this book. The ten strategies outlined in the book emerged from the years we worked together to share best practices and solutions for CND across the U.S. federal government.

Some sections of this book are based, in part, on material from other MITRE work. The following sections incorporate ideas and expertise from other MITRE staff members:

- Scott Foote, Chuck Boeckman, and Rosalie McQuaid: Cyber situational awareness
- Julie Connolly, Mark Davidson, Matt Richard, and Clem Skorupka: Cyber-attack life cycle
- Susan May: CSOC staffing
- Mike Cojocea: Security information and event management (SIEM) and log management best practices
- Joe Judge and Eugene Aronne: Original work on intrusion detection systems (IDS) and SIEM
- Frank Posluszny: Initial concept and development of material on Cyber Threat Analysis Cells
- Kathryn Knerler: CND resources and websites
- Therese Metcalf: Material on various government CSOCs, which were used throughout this book
- Katie Packard and Bob Martin: Technical editing and glossary
- Robin Cormier: Public release and project management
- Robert Pappalardo and John Ursino: Cover design
- Susan Robertson: Book layout and diagrams

The following individuals are recognized as peer reviewers for this book: Chuck Boeckman, Mike Cojocea, Dale Johnson, Kathryn Knerler, Eric Lippart, Rick Murad, Todd O'Boyle, Lora Randolph, Marnie Salisbury, Ben Schmoker, Wes Shields, and Dave Wilburn.

This book would not have been possible without the funding and mentorship provided by MITRE's cybersecurity leadership: Gary Gagnon, Marnie Salisbury, Marion Michaud, Bill Neugent, Mindy Rudell, and Deb Bodeau. In addition, Lora Randolph's and Marnie Salisbury's advice and support were instrumental in making this book possible.

This book was also inspired by the excellent work done by the Carnegie Mellon University (CMU) Software Engineering Institute (SEI) Computer Emergency Response Team (CERT®), whose materials are referenced herein. Their copyrighted material has been used with permission. The following acknowledgement is included per CMU SEI:

This publication incorporates portions of the "Handbook for Computer Security Incident Response Teams, 2nd Ed.," CMU/SEI-2003-HB-002, Copyright 2003 Carnegie Mellon University and "Organizational Models for Computer Security Incident Response Teams," CMU/SEI-2003-HB-001, Copyright 2003 Carnegie Mellon University with special permission from its Software Engineering Institute.

Any material of Carnegie Mellon University and/or its Software Engineering Institute contained herein is furnished on an "as-is" basis. Carnegie Mellon University makes no warranties of any kind, either expressed or implied, as to any matter including, but not limited to, warranty of fitness for purpose or merchantability, exclusivity, or results obtained from use of the material. Carnegie Mellon University does not make any warranty of any kind with respect to freedom from patent, trademark, or copyright infringement. This publication has not been reviewed nor is it endorsed by Carnegie Mellon University or its Software Engineering Institute.

CERT is a registered trademark of Carnegie Mellon University.

I would like to recognize the tireless efforts of the many operators, analysts, engineers, managers, and executives whose contributions to cyber defense have helped shape this book.

This book is dedicated to Kristin and Edward.

Contents

List of Figures

List of Tables

Executive Summary

This book presents an overview of how to organize and consider the many functions in cybersecurity operations centers (SOCs). It describes strategies that can be applied to SOCs of all sizes, from two people to large, multi-national centers with hundreds of people. It is intended for all cybersecurity operations center personnel, from new professionals just starting in a SOC to managers considering capability expansion of the SOC. Starting with a Fundamentals section table which summarizes functional categories and areas, the book guides cyber professionals through applying mission context to 11 strategies of a world-class SOC:

Strategy 1: Know What You Are Protecting and Why

Develop situational awareness through understanding the mission; legal regulatory environment; technical and data environment; user, user behaviors and service interactions; and the threat. Prioritize gaining insights into critical systems and data and iterate understanding over time.

Strategy 2: Give the SOC the Authority to Do Its Job

Empower the SOC to carry out the desired functions, scope, partnerships, and responsibilities through an approved charter and the SOCs alignment within the organization.

Strategy 3: Build a SOC Structure to Match Your Organizational Needs

Structure SOCs by considering the constituency, SOC functions and responsibilities, service availability, and any operational efficiencies gained by selecting one construct over another.

Strategy 4: Hire AND Grow Quality Staff

Create an environment to attract the right people and encourage them to stay through career progression opportunities and great culture and operating environment. Plan for turnover and build a pipeline to hire. Consider how many personnel are needed for the different SOC functions.

Strategy 5: Prioritize Incident Response

Prepare for handling incidents by defining incident categories, response steps, and escalation paths, and codifying those into SOPs and playbooks. Determine the priorities of incidents for the organization and allocate the resources to respond. Execute response with precision and care toward constituency mission and business.

Strategy 6: Illuminate Adversaries with Cyber Threat Intelligence

Tailor the collection and use of cyber threat intelligence by analyzing the intersection of adversary information, organization relevancy, and technical environment to prioritize defenses, monitoring, and other actions.

Strategy 7: Select and Collect the Right Data

Choose data by considering relative value of different data types such as sensor and log data collected by network and host systems, cloud resources, applications, and sensors. Consider the trade-offs of too little data and therefore not having the relevant information available and too much data such that tools and analysts become overwhelmed.

Strategy 8: Leverage Tools to Support Analyst Workflow

Consolidate and harmonize views into tools and data and integrate them to maximize SOC workflow. Consider how the many SOC tools, including SIEM, UEBA, SOAR, and others fit in with the organization's technical landscape, to include cloud and OT environments.

Strategy 9: Communicate Clearly, Collaborate Often, Share Generously

Engage within the SOC, with stakeholders and constituents, and with the broader cyber community to evolve capabilities and contribute to the overall security of the broader community.

Strategy 10: Measure Performance to Improve Performance

Determine qualitative and quantitative measures to know what is working well, and where to improve. A SOC metrics program includes business objectives, data sources and collection, data synthesis, reporting, and decision-making and action.

Strategy 11: Turn up the Volume by Expanding SOC Functionality

Enhance SOC activities to include threat hunting, red teaming, deception, malware analysis, forensics, and/or tabletop exercises, once incident response is mature. Any of these can improve the SOCs operating ability and increase the likelihood of finding more sophisticated adversaries.

Introduction

This book aims to help those who have a role in cybersecurity enhance their ability to find, analyze, and respond to cyber threats proactively and reactively. To do so, we gathered observations and proven approaches across people, process, and technology. The structure of this book includes 11 strategies that, when followed, can lead to an enhanced security operations capability. We address common questions often pondered by SOCs in the context of their organization's priorities.

This book's objectives are to help individuals and organizations:

- Articulate a coherent message of "this is the way security operations is done."
- Share observations and lessons learned from both mature SOCs and new SOCs.
- Provide context and options for critical SOC architecture, tools, and process decisions.
- Rationalize mission demands and requirements against how the SOC may be resourced and enabled.
- Differentiate the roles of different SOCs, given various constituency sizes and missions.
- Maximize the value of SOC staff and technology investments.
- Highlight not only internal SOC success factors but also shed light on conditions for success outside the SOC.

Audience

If you are part of, support, frequently work with, manage, or are trying to stand up a SOC, this book is for you. Its audience includes SOC managers, technical leads, engineers, and analysts. Portions of this book can be used also as a reference by those who interface with SOCs on a routine basis to better understand and support security operations. These include chief information security officers (CISOs); cybersecurity professionals working in other areas of cyber such as governance, risk, compliance, security architecture and engineering; network operations center (NOC) personnel; IT, cloud, mobile technology, and operational technology (OT) system administrators (sysadmins); counterintelligence and law enforcement personnel who work cyber cases; and people approaching cybersecurity anew, such as college students and individuals transitioning in from other fields.

Anyone reading this book is assumed to have a general understanding of IT and security concepts and a general awareness of cyber threats. A background in computer science, engineering, networking, or system administration is especially helpful, but is not required.

How to Use This Book

This book has been written as if you, the reader, have been given the task of operating a SOC. We take this approach to emphasize the operational and practical "real-world" nature of defending an enterprise. Regardless of your background, the book is intended to convey advice and best practices culled from a number of SOCs that MITRE and the book's authors have supported over the years. As a result, this book adopts two key conventions. First, tangible, concrete guidance is provided wherever possible. Second, many other excellent resources exist for a variety of cybersecurity technology, process, and organizational topics. This book leverages or references those works wherever possible.

Throughout the book, important key points will be designated as such:

This is a really important point, worthy of your consideration.

The book is organized as follows:

Strategy 0: Discusses the fundamentals of SOCs. It targets readers who do not have a strong background in cybersecurity. It includes a SOC's basic mission, capabilities, and technologies.

Strategies 1 through 11: Describe the 11 strategies of an effective SOC. Each strategy covers key design, architecture, and procedural issues. These sections house the main body of material covered in this book and crosscut issues of people, process, and technology.

Appendices and additional matter: Expand upon topics covered in the main strategies. These are relevant to supporting a world-class SOC but often include large tables that were better viewed outside of the structure of a specific chapter. The book also contains a Glossary and List of Abbreviations.

Scope

This book integrates subjects cutting across the people, process, and technology elements of security operations. However, the scope of cybersecurity operations is so broad that it is impossible to cover all potential topics. There are many excellent existing materials that explore specific cyber operations topics, so throughout the book the reader will find numerous references that can guide them towards further understanding of a given strategy area.

The topics chosen for this book were included because of their broad applicability to helping the reader understand how a SOC works and what types of decisions need to be made to stand up, execute, and improve on the security operations mission. As a result, this book does not dive deeply into the following areas:

- Technical details on how networks and systems operate
- Step-by-step use of SOC technologies or comparisons of specific SOC products
- Detailed description of how various cyber-attacks work

- Media, memory, and mobile forensics
- Malware analysis and exploits
- Tangentially related topics such as network and telecommunications monitoring and operations or in-depth big data architecture, beyond the scope of leveraging it for security operations
- Physical security operations (e.g., "gates, guards, guns")
- Legal aspects of monitoring such as privacy laws, details on chain of custody, and specific legal or regulatory requirements for retention of audit data
- Compliance with specific laws and regulations

Why the First Edition?

The MITRE Corporation supports a number of U.S. Government SOCs, which go by many names: Computer Security Incident Response Team (CSIRT), Computer Incident Response Team (CIRT), Computer Security Incident Response Capability (CSIRC), Network Operations and Security Center (NOSC), and, of course, CSOC. As a corporation that operates several federally funded research and development centers (FFRDCs), MITRE's support to these entities spans many years, with staff members who have worked across the full scope of the SOC mission, from standing up new SOCs to enhancing existing capabilities. Operational activities range from analyzing malicious executables to editing incident escalation procedures to architecting organizational sensor grids and providing SOC maturation guidance. We drew upon these experiences, and others, in developing the book.

Our firsthand experience, along with other trends and observations, motivated us to write and publish the first edition, *Ten Strategies of a World-Class Cybersecurity Operations Center*, in 2014. Leading up to the publication of the first edition of the book, the majority of recognized materials about SOCs were published between 1998 and 2005. A list of some of these books can be found in Appendix A. The first edition sought to build upon and advance those works by presenting a more modern reference for the cybersecurity community.

Why a Second Edition?

Although the official practice of IR is now over 30 years old [1], SOCs large and small, new and old, still wrangle with fundamental issues related not just to the technologies they must work with day-to-day but with larger issues related to people and process, from how to handle incident escalation, to where the SOC belongs on the constituency organization chart, to how to successfully integrate the various cybersecurity functions. Based on widespread feedback, many aspects of the material presented in *Ten Strategies* have endured since its publication in 2014. However, the industry and technology continue to evolve rapidly, and many new lessons have been learned since then.

Since the writing of first edition of this book, many more seminal references have been released. A few of these books are also included in Appendix A. Additionally, white papers, blog posts, podcasts, and conference presentations provide a continual stream of updated content. However, even with all this new content, MITRE still receives inquiries about the first edition of the book and downloads of the first edition continue. In the more crowded space there still appeared to be room for an updated book that took a holistic view of what makes a SOC a SOC, considers what strategies are especially important to a SOC's success, and puts into context all the technology trends and changes that have occurred since the first books was published.

Changes Since the Original *Ten Strategies of a World-Class Cybersecurity Operations Center*

The fundamental questions about the "hows" and "whys" of security operations has endured since the first edition. However, this second edition updates the strategies to encompass many industry and technological changes, along with new lessons learned. Originally, this book was intended to be a lightweight update to existing chapters, potentially with just the addition of some new sections within the existing strategies. As work began on this new edition, it quickly became clear that rethinking the whole structure would lead to a better product for the readers. In some cases, the changes below do appear as individual sections, but the challenges and opportunities they present go much deeper and are reflected throughout the entire approach to the content of this second edition of the book:

Evolved thinking

- Increased emphasis on understanding the mission or business context in which the SOC operates (Strategy 1)
- Maturation and recognition of the value of threat-oriented defense, including incorporation of cyber threat intelligence and threat hunting into routine security operations and the use of threat frameworks such as leveraging MITRE ATT&CK (Strategy 6)
- Prioritization of host sensors and instrumentation (such as through an EDR), along with log collection and cloud monitoring, decreased focus on standalone network IDS/IPS sensors, and packet capture (PCAP) collection (Strategy 7)
- Diversification of thinking around SOC operating constructs, such as the increasing use of partial or complete outsourcing of SOC functions to a managed services provider (Section 3.5), and tierless SOC models (Section 3.3.4)
- Recognition of the importance of communication and collaboration between the SOC and other parties such as executives, business/system/service owners, other elements of cybersecurity apparatus, and other SOCs (Strategy 9)
- Promotion of a more balanced approach to protecting SOC tools, data, and operations from the adversary while still being able to engage with constituents dynamically and routinely in furtherance of the SOC mission (Sections 3.7.5, 8.6, and 10.3.2; Strategy 9)

New and expanded material

- Monitoring and response in additional environments including cloud (Sections 5.6 and 7.5), mobile (Section 5.7), and operational technology (Sections 5.8 and 7.7)
- Cloud as a platform and as a venue for SOC operations, tools, and data (Section 8.7)
- Measurement of SOC effectiveness through metrics (Strategy 10)
- Expanded SOC functionality such as hunting, adversary emulation, and cyber deception (Sections 11.1 through 11.5)
- Updated technologies such as:
  - Threat intelligence platforms (TIP) (Section 6.6)
  - Endpoint detection and response (EDR) (Section 7.3.2)
  - User entity behavior analytics (UEBA) (Section 8.3)
  - Security orchestration, automation, and response (SOAR) (Section 8.5)
  - Use of big data and machine learning techniques to complement or wholesale replace traditional SIEM and to advance detection, investigation, and response (Section 8.2.8)
  - Breach and attack simulation (Section 11.4)

Fundamentals

Before jumping into the 11 strategies covered in this book, it is important to ensure a common understanding of the functions and practices of SOC operations so that readers of all experience levels can benefit from the content. Even if security operations and incident response (IR) are familiar, it may be helpful to review this section as it lays out foundational material and definitions that the rest of the book will reuse.

0.1 The Importance and Role of the SOC

Ensuring the confidentiality, integrity, and availability of the modern digital enterprise is a big job. It encompasses many parallel and related efforts, from robust systems engineering to effective cybersecurity policy and comprehensive workforce training. One essential element is cybersecurity operations: monitoring, analyzing, responding, and recovering from all measures of cyber attack. The operational focal point for incident detection, analysis and response is the cybersecurity operations center (CSOC, or simply SOC). Virtually all medium and large organizations have some form of a SOC, and their importance is growing as cyber becomes increasingly integral to every type of organization, including for-profit and nonprofit businesses, government, and academia.

Since SOCs began rising in prominence during the 1990s, several movements in computing have changed the way the SOC mission is executed, along with the SOC's importance to an organization. While many are cliched, they are worth observing:

- The rise of the advanced persistent threat (APT) [2] and an acceleration in the evolution of the adversary's tactics, techniques, and procedures (TTPs)
- Digital transformation, meaning the integration of information technology (IT) into nearly every aspect of business and government, even those not traditionally associated with computers
- The dissolution of organizational boundaries, with the onset of both mobile and cloud computing including the sudden shift to remote work/work from home as a result the COVID19 global pandemic
- The integration and proliferation of non-traditional IT, such as with embedded computing and Industrial Control Systems (ICSs)/Supervisory Control and Data Acquisition (SCADA)
- A transition from network-based buffer overflow attacks to client-side attacks
- The rise of cybersecurity from near obscurity to a daily top news topic
- The recognition by executives across industry verticals and across the world of the importance of cybersecurity to business operations and the integration of cybersecurity into organizational risk management calculations

0.1.1 The SOC Name

A SOC is defined primarily by what it does: cyber defense. Adapting the definition from the Committee on National Security Systems (CNSS) [3], cyber defense is "the practice of defense against unauthorized activity within cyberspace, including monitoring, detection, analysis (such as trend and pattern analysis), and response and restoration activities." Further, this book combines definitions of CSIRT from CNSS [3] and Request for Comments (RFC) 2350 Expectations for Computer Security Incident Response [4] to define a SOC as:

> *A SOC is a team, primarily composed of cybersecurity specialists, organized to prevent, detect, analyze, respond to, and report on cybersecurity incidents.*

There are many terms that have been used to reference a team of cybersecurity experts assembled to perform these functions. They include:

- Computer Security Incident Response Team (CSIRT)
- Computer Incident Response Team (CIRT)
- Computer Incident Response Center (or Capability) (CIRC)
- Computer Security Incident Response Center (or Capability) (CSIRC)
- Security Operations Center (SOC)
- Cybersecurity Operations Center (CSOC)
- Computer Emergency Response Team (CERT®) [5]

Common variations in the name for these teams include words such as "network," "computer," "security," "cyber," "emergency," "incident," "operations," or "enterprise."

There is no one universally agreed-to term for this set of cybersecurity specialists, and preference for terms has changed over time, may be country specific, and is often tied to the scope of the team's mission. However, for simplicity, this book uses the acronym "SOC" to refer to this team of cybersecurity specialists in all its forms and sizes.[1] As is explained throughout the book, there are various services and supporting functions a SOC can perform, such as engineering its own tools or curating cyber threat intelligence, although not all SOCs perform all functions equally. At its core, however, these functions again circle back to the core mission of finding, analyzing, and responding to intrusions, which is at the heart of the SOC mission.

0.1.2 The SOC Constituency

A SOC can take many forms as will be discussed throughout the book, but all SOCs provide services to a set of customers referred to as a constituency which is a bounded set of users, sites, IT/OT assets, cloud assets, data, networks, and organizations. Combining

---

[1] Some physical security operations centers also go by "SOC." Thus, the title of this book includes the term "Cyber" to disambiguate the topic even though we use the shorter term "SOC" throughout the text.

definitions from the *Request for Comments 2350 Expectations for Computer Security Incident Response* [4] and the *Handbook for Computer Security Incident Response Teams (CSIRTs)* [6], a constituency can be established according to organizational, geographical, political, technical, or contractual demarcations.

Once again borrowing from the historical definition of "CSIRT," *RFC4949, Internet Security Glossary, Version 2* [7] articulates three criteria that an organization must meet to be considered a CSIRT, which is also relevant to a SOC. For an organization to be considered an SOC, it must:

- Provide a means for constituents to report suspected cybersecurity incidents.
- Provide incident handling assistance to constituents.
- Disseminate incident-related information to constituents and external parties.

The SOC may also provide a set of services to the constituency that is related to the core mission of incident detection and response, such as engaging in vulnerability assessments, performing penetration testing, or supporting supply chain risk management efforts. The way a SOC serves its constituency can be compared to the way the medical system supports its community; there are primary physicians who work to prevent illness while also detecting and responding when it is found, but they are not the only line of defense. There are also emergency services for when the unexpected happens. And there are specialists that are called in as their expertise is needed. And behind the scenes, there are programs and policies that must be put into place to support the execution of the medical services and community outreach and training to ensure those services are as effective as possible in helping to prevent disease as well as detecting and responding to it. These different services work in concert with each other, with any given medical facility offering a range of services depending on their mission and constituency.

Similarly, all SOCs will be able to perform basic cyber defense services, but some SOCs will also have the skills and resources to perform more specialized activities, such as detailed forensics on compromised systems. Others, however, must call on partner SOCs or external resources when in-depth forensics must be performed. This is similar to how a primary care physician may need to engage with a specialist if a medical issue is not within their specialty. The services that will be available locally within a SOC will depend on the needs of the constituency.

0.1.3 Comparing and Contrasting the SOC to Similar Entities Within the Constituency

Other entities in the constituency are similar to SOCs, but often have different roles. A SOC is usually distinct from:

- A NOC or IT operations center because a SOC is primarily looking for cyber attacks, whereas a NOC (and typically other IT operations staff) is concerned with operating and maintaining network and other IT equipment [8].
- A chief information officer (CIO) or chief information security officer (CISO) because the SOC is a real-time operational capability, and its monitoring efforts are not usually

focused on other areas of cybersecurity like policy and governance, risk management, or secure system engineering (though some SOCs report directly to a CISO or CIO).

- An Information Security Continuous Monitoring (ISCM) program because the SOC is responsible for incident detection and response, whereas ISCM is generally focused on security compliance and risk measurement [9].
- An information systems security officer (ISSO) or information systems security manager (ISSM) organization (such as in the government) because the SOC is responsible for monitoring and responding to the full-scope cyber threat across the entire constituency, whereas ISSOs are often more focused on IT compliance and ensuring the security of specific systems.
- Physical security monitoring (e.g., "gates, guards, and guns") because a SOC is concerned with the cyber domain, whereas physical security monitoring is primarily concerned with protecting physical assets and ensuring personnel safety.
- Law enforcement because SOCs rarely hold legal investigative authorities. While SOCs may find intrusions that result in legal action, their primary duty is usually not the collection, analysis, and presentation of evidence that will be used in legal proceedings.

These delineations highlight the fact that SOCs must work with all these groups on a regular basis, and they must maintain skill sets in many areas of IT and cybersecurity.

0.1.4 The SOC Mission

SOCs can range from small operations with just a few people, each working part time, where each team member "does everything," to national coordination centers employing hundreds of analysts and responders, where each person typically has a very specialized role. A typical midsize SOC's mission statement typically includes the following elements:

- Preventing cybersecurity incidents through proactive measures, including:
  - Continuous analysis of threats
  - Scanning for vulnerabilities
  - Deploying coordinated countermeasures
  - Consulting on security policy and architecture
- Monitoring, detection, and analysis of potential intrusions in real time and through adversary hunting, utilizing a variety of security-relevant data sources
- Responding to confirmed incidents, by coordinating resources and directing use of timely and appropriate countermeasures
- Providing situational awareness and reporting on cybersecurity status, incidents, and trends in adversary behavior to appropriate organizations
- Engineering and operating SOC technologies, such as host sensors, network sensors, log collection, and analysis systems

The SOC mission can also be thought of in terms of the infrastructure and data the SOC defends. For a few small SOCs that may only consist of local on-prem digital assets and data. However, for the majority of SOCs, their responsibilities will also include monitoring, detecting, and responding to potential incidents on remote systems, systems and data in the cloud, and the constituency's mobile infrastructure. As of this book's publication, some SOCs are starting to integrate OT monitoring and defense into the SOC mission as well. Throughout the book the assumption is that the SOC will be responsible for all these aspects: IT (on-prem and remote), OT, cloud, and mobile. When there are unique considerations for a particular environment, such as with data collection and monitoring, additional context is provided.

0.2 SOC Functions

A SOC satisfies the constituency's cyber monitoring and defense needs by performing a set of functions for its constituency. Table 1 provides a list of those functions that tend to find a home in the SOC. However, it is very rare for a single SOC to fulfill every single one. Consequently, "Strategy 3: Build a SOC Structure to Match Your Organizational Needs" includes details on how to make careful choices when picking from this set of services based on factors such as constituency size, SOC resourcing, and maturity. In addition, when looking at the SOC Service Areas and Service Descriptions table, consider the following:

- The vulnerability management (VM) functional area is included in this list of services for completeness. Although many constituencies have a separate team that performs some or all of this function, there are also SOCs that find themselves tasked with some or all of these duties.
- For the purposes of this book, the attack simulation and assessments functional area includes penetration testing (pen testing). However, pen testing is often closely associated with a vulnerability management program and may find its home in the organization responsible for VM functions if outside the SOC.
- Functional areas like situational awareness and communications, metrics, training, and process improvement will usually be included in the execution of many of the other services. They are called out as unique functional areas to show the need to have a mechanism to integrate these efforts across the SOC.
- Security Architecture, Engineering, and Administration includes deployment and management for different technical environments. As with other functional areas, not all environments are applicable to all SOCs, and new environments may emerge over time.
- Insider Threat programs usually combine physical security, personnel awareness, and information-centric principles [10]. Depending on the constituency, the SOC may have a larger or smaller role to play in identifying and protecting against this threat.

Table 1. SOC Functional Categories and Functional Areas

Incident Triage, Analysis, and Response	
Real-Time Alert Monitoring and Triage	Performing triage and short-turn analysis of potential security incidents generated by near-real-time security alert feeds.
Incident Reporting Acceptance	Receiving and processing reports of potential security incidents from constituents, other SOCs, and third parties. These reports may come through written (e.g., email) or verbal means.
Incident Analysis and Investigation	Performing in-depth, detailed analysis of suspected incidents. This includes identifying details such as the origin, extent, and implications of an incident, and characterizing the confidence of these conclusions.
Containment, Eradication, and Recovery	Performing activities supporting incident/adversary containment, damage management, adversary eviction, and system recovery to reduce current impact and move to a state that will prevent future incidents.
Incident Coordination	Performing information gathering, information distribution, and notification in support of an ongoing incident. Directing and/or coordinating response in partnership with constituents, incident response stakeholders, other SOCs, and third parties.
Forensic Artifact Analysis	Examining media samples and digital artifacts (hard drives, files, memory) to draw detailed observations and conclusions about suspected activity, such as content analysis and timeline reconstruction.
Malware Analysis	Examining suspicious files to understand the provenance, pedigree, functions, and intent of suspected malware samples. This includes utilizing various methodologies, including static code reverse engineering and dynamic runtime analysis.
Fly-Away Incident Response	Tools, procedures, and coordination practiced to rapidly relocate and provide onsite incident response services for constituents at physical locations where SOC analysts do not routinely reside.
Cyber Threat Intelligence, Hunting, and Analytics	
Cyber Threat Intelligence Collection, Processing, and Fusion	Collecting cyber threat intelligence products, including CTI feeds and reports. Processing and integrating CTI into SOC systems and parsing and filtering information for further consumption by the SOC and its constituency.
Cyber Threat Intelligence Analysis and Production	Utilizing analytic techniques to track, trend, and correlate adversary behavior over time, and support risk decision making. This includes creating and producing CTI reports describing specific adversaries, their TTPs, and campaigns. This may include using a cyber threat intelligence platform or other tools to enhance analysis.
Cyber Threat Intelligence Sharing and Distribution	Sharing CTI and incident reports with parties outside the SOC, including partners, other SOCs, and the broader cybersecurity community.
Threat Hunting	Performing proactive operations to identify potentially malicious activity, outside the scope of established SOC alerts, based on hypotheses that the adversary is operating in or against the constituency. This includes developing and refining custom analytic capabilities.
Sensor and Analytics Tuning	Performing curation, tuning and optimization of detections, analytics, signatures, correlation rules, and response rules deployed on SOC detection and analytics systems, such as EDR, SEIM, and SOAR.
Custom Analytics and Detection Creation	Using knowledge of adversary TTPs and constituency systems to create detections and analytics to detect and understand various activity in SOC sensors and analytic systems, usually from scratch.

Data Science and Machine Learning	Defining, implementing, and curating data science and machine learning techniques to support SOC functions, such as bespoke machine learning models tailored to the constituency.
Expanded SOC Operations	
Attack Simulation and Assessments	Performing red teaming, pen testing, adversary emulation, purple teaming, breach and attack simulation, or other testing detections with the goal of improving SOC operations and the constituency's overall defensive posture.
Deception	Performing actions to conceal networks and assets, create uncertainty and confusion, and/or influence and misdirect adversary perceptions and decisions.
Insider Threat	Supporting detections, analytics, and investigations focused on finding malicious or anomalous activities carried out by users with legitimate access to constituency systems.
Vulnerability Management	
Asset Mapping and Composite Inventory	Collecting and curating knowledge of constituency assets, networks, and services, mapping their interdependencies, and calculating criticality and risk.
Vulnerability Scanning	Interrogation of constituency assets for vulnerability status, including patch level and installed software, and security-relevant configuration, for purposes of calculating security risk and compliance status.
Vulnerability Assessment	Performing the "systematic examination of an information system or product to determine the adequacy of security measures, identify security deficiencies, provide data from which to predict the effectiveness of proposed security measures, and confirm the adequacy of such measures after implementation." [11]
Vulnerability Report Intake and Analysis	Accepting, triaging, and analyzing vulnerability reports from vulnerability researchers to understand them and find mitigations for constituency products or assets. This is sometimes known as a responsible vulnerability disclosure program.
Vulnerability Research, Discovery, and Disclosure	Performing proactive discovery of security vulnerabilities not previously known to the SOC (e.g., "0 days"), through reviewing internal incidents, cyber threat intelligence collection, and software reverse engineering. This includes sharing vulnerability information with vendors, other SOCs, and constituents such that they can act on that information.
Vulnerability Patching and Mitigation	Addressing vulnerabilities through applying patches or mitigating the risk of vulnerability exploitation through minimizing vulnerability exposure to adversaries such as system or service configuration changes.
SOC Tools, Architecture, and Engineering	
Sensing and SOC Enclave Architecture	Defining the overall architecture for the sensing, analytic and SOC operating environments, including the integration between various components and data management planning. This may include evaluating commercial products for intended use.
Network Security Capability Engineering and Management	Engineering, deploying, operating, and maintaining network and enterprise services detection and protection capabilities including firewalls, web proxies, email proxies, and content filters. This includes creating and maintaining the connections and data flows that interconnect tools and data.
Endpoint Security Capability Engineering and Management	Engineering, deploying, operating, and maintaining endpoint detection and protection capabilities such as EDR. This includes creating and maintaining the connections and data flows that interconnect tools and data.

Cloud Security Capability Engineering and Management	Engineering, deploying, operating, and maintaining cloud detection and protections capabilities such as CASB and other cloud-native tools that protect cloud services. This includes creating and maintaining the connections and data flows that interconnect tools and data.
Mobile Security Capability Engineering and Management	Engineering, deploying, operating, and maintaining enterprise and endpoint detection and protection capabilities for mobile devices. This includes creating and maintaining the connections and data flows that interconnect tools and data.
Operational Technology Security Capability Engineering and Management	Engineering, deploying, operating, and maintaining cyber detection and protection capabilities for operational technologies. This includes creating and maintaining the connections and data flows that interconnect tools and data.
Analytic Platform Engineering and Management	Engineering, deploying, operating, and maintaining SIEM, SOAR, CTI platforms, UEBA, Big Data Platform, and other technologies. This includes creating and maintaining the connections and data flows that interconnect tools and data.
SOC Enclave Engineering and Management	Deploying, operating, and maintaining technologies, outside the scope of SOC and sensor capabilities, that support SOC operations including research environments, servers, workstations, printers, file shares, and enclave network systems.
Custom Capability Development	Creating the custom tools and systems necessary to fulfill various SOC requirements when no suitable commercial or open-source capability fits the need.
Situational Awareness, Communications, and Training	
Situational Awareness and Communications	Synthesizing and redistributing the SOC's knowledge of constituency assets, risks, threats, incidents, and vulnerabilities to constituents, supporting improvement of constituency cybersecurity posture and practices. Engaging within the constituency, and with other external organizations, to inform and be informed, collaborate, and share information.
Internal Training and Education	Gathering, formulating, and delivering training to SOC analysts to increase their proficiency in SOC functional areas.
External Training and Education	Gathering, formulating, and delivering training to constituents, to increase their knowledge of various cybersecurity topics.
Exercises	Formulating and facilitating cybersecurity scenario-based simulations and exercises, such as mock critical severity incidents.
Leadership and Management	
SOC Operations Management	Executing the day-to-day functions of running a SOC including financial and personnel management.
Strategy, Planning, and Process Improvement	Identifying the future state of the SOC and guiding the organization towards those outcomes. This includes looking at and learning from the past, performing assessments of the current state, and identifying new opportunities.
Continuity of Operations	Creating and evaluating plans designed to help the SOC sustain mission and business processes during and after a disruption.
Metrics	Defining, measuring, and reporting on key performance indicators of operational processes, the output of operations, and/or situational awareness of the constituency.

One way to use this table is as a resource to account for the various activities of the SOC in its aggregate. This can be helpful when thinking about all the personnel and functions that need to be budgeted for. For additional thoughts on SOC services consider [6] and [12].

0.3 SOC Basics

The SOC has a core mission of identifying and responding to potential cyber threats. This includes understanding how the SOC moves from tip-offs to response, along with understanding how knowledge of the adversary shapes those activities. This section presents a basic SOC workflow to show how data and tools come together to support the fundamental processes performed within the SOC.

0.3.1 From Tip-Offs to Response

SOCs accomplish their mission in large part by being purveyors and curators of copious amounts of security-relevant data. They must be able to collect and understand the right data at the right time in the right context. Virtually every mature SOC employs several different technologies, along with automation processes, to generate, collect, enrich, analyze, store, and present tremendous amounts of security-relevant data to SOC members. More details on both data and tools will be found throughout the book, specifically in "Strategy 7: Select and Collect the Right Data" and "Strategy 8: Leverage Tools to Support Analyst Workflow."

Among the data sources a SOC is likely to ingest, the most prominent are host sensors such as endpoint detection and response (EDR) capabilities, network traffic metadata, and various log sources such as application or operating system (OS) logs from on-prem devices, the cloud, or OT. These sensors are placed on either the host or network, or cloud to detect potentially malicious or unwanted activity that warrants further attention by a SOC analyst. Combined with security audit logs and other data feeds, this data will then be sent to a variety of systems within the SOC such as security information and event management (SIEM) or security orchestration, automation, and response (SOAR) technologies or specialized capabilities for performing functions such as malware analysis. Figure 1 illustrates the high-level flow of data into the tools and technologies a SOC might use.

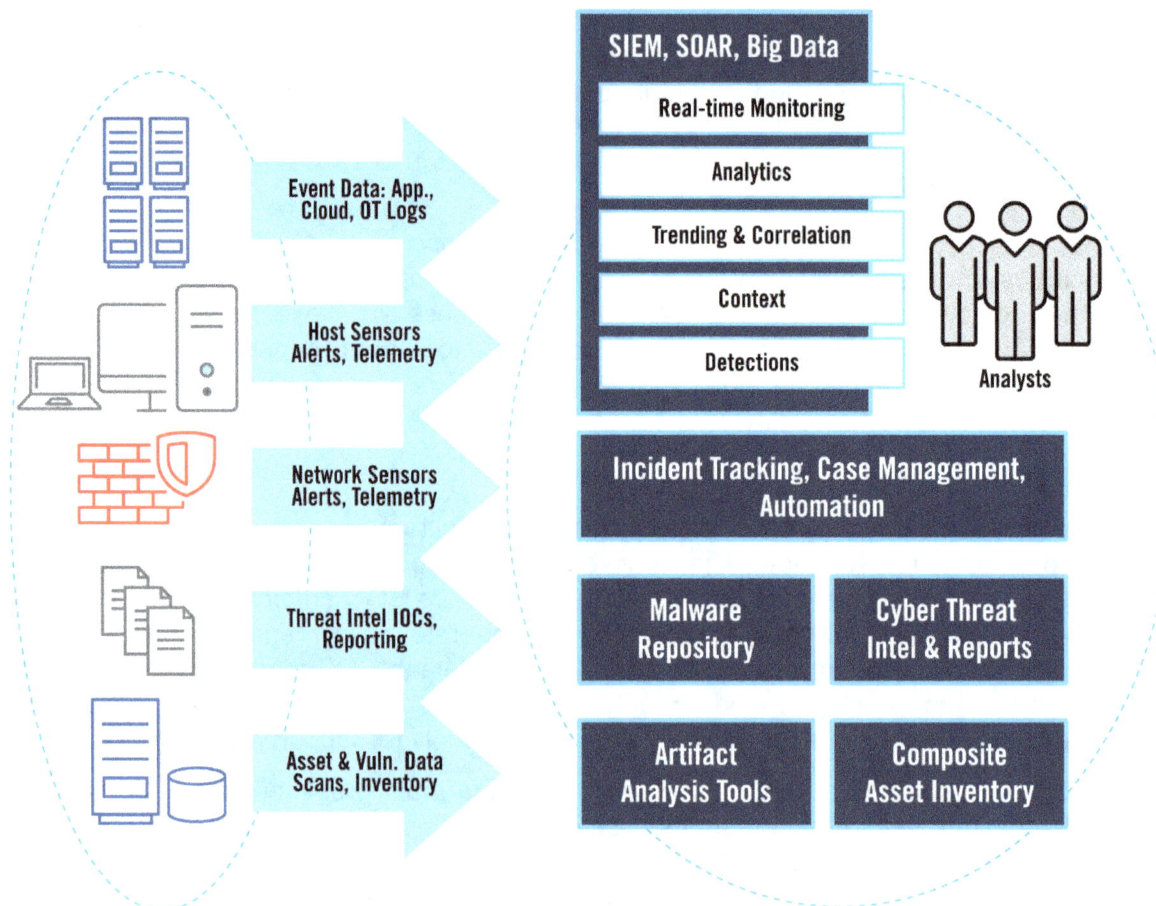

Figure 1. Typical SOC Data and Tools

Tip-Offs

A typical SOC will collect, analyze, and store anywhere from millions to tens of billions of security events every day. Events include a user connecting to a file share, a server receiving a request for a web page, a user sending email, and a firewall blocking a connection attempt. Events do not necessarily indicate good or bad behavior; they simply are things that happened.

> *An event is "any observable occurrence in a*
> *system and/or network" [3].*

In contrast, the term alert is typically used to reference an event that generated with the implication it may be a potential attack. Intrusion detection systems (IDS) and SIEM systems are typical generators of alerts.

> *An alert is a technical notification that a particular event,*
> *or series of events, has occurred.*

Alerts will often come in two forms, signature-based and anomaly detections. Misuse or signature-based detection which is where the system has prior knowledge of how to characterize and therefore detect malicious behavior, such as with an indicator of compromise (IOC) matching. IOCs are forensic artifacts from intrusions that are identified on constituency systems at the host or network level [11]. They are discrete pieces of information, such as IP addresses, hashes/checksums, or malware characteristics. See Section 6.2.1 for more discussion on IOCs. Anomaly detection is where the system characterizes normal or benign behavior and alerts whenever it observes something that falls outside the scope of that behavior.

Both events and alerts are nothing more than data; both must be evaluated within the context of the system(s) they occurred on, the surrounding environment, supported mission, relevant cyber intelligence data, and other sources of data that can confirm or repudiate whether there is any cause for concern. Just because the SOC receives an alert, that does not necessarily mean something bad happened. It just that a pre-defined set of criteria was met.[2] It takes human analysis, the process of evaluating the meaning of a collection of security-relevant data to establish whether further action is warranted. This is typically performed with the assistance of specialized tools and automation.

In addition to its own sensing, some of the best tips a SOC receives come from constituents and partners. These partners include:

- Constituents with varying levels of access and expertise, from ordinary unprivileged users to system and network administrators
- Constituency IT help desk
- Peered, subordinate, coordinating, or national SOCs
- Law enforcement or other investigatory entities

These tip-offs can be delivered through a variety of methods, typically email messages, phone calls, real-time text chat, walk-in reports, an incident reporting form on SOC website, and tips from other SOCs.

Context
Context is an incredibly important factor in SOC decision making, as it is in many endeavors. For example, on the morning of 7 December 1941, a U.S. Army radar station in Oahu, Hawaii picked up a huge blip on its instruments. Vigilant radar operators sent news of their finding to another unit on the island; however, they could not understand what was on their radar because they had no other observables to confirm what they were seeing or provide context [13]. They speculated that the blip was caused by friendly B-17s flying a mission that same morning. Unfortunately, what was actually on their radar were Japanese warplanes on their way to bomb Pearl Harbor. Had the radar operators been able to confirm that the blip on their instruments was the enemy, they could have taken preemptive measures to prepare for the attack.

---

[2] Section 5.3.4 will discuss the idea of true and false positives and negatives which are ways of describing if an alert is acting as expected.

The same can be said of many alerts seen by SOC analysts, even high severity ones. Without supporting context, the alert is worth little.

Context can come in many forms. It can include business related information, such as knowing if the constituency should be expecting connections from foreign countries, or technical context such as the details contained in media images. For more on overall context see "Strategy 1: Know What You Are Protecting and Why." Technical context helps establish a complete accounting of what happened on the host or in a cloud service. SOC's monitoring should cover both contextual sources and actionable alerts, so that its analysts will be able to spot and analyze suspected incidents more effectively. For example, system memory images and media artifacts (such as hard drives) are undisputable sources of ground truth in a severe incident. On the other hand, output of custom detections and analytics such as the SIEM or a ML architecture, if implemented properly, should be alerts that analysts can act upon, but they will still need additional context to fully understand what happened. There are some sources of data that serve both needs; for example: a good EDR will provide both robust host detections as well as rich host telemetry, including process execution, network connections, and authentication events. With this information an analyst can both be alerted to what has happened and have the context needed to decide what to do next. Example data feeds that complete this picture are depicted in Figure 2.

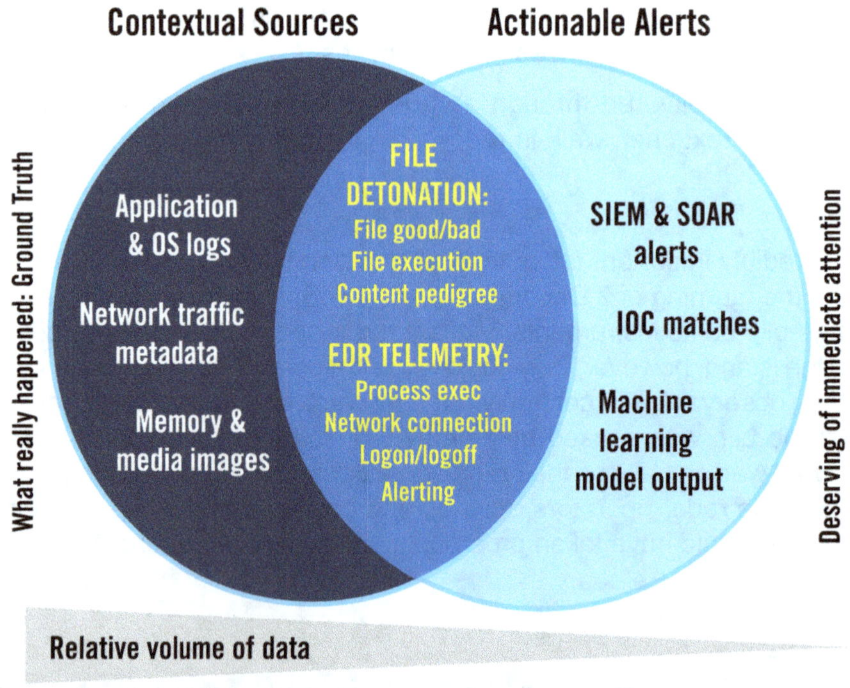

Figure 2. Context to Tip-Offs

Ground Truth

Analysts will start with initial indicators (such as a high-priority alert or analytic trigger) and use a combination of automation, rote process, and their own experience to gather additional contextual data to establish the ground truth disposition of the event(s) they are evaluating. Handling the constant influx of alerts and referrals is analogous to triaging patients in an emergency—there are hundreds of wounded people, so who requires urgency or prioritization?

In cybersecurity operations, triage is the process of sorting, categorizing, and prioritizing incoming events and other requests for SOC resources [5].

Triaging incoming events, alerts, and other tip-offs is benefited by high degrees of automation (implemented by the SOC using tools discussed later in this book), allowing a relatively small group of analysts to handle alerts derived from millions of events. Confidence in this data can be enhanced and volume lowered through techniques such as filtering, deduplication, alert enrichment, cyber threat intelligence fusion, ML-based prioritization with the SOC's big data analytic platform, SIEM and SOAR. This is true not only for the alert lifecycle but also for processing CTI.

Leveraging automation early in the alert lifecycle is critical to the SOC's maturity and keeping pace with expanding and changing mission demands.

The threshold at which an event or alert is escalated can be defined according to various types of potential "badness" (type of incident, targeted asset or information, impacted mission, etc.). In such an arrangement, the time span the triage analyst examines each alert is usually measured in minutes; this depends on the SOC's escalation policy, concept of operations (CONOPS), number of analysts, size of constituency, and alert volume.

A number of technologies enable the SOC to comb through hundreds of alerts and billions of events every day, supporting the incident life cycle from cradle to grave. This includes tools such as SIEM, SOAR, and big data analytic platforms tools that allow the SOC to collect, store, correlate, and display myriad security-relevant data feeds, supporting triage, analysis, escalation, and response activities. The variety of data feeds combined by these platforms support both the tip-off and contextual "ground truth" analysis needed to support incident identification and investigation. Any one source of security events has little value; the whole is greater than the sum of the parts. That said, there is an enduring need for human analysts to reason about alerting, events and analytics that cannot otherwise be fully automated. Until a SOC analyst has evaluated the disposition of an alert, the SOC cannot be certain there is a confirmed incident or not. Furthermore, no one analyst can know all the technologies or all the behaviors of the enclaves and hosts they are watching; multiple sets of eyes and analytics usually are better than one. While alert enrichment, prioritization, and automation will save the SOC from drowning in a sea of raw data, do not forget the following truth:

Once an event is escalated it may take days or weeks to collect and carefully inspect all the necessary data to determine the case's extent and severity. A single event can spawn an incident, but, for every incident, there are millions of events that are simply benign.

A cyber incident is defined as: "actions taken through the use of an information system or network that result in an actual or potentially adverse effect on an information system, network, and/or the information residing therein." [3] [14]

Response

The SOC is responsible for leading and coordinating activities that support adversary identification, containment, eradication, recovery, and reporting. During this process, the SOC will typically leverage internal and external resources when responding to the incident to ensure the response matches with organizational needs. Yet even with all the techniques at their disposal, the SOC cannot always be 100% certain they have the complete picture of what occurred. This is often due to incomplete, inconclusive, or ambiguous data. Given this, the SOC may not always deploy countermeasures at the first sign of an intrusion. There are three reasons for this 1) the SOC wants to be sure that it is not blocking legitimate activity 2) to determine the nature and full scope of the attack or 3) sometimes, forensic evidence must be preserved, collected, and analyzed, and stored in a legally sound manner.

Of course, there are cases where a set of indicators is correct often enough that certain response actions can be automated, leading to use of automated response at the asset and network level, or orchestrated through SOAR tools. Or the SOC may be able to create an analytic or detection to automatically prevent future activity of the specified nature in-line as with an EDR capability or Network-based Intrusion and Prevention System (NIPS). High-maturity SOCs typically build several layers of alert correlation, enrichment, and automation on top of alerts generated from such platforms. However, there is always the chance that a positive indicator will turn out to be incorrect. Given this, the decision to automate responses must include the risk calculation of the increase in the speed of the response vs the risk of taking the wrong action. For more on incident response see "Strategy 5: Prioritize Incident Response."

0.3.2 Understanding the Adversary

In support of the identification, protection, detection, response, and recovery efforts, the SOC folds in information from a variety of external sources that provides insight and context into threats, vulnerabilities, motivations, and interests. This information is usually referred to

---

[3] Other authoritative definitions for an incident such as those from CNSS or FISMA, also exist. The CNSS definition for "incident" can be found here: [3] ; FISMA law definition is here: [519].

as cyber threat intelligence (CTI) and it may include cyber threat intelligence feeds; tactics, techniques, and procedures (TTPs) such as correlated and curated IOCs, malicious domains and internet protocol (IP) addresses with context; incident reports; and adversary campaign and reporting.

Cybersecurity operations practices have matured since the very first CSIRTs in the late 1980s and early 1990s. Since that time, the structure and focus on how the cybersecurity community understands and articulates knowledge of the adversary has increased tremendously. Historically, SOCs focused their efforts on detecting an incident while the adversary is performing reconnaissance, or during direct attack. By contrast, today the SOC must expand its situational awareness far beyond this focus. Indeed, in an ideal circumstance, the SOC will be part of a larger cybersecurity effort that understands the adversary well enough so that they can prevent or mitigate attacks before they occur, or at least detect an incident before significant damage is done.

As the defender, the SOC is in a constant race to maintain parity with the changing environment and threat landscape. Continually feeding timely CTI into SOC monitoring tools is key to keeping up with adversaries. In a given week, the SOC likely will process dozens of pieces of CTI that can drive anything from sensor detection updates to emergency patch pushes. A SOC must discriminate among the data that it harvests; CTI must be actionable, timely, relevant, and accurate about the incident, vulnerability, or threat it describes. In addition, CTI may sometimes be accompanied by actual digital artifacts, such as suspect files or code snippets. CTI may have handling restrictions depending on the originating organization, so proper usage is paramount to ensuring adversaries are not tipped off.

It is important for the SOC to comprehend the adversary and instrument the constituency across the entire cyber-attack life cycle, which is facilitated through the use of CTI. The SOC strives to detect and respond to adversaries, not just when they deliver their attack to a target, but also "left of hack" and "right of hack." Left of hack includes actions the adversary performs prior to trying to get into the environment. These actions help an adversary prepare for an attack and potentially increase their chances of success. Examples include searching for information on a potential victim, developing technical capabilities such as malware, and active scanning of targeted victim environments. These left of hack actions can be more difficult to observe but are critical to a proactive defense. Right of hack includes all the actions an adversary might take after they have a foothold. This may include trying to gain additional access, collecting data and exfiltrating it, or taking actions to create an impact such as destroying information. CTI can help the SOC gain insights both left and right of hack.

Left of hack are the actions the adversary performs prior to trying to get into the environment. Right of hack is all the actions an adversary might take after they have a foothold.

Using knowledge of the entire cyber-attack life cycle the SOC can take a more holistic approach to sensing and analytics. This is often done in the context of a cyber threat framework such as MITRE ATT&CK which assists the defender in understanding potential adversary activity. For instance, sensor instrumentation of constituency networks and hosts should not only provide indications of reconnaissance and exploit activity but should also reveal the presence of remote access tools (RATs), credential theft, and lateral movement. Furthermore, evidence of any one technique may be elusive; by looking at many different adversary TTPs the likelihood of detections is increased substantially.

> *The SOC has the best chance of preventing or catching the adversary by equipping the constituency with capabilities covering the entire cyber-attack life cycle.*

Just as important, it is necessary to for the SOC to understand the adversary at each stage of the lifecycle, because many adversaries and incidents will share the same traits for a single step of the lifecycle.

Some SOCs also allocate resources to look for all the unstructured indicators of incidents in addition to the routine detections and alerting that are processed every day. This is usually referred to as threat hunting–starting with different hypotheses of adversary presence in the constituency and using various analytical techniques to prove or disprove that hypothesis. Indeed, many cases stem from the non-routine indicators and analytics that do not show up as routine detections set in place by the SOC. In larger SOCs, these teams work in concert to find and evaluate the disposition of suspicious or anomalous activity on constituency digital assets. A mature, structured hunt program builds on, and further enhances its CTI, routine alert handling, and response functions. "Strategy 6: Illuminate Adversaries with Cyber Threat Intelligence" is dedicated to the topic of CTI and more can be found about threat hunting in "Strategy 11: Turn Up the Volume by Expanding SOC Functionality."

0.3.3 The Basic SOC Workflow

Figure 3 shows a basic SOC workflow. The terms, concepts, and flows will be discussed in more detail throughout the book but are presented here to provide initial context for further reading. In this figure there are many sources of information coming into the SOC including security-relevant events from constituency assets, information from constituents themselves, and cyber threat intelligence. These inputs are filtered and assessed by both humans and machines with the goal of being able to take a response action or deciding that no action is needed. Throughout the process the SOC will coordinate and consult with many others such as system administrator and service owners to ensure that any response actions taken are done in the context of the business environment the SOC supports.

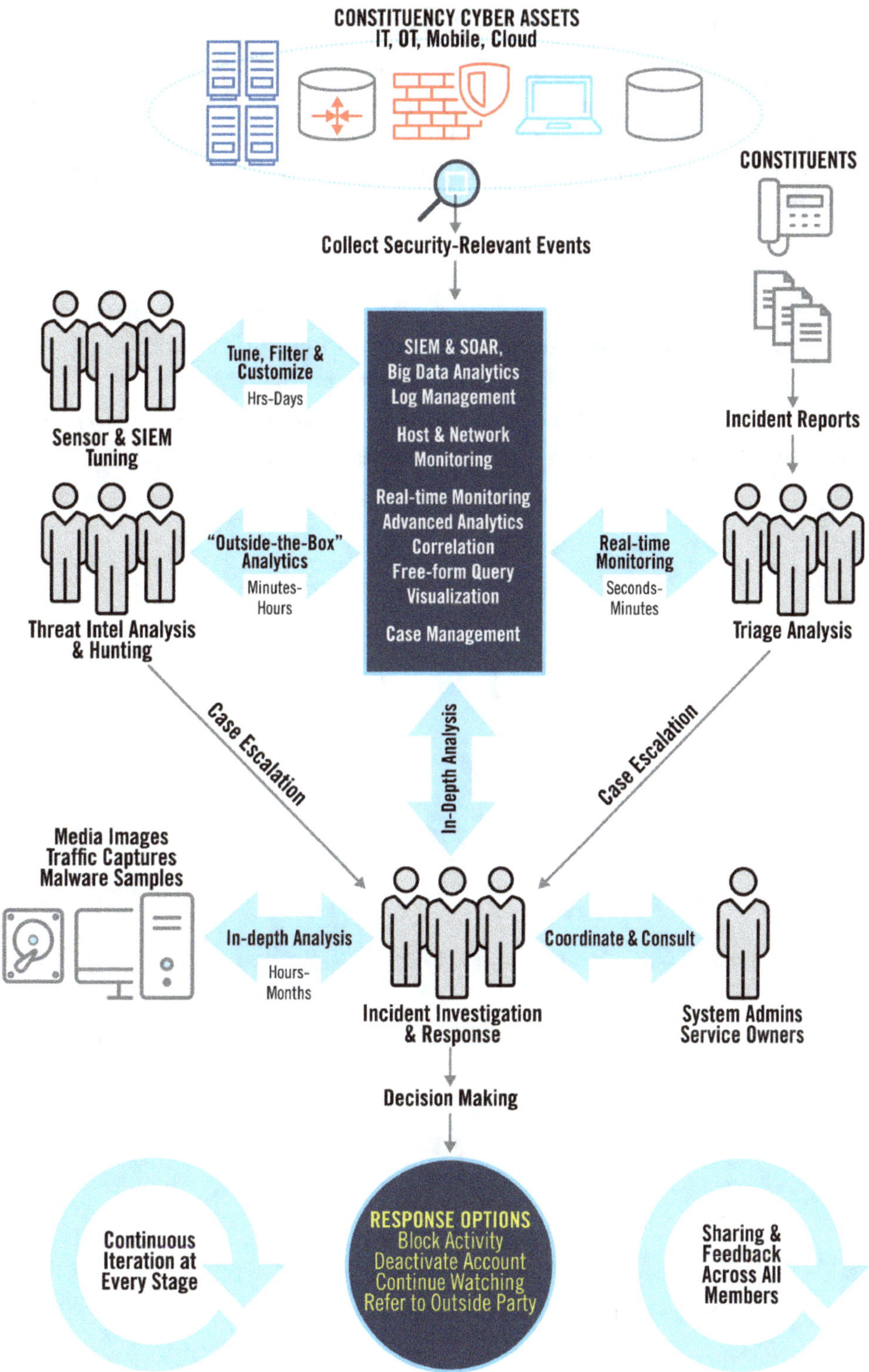

Figure 3. Basic SOC Workflow

0.4 People, Process, and Technology… at Speed

The best SOCs stand out in a number of ways, but a high operations (ops) tempo is one of the most prominent. Specifically, it is the SOC's ability to both comprehend its constituency, and act in timescales relevant to the timescales of the adversary that will set it apart. Both require skill and sophistication in analytic tooling and tradecraft, and the right people and processes to act decisively.

> *The key to effective security operations is having the people, process, and technology to enable the SOC to detect, understand, and respond to the adversary rapidly, both proactively and reactively.*

Throughout this book, the reader is encouraged to routinely reflect on how to best orient their SOC to constantly improve efficiency and effectiveness in various ways.

Strategy 1: Know What You Are Protecting and Why

Very few businesses, government agencies, or private sector organizations have cybersecurity operations as their primary business function. Even for those that do, such as a company that offers security services like a Managed Security Services Provider (MSSP), the protection of their own systems and data is not their core mission. Instead, cybersecurity operations must serve as mission enablers supporting the goals of their constituency. For the SOC, that means having the needed context for the data that it sees and the actions it takes. This is especially true given that most SOCs receive more data than they can possibly act upon and will need to prioritize their decision making. This first strategy therefore discusses the importance of understanding what the SOC is defending and why.

1.1 Situational Awareness

For a SOC to effectively provide a set of capabilities to constituents, it must understand maintain and share situational awareness (SA). "Situational awareness is the perception of the elements of the environment within a volume of time and space, the comprehension of their meaning, and the projection of their status in the near future" [15, pp. 32-64].

The idea of SA grew out of aviation in the latter half of the 20th century [15, p. 49]. Imagine a fighter pilot in their aircraft. For the fighter pilot to effectively defend themselves, their aircraft, and their fellow service members against attack, they must be an expert at comprehending a variety of sensory inputs, synthesizing their meaning in the aggregate, and then acting on that understanding. The SOC must achieve the same outcomes as the pilot, although the cyber realm provides unique challenges.

While a pilot has one aircraft to control and perhaps no more than a few dozen friendlies or foes around them to keep track of, a SOC may have hundreds of sensors, tens of thousands of assets, and hundreds of potential adversaries. Aviators operate in kinetic space, where instruments normally can be trusted and the results of one's actions are usually obvious. In the cyber realm, analysts must frequently cope with ambiguity. The confidence a pilot places in their instruments is high; SOC analysts must drill down to raw data to establish the ground truth of an incident. In fact, sometimes the analyst cannot fully understand exactly what happened, due to incomplete or inconclusive data. Whereas aviation is a topic that has been understood and practiced by over a million people for more than 100 years, cyber defense is still rapidly evolving and has been understood and practiced for far less time.

The general definition of SA can be extended to cyberspace:

For the SOC, gaining and using SA follows the observe, orient, decide, and act loop (OODA Loop). As shown in Figure 4, the OODA Loop is a self-reinforcing situational awareness decision cycle.

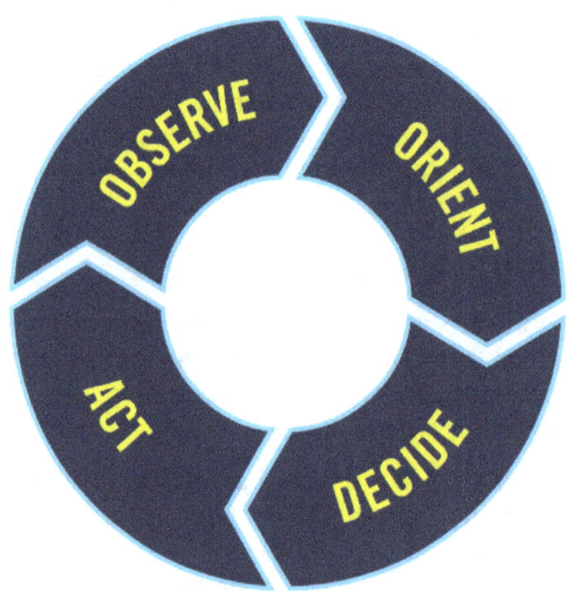

Figure 4. OODA Loop [16], [17]

The analyst is constantly making observations about the constituency, orienting that information with previous information and experience, making decisions based on that synthesis, taking action, and then repeating the process [15]. Over time spans varying from minutes to years, SOC analysts build their familiarity with their constituency and relevant cyber threats. As that SA is enhanced, they become more effective operators. Whether the analyst realizes it or not, they follow the OODA loop while carrying out various elements of the SOC mission, from the tactical to the strategic: when performing alert analysis, during hunting, while assessing the impact of a vulnerability incident, and while iteratively executing growth in the SOC's tooling and data collection activities. The timescales at which the OODA loop are executed can vary from minutes to months.

1.2   SOC Operating Context

This book organizes Cyber SA into five related, deeply coupled, and equally important areas:

- **Business/mission:** This area is focused on understanding a constituency's reason for being and how it operates. This includes the products and/or services offered, primary customers, the geographical location(s) in which the constituency operates, and relationships with other external parties such as suppliers or distributors.
- **Legal and regulatory environment:** This area includes government laws and industry regulations that are pertinent to cybersecurity operations such as reporting requirements or privacy regulations.
- **Technical and data environment:** This area includes understanding the number, type, location, and network connectivity of IT and OT assets along with the status of those assets (e.g., patch status, vulnerability status, or up/down status). This also includes knowing the constituency's critical systems and data, the connection and value of that data to the business, and the location of that data (on-prem systems, cloud, partner IT, etc.). This can include not only the business's intellectual property but also data necessary to its processing, such as billing transactions or OT control system status.
- **Users, user behaviors, and service interactions:** This area includes understanding typical patterns of behavior, including user to service and service to service interactions. The focus is on understanding normal behavior and then looking for deviations from that baseline.
- **Threat:** This includes understanding the various types of threats (hacktivists, criminal, nation state, etc.) likely to be of particular concern to the constituency, how they operate, and how that should affect the constituency's defensive posture.

Understanding these areas will help the SOC to better understand the consequences of a security breach and therefore improve the SOCs ability to prioritize actions. It will also enhance a SOC's ability to support its constituents effectively and efficiently. Figure 5 shows these five areas in context of each other.

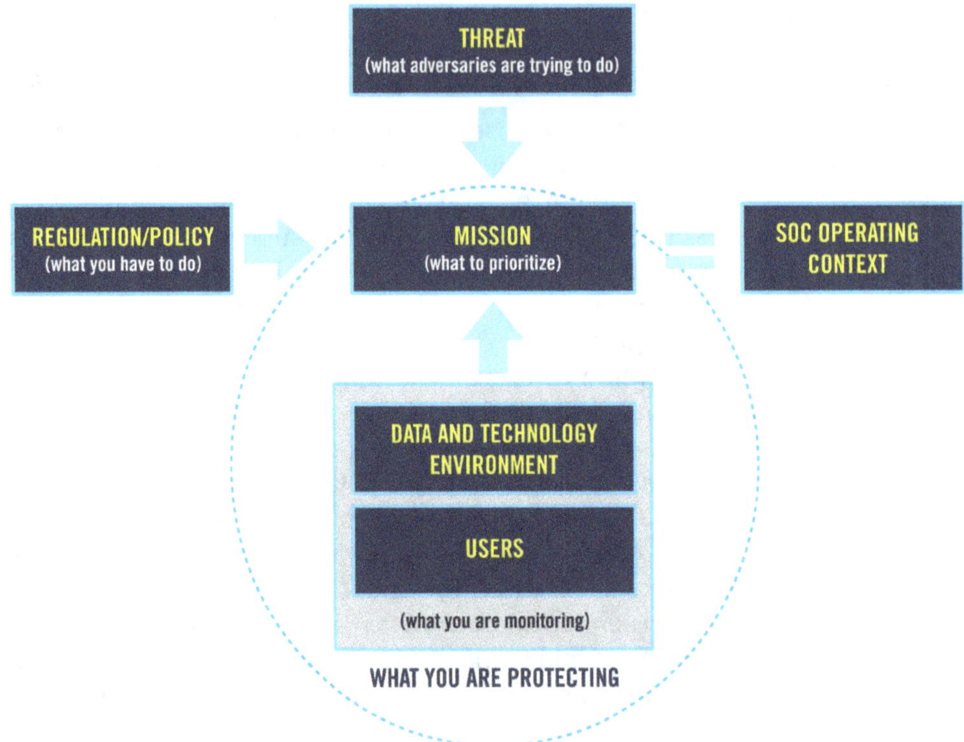

Figure 5. SOC Operating Context

Building on an understanding of these five areas, SA takes on different forms, depending on the level at which cybersecurity decisions will be made. At the lowest tactical level, the cyber analyst sees out to the end asset and enclave, with a direct understanding of hosts and users. Above that, at the operational level, are lines of business and large networks. At the top is the strategic level, where long-term campaigns are waged by the adversary and where entire enterprises exist. As a result, the need to understand the constituency and actors varies widely, depending on what level of SA is appropriate—from the junior SOC analyst to the CIO and beyond. Additionally, cyber SA can be presented in a variety of ways with different levels of analytical depth. Strategies 8 and 9 discuss the data and tools that form much of the analyst's situational awareness.

The constituency, especially its executives, naturally looks to the SOC to answer questions such as, "What cyber threats is the constituency currently facing?" A SOC can level-set its priorities by providing details to constituents and other SOCs, during both normal operations and a critical incident, and soliciting their feedback in turn. Without proactively providing enough detail, the SOC either will be marginalized or will constantly field ad hoc data calls. If the SOC provides too much detail, its resources will be overcommitted to answering questions from constituency management and thus unable to adequately spot and analyze intrusion activity. Many SOCs participate in and curate strong metrics and reporting to their constituents, usually as a member of the large cybersecurity apparatus.

Effective reporting will also spur partner organizations—especially other cybersecurity stakeholders—to provide feedback, reinforcing and growing the SOC's SA.

Having mature cyber SA allows the cyber defender to answer questions like:

- What is our security posture?
- What are the consequences of a successful attack, including the consequences for upstream or downstream services?
- From which adversaries is the constituency facing imminent threat of attack?
- How are they attacking us and what are they after?
- What is our best course of action in response to these attacks?
- What is the patch status of the constituency? Which patches need to be prioritized?
- To which systems should I apply a given set of security controls, thereby rendering the best mitigation?
- What is changing about the threats faced by the constituency? How are the adversaries' TTPs changing, and what do I have or need to detect and defend against those new threats?
- Who is acting outside their typical lines of behavior, and is this cause for concern?

To develop a picture of all these systems, the SOC partners with other parts of the constituency to draw upon existing materials or collaboratively build needed insights. Additionally, saying "the SOC understands X," does not mean everyone within the SOC has every piece of context. Rather, it means that the context is institutionalized in such a way that it becomes part of the SOC process. This could range from SOC leadership understanding constituency leadership priorities, and using that to guide decision making, to having asset inventories that are automatically matched with incoming alerts.

*The SOC should be able to put any cyber event it observes into
constituency context so that it can effectively prioritize its actions.*

However, as vital as situational awareness is, SOCs have long struggled to achieve this understanding and context and to encode it in a way that is both durable and not siloed to specific individuals in the SOC. The following sections discuss the importance of the area, along with ideas about the types of information the SOC might want to collect or develop, and how to utilize that knowledge in SOC operations. Some of these areas are also covered in more detail in later strategies. Most importantly, each subsection provides concrete actionable advice on how to make progress in that area of situational awareness.

These areas are not independent of one another and gaining understanding of one piece of information will inform the SOC's understanding in multiple areas. For example, understanding the geographic location in which the constituency operates will inform the legal regulations that apply, affect the IT environment, and speak to the users and user behaviors that can be expected. As will be discussed at the end of this section, this is not something the SOC is likely to have available on day one, it is a continual process that the SOC can build upon over time.

1.3 Understand the Constituency's Mission

Understanding the constituency's mission is at the heart of a SOC optimally performing its functions. Understanding the constituency mission means knowing what functions need high confidentiality, integrity, or availability, even during a cyber attack. Understanding the mission means knowing if the SOC is primarily trying to protect against the theft of intellectual property, support secure financial transactions, enable commerce, or something else. Understanding the mission means knowing what keeps the senior leaders up at night worrying so that the SOC can pass along right information to allow its leaders to make informed decisions.

For example, for some constituencies maximum confidentiality is the most critical. This might include a high-tech research or a legal office. Other constituencies will be focused primarily on maximum integrity. An example would be the stock market where accuracy is by far the most important aspect of their data. Other constituencies will prioritize maximum availability and care about keeping their infrastructure running above all else. This might include critical infrastructure organizations such as power or water companies. A utility that runs control systems for electricity will prioritize SOC activities based on prioritizing safety and delivery of electricity to customers, for example.

Of course, most constituencies are more balanced in their priorities but, when possible, it helps both the SOC and the constituency to understand which aspect of confidentiality, integrity, and availability is most critical. It can be a challenge to translate strategic business plans into practical context the SOC can use to execute its mission. To address this challenge the SOC should focus on identifying information which supports prioritizing monitoring and response decisions and informs what 'normal' looks like across the environment. Mission context that the SOC will want to be aware of includes:

- The lines of business and mission the constituency engages in, including their subordinate services and functions, and corresponding value expressed in revenue, expenditures, or lives
- Geographic/physical location where different parts of the mission and business occur
- How constituency mission, lines of business, and services map to constituency digital assets, services, enclaves, data, and the dependencies among them
- The business relationship between the constituency and external parties such as other businesses, government entities, educational institutions, and not-for-profit organizations

To identify this information the SOC should directly engage with the constituency's leadership and other stakeholders on a routine basis. If possible, a capability should be put in place that allows parts of the constituency to register their services and associate their assets with a particular part of the business. An advanced approach to this is known as mission mapping or mission dependency modeling.

A SOC with mature and sustained understanding of the constituency mission will most likely leverage some of the following techniques and practices, in partnership with the larger cybersecurity apparatus:

- Recruit members to the SOC who have prior hands-on expertise in constituency mission and business systems and operations
- Routinely engage with security champions in other organizations, and potentially include them as associate members of the SOC, as detailed in Section 3.7.4
- Routinely engage with business executives and system owners, providing both a window into the latest efforts by the SOC to support them, and insight into how their business and mission are evolving
- Facilitate routine tabletop and hands-on technical exercises with major system owners, as detailed in Section 11.2
- Support and leverage a service and asset inventory system, as discussed in Section 1.5.1
- Engage and leverage with the larger security apparatus around service risk and criticality scoring
- Perform routine internal SOC briefings on key mission areas, either by asking SOC team members to research the area and/or invite mission partners into brief
- Record mission and service information as part of routine service onboarding to SOC services.
- Curate links to other mission groups' information, such through a service registry

1.4 Understand the Legal, Regulatory, and Compliance Environment

Just as the SOC operates within the context of a constituency, it also operates within a legal and regulatory environment. Much of this context will be specific to the country or countries in which the SOC operates and will be driven by the type of legal system of the country and the types of laws in place. Additionally, the industry that the constituency operates in may add on regulations that the SOC has to abide by. Things that the SOC will want to be aware of include:

- Type of legal system the SOC operates under (e.g., civil or common law)
- Types of laws the SOC is subject to upholding (e.g., criminal, intellectual property, or data privacy)
- Role of the SOC in legal and regulatory compliance
- Industry standards (e.g., incident reporting requirements or data standards for healthcare providers)

Many regulations or standards only apply to certain countries or industries. Some of these will impact how the SOC helps protect the systems and data of the constituency and some will define how the SOC executes in own mission through actions such as reporting requirements. The SOC is not expected to be an expert in this area, rather they should build their operations with the support of legal counsel to ensure compliance.

- Common regulations that may impact the SOC in the United States, particularly those supporting the federal government:
  - 1996 Health Insurance Portability and Accountability Act (HIPPA) [18]
  - 1999 Gramm-Leach-Bliley Act [19]
  - 2002 Homeland Security Act which includes the Federal Information Security Management Act (FISMA) [20]
- Common standards and regulations that may impact the SOC in the European Union:
  - 2016 Directive on Security of Network and Information Systems (the NIS Directive) [21]
  - European Union General Data Protection Regulation (GDPR) [22]
- Global standards:
  - Payment Card Industry Data Security Standards (PCI DSS) [23]
  - International Organization for Standardization (ISO) [24]

A SOC with mature and sustained understanding of the legal, regulatory and compliance environment will probably leverage some of the following techniques and practices, in partnership with the larger cybersecurity apparatus:

- Designated external points of contact (POCs) for each regulation of particular interest, such as audit coordinators and legal counsel, that serve as focal points
- References to regulations, and if possible, specific Memorandum of Agreement (MOA)/Memorandum of Understanding (MOU) that carve out SOC's responsibilities in the context of each regulation or standard
- Guidelines for specific responsibilities around forensic data integrity and handling
- Internal POCs responsible for support to routine (annual) audits, including evidence preparation and representation

1.5 Understand the Technical Environment, Especially Critical Systems and Data

When it comes to understanding what the SOC is protecting, systems and data are often the first thing that come to mind. A 2019 Ponemon Institute study identified that "The top barrier to SOC success, according to 65 percent of respondents, is the lack of visibility into the IT security infrastructure and the top reason for SOC ineffectiveness, according to 69 percent, is lack of visibility into network traffic" [25]. Ideally the SOC will have access to robust information about all constituency systems and data. However, this may serve as a daunting, seemingly unachievable goal to many cybersecurity professionals. Therefore, the importance of understanding this information is truly driven by constituency priorities and not all information is equal. Ideally, the SOC will want insight into:

- Location of constituency digital assets:
  - The geographic footprint of the IT and OT environment
  - Number, type, location, and network connectivity of IT and OT assets, including laptops, servers, network devices, mobile devices, and internet of things (IoT) devices

- ○ Network topology, including physical and logical connectivity, boundaries that separate differing zones of trust, and external connections
  - ○ Asset, network, and application architecture (including authentication, access control, and audit)
  - ○ Where data is stored relative to system assets such as in a closed network, in the cloud, or on mobile devices
- The relative importance of constituency digital assets:
  - ○ The most important types of data and where are they located and processed
  - ○ What systems perform essential functions for the constituency and what data requires high confidentiality, integrity, and/or availability
- The state of constituency digital assets:
  - ○ What normal state looks like across major network segments and hosts
  - ○ Changes in that state, such as changes in configuration, host behavior, ports and protocols, and traffic volume
  - ○ The vulnerability of hosts and applications, and countermeasures that mitigate those vulnerabilities

The SOC should start by understanding the big picture of how the constituency operates. For example, is it an academic institution with students bringing in new outside equipment each year? Is it a business with a significant number of interconnected supplier networks? Does the company have a large number of remote users? Are mobile devices owned and operated by the constituency or is there a bring your own device practice in place? All these approaches will define how the SOC needs to think about the IT environment.

Next, the SOC should work to gain insights into the relative importance of various systems and data. If a single laptop gets compromised, and it does not process or contain any sensitive data, that would be a much lower priority than a system that contains trade secrets or financial information. Finally, the SOC needs to identify the types of information it needs about the systems and data. The SOC can gain insight into the environment by engaging with IT, OT, and network operations routinely such as through regular ops syncs, participation and routine change management boards, and potentially physical co-location with NOC, IT ops center, and OT ops center.

A SOC with mature and sustained understanding of the technical environment and constituency data will probably leverage some of the following techniques and practices, in partnership with the larger cybersecurity apparatus:

- Direct support to and/or engagement in a comprehensive IT/OT composite inventory detailed in Section 1.5.1:
  - ○ Curate and/or leverage a knowledge repository of network maps and major system diagrams
  - ○ Collect and leverage vulnerability scanning data
  - ○ Curate and/or leverage a set of points of contact for major systems and services, potentially as part of composite inventory
- Have access to major constituency IT change management tracking systems and trouble ticketing

- Maintain access to or collect security-relevant telemetry, including application and database level telemetry for major applications and data stores
- Network sensors, host sensors, big data analytics, log management, and SIEM, as detailed in "Strategy 7: Select and Collect the Right Data" and "Strategy 8: Leverage Tools to Support Analyst Workflow"

1.5.1 Composite Inventory

The SOC's mission, and that of the larger cybersecurity apparatus, can be greatly enabled by an accurate, comprehensive, and current accounting of the constituency's cyber assets: on-prem, cloud, mobile, and so forth. This is essential to any sort of activity where the security org wishes to drive cybersecurity hygiene and compliance in a consistent and scalable manner. Asset inventory serves as the denominator for which any and all hygiene metrics are calculated.

Most of the time, cybersecurity organizations, including SOCs, are not responsible for asset management beyond the assets itself owns (e.g., the SOC's own monitoring and scanning tools). However, feeding asset data into the SOC is a wise move to provide a complete picture. This is especially true when the cybersecurity organization and SOC serve distributed or disjoint IT services and lines of business.

No single source of asset inventory is likely an authoritative and complete source for all the constituency cyber landscape. Instead, there must be a composite or synthesis of disparate asset data. The larger and more complex the constituency, the more sources of asset data are likely applicable.

To complicate matters, tracking ordinary hosts alone is no longer sufficient. Asset inventory must also encompass digital assets that may not be tied to a host, such as cloud Platform as a Service (PaaS) & Software as a Service (SaaS) resources: databases, cloud storage, message buses, secrets vaults, serverless compute, and so forth. Finally, asset management must compensate for ephemeral cloud compute such as elastic workloads, endpoint systems that are not always on, and IoT devices residing in disconnected or low bandwidth environments.

Thoughtful planning is needed for computing the number of assets in a cloud environment. For example, consider a SOC with a constituency of 10,000 on-prem hosts, 5,000 IaaS cloud resources, and a varying number of elastic compute nodes. On average, the pool of elastic compute resources is running about 2,000 nodes, with each node living an average of 24 hours. Therefore, the correct and appropriate way to size this SOC's constituency is 17,000 assets, on average. However, without the correct controls, the SOC might claim it is responsible for 75,000 assets or perhaps even more if its asset tracking system keeps a month of data. Why? Each cloud elastic compute node lives for 24 hours; 2000 unique nodes * 30 days in a month (roughly) = 60,000 unique systems seen, if measured over the course of an entire month. Different cloud providers and virtualization platforms have different means of tracking unique hosts; the data processing framework that aggregates asset data must respect and compensate for this in a deterministic manner.

In a large enterprise, the SOC is likely to leverage several sources of asset data:

- Machine and user directory services such as Lightweight Directory Access Protocol (LDAP) and Windows Active Directory
- Dynamic Host Configuration Protocol (DHCP) logs and lease databases
- Inventory databases owned and operated by parties outside the SOC, ranging from simple spreadsheets of on-prem hardware to commercial and proprietary asset inventory, lifecycle, and operations management systems
- Cloud resource/asset inventory
- Mobile device management (MDM) and EDR, including features that passively detect the presence of unmanaged devices on adjacent networks
- Network scanners and mapping
- Vulnerability scanners
- Security correlation and analytic platforms that automatically generate asset lists based on hosts "seen" through event feeds consumed by the tool, in particular: SIEM, user entity behavior analytics (UEBA) tools, and some firewalls
- System management, patch management and software distribution such as Microsoft System Center Configuration Manager (SCCM), HCL BigFix, or Ivanti Unified Endpoint Manager

If the cybersecurity organization believes its digital landscape can only be understood by synthesizing multiple sources of asset data, such as the above, it may consider a solution with the following qualities:

- Features automated data processing with measurable end-to-end latency that minimizes latency between when data is updated from source systems, to when it is viewable by the user
- Allows administrators to deconflict, reconcile, and merge differences in overlapping data
- Can cope with overlap in IP addresses and hostnames, such as due to RFC1918 private address space and reprovisioned, rebuilt, and recycled assets
- Support connecting with risk management, cybersecurity compliance, risk scoring and mission mapping systems as applicable
- Includes fail-safes such that when an asset data sources break, the system reverts to the last known good data
- Retains asset history metadata over time

Aggregating and evolving a picture of digital assets can be one of the most tedious of endeavors for the SOC or the larger cybersecurity organization. In fact, the resourcing and complexity for this may rival that of some of the most complex SIEM and analytic architectures described in "Strategy 8: Leverage Tools to Support Analyst Workflow." However, it serves as the foundation for so much of what the SOC does, and for those who do it well, it serves as the foundation of almost every cybersecurity function performed [26], [27], [28], [29], [30], [31].

1.6 Understand the Users, User Behaviors, and Service Interactions

The IT environment and user behaviors are very closely intertwined and defining "normal" needs to incorporate both pieces. As with the IT environment the SOC should both understand the general context in which users within the constituency work as well as having more detailed knowledge of specific actions users might take. For example, is the constituency a research center with visiting scientists from all over the world needing access? Or are they a small company where any access from outside a local network would be suspicious? Things that the SOC will want to be aware of:

- The meaning of activity on constituency networks and hosts in the context of the mission
- The role, importance, and public profile of major user groups, such as:
  - System administrators
  - Executives and their administrative staff
  - Those with access to sensitive information (intellectual property, finance)
  - General constituency user population
  - Users external to the constituency
- Baseline metrics for how various systems and data, particularly critical ones, are accessed by users over time
- App authorization and delegation, both for user to service and service to service interactions
- Inter-organizational and Inter-business zones of trust and trust dependencies

Developing a baseline of user behavior can be done through user entity behavior analytics (UEBA). This can be as straightforward as looking at access logs for suspicious connections to utilizing advance machine learning algorithms to identify unexpected activity. Common areas to focus on include tracking administer access and behaviors, monitoring file shares, and monitoring access and use of critical systems. Additionally, UEBA can be expanded to include devices which is known as user entity behavior analytics. UEBA tracks not only user activity but communications between and among servers, routers, endpoints, and IoT devices. As mentioned before, each of the five operating context areas is closely intertwined with the others and concept of UEBA brings together the awareness of the technology and the users. For more on UEBA see Section 8.3.

A SOC with mature and sustained understanding of users, user behaviors and service/machine behaviors will probably leverage the following techniques and practices in partnership with the larger cybersecurity apparatus:

- Support and leverage service and asset inventories, as stated above
- Maintain access to application and user authorization directories, in particular those supported by federated app registration and delegated access providers
- Have points of contact inside the SOC who are conversant in the Identity, Credential, and Access Management (ICAM) and zero trust protocols and platforms being used, such as Open Authorization (OAUTH), Web Service Security (WSS), Security

Assertion Markup Language (SAML), JSON Web Token (JWT), Azure Active Directory (Azure AD), Okta, or Active Directory Federation Services (ADFS)

- Curate or have access to knowledge repositories that depict design and architecture of ICAM solutions in use, such as AD forests and trusts
- Maintain access to, instrument, collect, and/or curate ICAM login/logoff, user access, directory object modification from identity planes in use
- Operate and maintain a UEBA capability proportionate to the SOC's own sophistication and resourcing, as mentioned above

1.7 Understand the Threat

Understanding the SOCs operating context would not be complete without understanding the threat to the constituency. If there was not a threat the SOC would not need to exist! "Strategy 6: Illuminate Adversaries with Cyber Threat Intelligence" is dedicated to discussing cyber threat intelligence and how it can be used to inform the SOC. This section simply notes that understanding the constituency's mission can provide context for understanding what types of adversaries are most likely to be higher risk. To help contextualize the threats to the constituency the SOC will want to be aware of:

- **What about the constituency might be of interest to an adversary:** It could simply be unsecured systems that allow an adversary to gain a foothold they can use to further their activities or that allow an adversary to hold a constituency hostage such as with ransomware. Or the adversary might be interested in something very targeted such as specific intellectual property or financial information.
- **What are the types or groups of adversaries likely to target the constituency:** Understanding if the constituency is more likely to be targeted by random hackers, hacktivists, criminals, or APTs can help the SOC bring in the right cyber threat intelligence to inform cyber defense practices.
- **What historical cyber incidents have happened within the constituency:** Understanding both the type of systems and data targeted and the impact and consequences of previous incidents can help the SOC prioritize efforts.

1.8 Building Awareness over Time

Understanding everything in context is a lofty, and incredibly challenging, goal. Developing this picture will take time and will require constant updating as the situation changes. Ideally the SOC can map mission to systems and data accessed by users, but that is hard for even the most advanced organizations. So how to get started?

- Build over time, do not try to gather everything on the first pass.
- Start small—talk to users. Reach out to mission experts to learn what they do and what is important to them. Talk to constituency leadership to get their inputs on what to prioritize.
- Gather the laws and regulations that apply to SOC operations.

- Assess the IT environment for which the SOC is responsible.
- Partner with other parts of the constituency to bring this information together. Senior management, legal, and IT operations will all have an important role to play in helping the SOC gain awareness.
- Identify high priority systems and data and begin with base lining everything about those systems and data. Understand these systems and data in the context of the mission. Look to "Strategy 7: Select and Collect the Right Data" to learn more.
- Use tools to store information and make it available for context. Reference "Strategy 8: Leverage Tools to Support Analyst Workflow" for tool ideas.
- Continually validate with leadership that the SOC understands the constituency's priorities.
- Build standard operation procedures (SOPs) and TTPs for SOC operations around and understanding of the constituency's priorities and run exercises to make sure the data collection, alerting, incident coordination, and reporting all support those priorities.

1.9 Summary – Strategy 1: Know What You Are Protecting and Why

1.1 The iterative process of observing, orienting, deciding, and acting in cyberspace is at the core of the SOC's operations, at both the tactical and strategic level.

1.2 The SOC must frame what it knows and sees in the operating context for its constituency. That situational awareness and operating context will enable it to answer questions ranging from "what is this adversary trying to achieve by attacking us" to "what are the greatest areas of security concern across the constituency." SOC operating context is organized into five pillars.

1.3 The first is the constituency mission, what the SOC is there to protect, and the connection to the underlying assets, mission, and data.

1.4 The second is the legal and regulatory environment, including things it is compelled to do or protect due to external requirements.

1.5 The third is the technical environment, especially the constituency's critical systems and data. Forming a strong understanding of the technical environment may compel the SOC to form a composite picture of the overall constituency digital asset landscape by partnering with other stakeholders and synthesizing a picture from disparate sources.

1.6 The fourth are the users, their behaviors and service interaction across the constituency.

1.7 The fifth and final are the threat, both internal and external.

1.8 The SOC does not have to develop awareness in all the operational context areas on its own, or all at once. Senior management, legal, and IT operations will all have an important role to play in helping the SOC gain awareness. The SOC should prioritize gaining insights on mission critical systems and data and then iterate over time to add to their knowledge.

Strategy 2: Give the SOC the Authority to Do Its Job

The SOC must execute its mission against constituency digital assets that almost always belong to someone else. Even though the SOC is usually a member of the constituency it serves, primary ownership and operation of hosts and networks is vested in another member of the constituency that the SOC must interact with. As a result, the SOC's ability to assert proactive and reactive authority must either be codified through written authority or inherited through the SOC's parent organization. This second strategy addresses both issues: (1) written authorities the SOC needs and (2) how organizational alignment supports, or does not support, the SOCs mission.

2.1 Written Authorities

Written guidance that grants a SOC the authority to exist, procure resources, and enact change is an important component of building and operating of a SOC. In addition, there are hosts of supporting constituency IT and cybersecurity policies which further enable a SOC to execute its mission. SOC related policies and authorities should tie into overall constituency policies and authorities. The SOC may wish to consult with their constituency's General Counsel in framing relevant authorities, particularly those related to its charter and digital artifact handling. When crafting policy, the SOC may want to leverage the free policy templates available on the SANS website [32].

2.1.1 Charter Authority

While the SOC will develop working r elationships with various cybersecurity stakeholders, it also must be able to point to a formal document if there are any questions about its mission and responsibilities. The most common way to do this is through the development of a charter. In addition to mission and responsibilities, the charter may also include the services expected of the SOC. In this case the charter should not only describe what the SOC should be doing in the future, but also what it is currently capable of doing. Even in cases where a SOC is not expected to fulfill the entire scope of its charter from the beginning, having the future state well defined helps the SOC grow into such a role. It is important also to recognize that the charter does not describe *how* a SOC fulfills its mission, only *what* it does and who has supporting responsibilities.

A charter also helps eliminate misconceptions about what the SOC is and what it must do. Conversely, SOCs that lack such written authority often spend more time in conflict with other organizations, and/or a lot more energy begging for help—time and energy not spent on making a positive impact.

Every organization has a different approach to writing IT/cybersecurity policy. With that in mind, this strategy discusses SOC related policy and authorities, or what will enable a SOC to execute its mission, without overly focusing on how the SOC will do it. How these elements are allocated among a charter and other policy documents may vary. The main distinction is that the core scope of the mission should always get the signature of the constituency's chief executive. Other items may be codified elsewhere and, therefore, updated with greater frequency.

2.1.2 Elements of a SOC Charter

The following elements should be codified in the charter of a SOC that is the sole cyber operations provider for a given constituency or that sits at the lowest tier in a multitiered arrangement. As appropriate, the SOC should modify what is included in the charter to match their own environment.

- The SOC's function as the operational center and head of cyber intrusion monitoring, defense, and incident response for the constituency
- The scope of the SOC's responsibilities including the organizations, data, systems, and users the SOC is responsible for defending, preferably in a way that is unambiguous and easy to interpret and thus relatively future-proofed against specific new technology or asset types
- Within its constituency, the SOC's authorities to:
  - Deploy, operate, and maintain active and passive monitoring capabilities, both on the network and on end hosts
  - Proactively and reactively scan hosts and networks for network mapping, security configuration, and vulnerability/patch status
  - Coordinate or directly apply active or passive countermeasures, including but not limited to disabling or suspending network connections, hosts, user accounts, and networks
  - Respond directly to confirmed incidents, in direct communication and cooperation with appropriate parties
  - Gather, retain, and analyze digital artifacts (including media, logs, and network traffic) in order to facilitate incident analysis on both an ad hoc and sustained basis (complying with applicable laws, regulations, policies, or statutes)

- The support expected from help desk staff, security officers, and NOC, OT, and IT operations staff when reporting, diagnosing, analyzing, or responding to misconfiguration issues, outages, incidents, or other problems that the SOC needs external support to resolve
- The SOC's role in architecting, acquiring, engineering, integrating, operating, and maintaining monitoring systems and the SOC functions or enclave
- The SOC's level of control over funding for tool engineering, maintenance, staffing, and operational costs related to SOC functions
- The SOC's responsibilities for supporting any other capabilities it intends to offer, such as security awareness building, cybersecurity education/training, or audit collection

Some readers might question why a charter authority includes a statement regarding the SOC's authority to collect digital artifacts or other specific details. A simple one or two sentence statement in the charter affirming the SOC's role will save countless hours and days of discussion down the line when service owners or users express reluctance at providing logs or other digital artifacts, either during an incident or during ordinary monitoring rollout. This type of statement will then be augmented by lower-level guidance, usually measured in several pages, which will say a lot more about "how" that data is collected, exactly what data is required, and so forth.

2.1.3 Central Coordination SOC

If the SOC follows a tiered model, the central coordination in a SOC will likely need the following authorities in addition to those of the lower tier SOC:

- Serve as an entity with higher authorities to subordinate SOCs
- Have access to security-relevant data from all subordinate SOCs which may come in the form of aggregate metrics, summarized data, a view into the data, or situation specific content rather than all raw logs
- Coordinate response actions among subordinate SOCs
- Direct improvements to subordinate SOC capabilities and operations, in accordance with fulfilling incident response requirements across the greater constituency
- Manage devices that aggregate security-relevant data from subordinate SOCs and sensors directly placed on hosts and networks, especially when the subordinate SOCs do not have the engineering skills to do this on their own
- Act as the focal point for constituency-wide security information sharing and SA through common practices, SOC–provided and developed tools, and preferred technologies or standards
- Propose standards such as constituency-wide preferred standard network and security monitoring technologies and practices
- Negotiate constituency-wide licensing/pricing agreements of monitoring technologies that may benefit subordinate SOCs, where possible

2.1.4 Other Enabling Policies and Agreements

Apart from policies that directly enable a SOC to function, there are other IT and cybersecurity policies that enable effective security operations. In coordination with the IT organization the SOC should consider influencing or providing input to these policies or seeing that they are created, if they do not already exist:

- **User consent to monitoring:** Giving the SOC and auditors the unambiguous ability to monitor and retain any and all activity on all systems and networks in the constituency
- **Acceptable use policy:** IT system usage rules of behavior, including restrictions on Internet and social media website use, authorized software on constituency systems, and any travel or remote use restrictions
- **Privacy and sensitive data handling policies:** Instructions for managing and protecting the types of information flowing across the monitored network, including personal, health, financial, and national security information
- **Internally permitted ports and protocols:** Enumeration of the ports and protocols allowed within the constituency, across the core, and through enclave boundaries
- **Externally permitted ports and protocols:** Enumeration of ports and protocols allowed by devices through external boundaries such as through a demilitarized zone (DMZ), to business partners and to the Internet
- **Host naming conventions:** Describing conventions for naming and understanding the basic type and role of host compute assets based on their Domain Name System (DNS) record
- **Other IT configuration and compliance policy:** Everything from password complexity to how systems should be hardened and configured
- **Bring your own device and mobile policies (if applicable):** Rules that govern how employees may interface with constituency networks, applications, and data with personally owned IT equipment and mobile devices
- **Approved OSes, applications, and system images:** The general approved list of OSes, applications, and system baselines for hosts of each type—desktops, laptops, servers, routers/switches, and appliances
- **Authorized third-party scanning:** Rules for notifying the SOC when another organization wishes to perform scanning activity such as for vulnerabilities or network discovery
- **Audit policy:** High-level description of the event types that must be captured on which system types, how long the data must be retained, who is responsible for reviewing the data, and who is responsible for collecting and retaining the data—with recognition of the performance impact value of the data gathered
- **Roles and responsibilities of other organizations:** Particularly with respect to incident response
- **Legal policies:** Including those concerning classifications of information, privacy, information retention, evidence admissibility, and testifying during investigations and prosecutions of incidents

In addition to the polices mentioned above, service provider agreements and service level agreements should exist. For capabilities hosted outside the organizational boundary, such as by cloud providers, service provider agreements should detail what and how information needs to be shared in the case of an incident. This is a very good opportunity for a representative of the SOC (or security generally) to be involved in developing contracts with external IT providers, and in particular, cloud providers. The SOC's requirements for eDiscovery, breach notification, digital artifact recovery, monitoring, and response should be represented in these agreements. Service level agreements (SLAs) should be created for both services the SOC receives and services the SOC provides. Elements of these agreements could include:

- Network capacity and availability requirements
- Contingency planning if contracted network services fail
- Network outage (incident) alerts and restoration and escalation/reporting times
- Security incident alerts and remediation procedures and escalation/reporting times
- Clear understanding of each party's responsibilities for implementing, operating, and maintaining the security controls or mechanisms that must be applied to the network services being purchased

2.2 Organizational Alignment

The SOC draws its authorities, budget, and mission focus from the constituency to which it belongs. Therefore, the decision about where to align the SOC in the organization chart should be deliberate.

The SOC's success is keenly influenced by the following factors:

- **Organizational depth:** How far down in the organization chart is the SOC is placed? Does it report directly to the CIO / CISO or is it located further down in the organizational structure? If the latter, what policy and process can be put in place to mitigate the lack of visibility and authority that the SOC is likely to experience?
- **What authorities the parent organization has:** This includes authorities on paper and in practice.
- **The power and influence wielded by parent organization executives:** How are they attuned to the cyber operations mission and are they likely to support SOC staff in their execution of the day-to-day security operations mission?
- **Established funding lines and budget of the parent organization:** Are they able to fund tools for comprehensive monitoring and people who can staff all the capabilities implied by the SOC's charter?
- **What capabilities the SOC will offer:** See Section 0.2 for a list of potential SOC functions.
- **What organizational model the SOC features:** If the SOC is tiered, can the subordinate SOC live within business units while the coordinating SOC sits under the CIO? Or, if the SOC is a centralized SOC can it sit near the NOC or IT operations and preside over the entire constituency?

The choice of organizational placement is intertwined with another, perhaps more interesting question: "Who is in control of defending the enterprise?" There are often multiple executives who feel they are the protectors of the mission. Under which executive would the SOC flourish, and how far down the command chain should it sit?

In order to direct defense of an entire organization, the SOC or its parent organization needs two things: (1) SA over the constituency, down to specific incidents and the systems and mission they impact, and (2) the authority and capability to direct changes to IT systems proactively or in response to an incident—such as changing domain policies, routers, or firewalls or pulling systems off the network.

These two needs can be at odds with each other, creating tension in where the SOC should sit and to whom it should report. Several executives—chief executive officer (CEO), chief operating officer (COO), chief security officer (CSO), CIO, CISO, chief technology officer (CTO), and their subordinates— have at least some sense of ownership over cybersecurity and have a legitimate need for the SA a SOC can provide. When there is a serious incident, it is likely that many of these parties will want to be informed. It is important to minimize the likelihood that multiple parties will assert (potentially conflicting) roles in directing, implementing, or approving response to an incident.

During a critical cyber incident, the roles and decision-making authorities of an organization's senior leaders should be as clear as possible.

Clarifying who truly has decision authority to act through policy signed by the chief executive of the constituency is critical. This is true regardless of the authorities delegated to the SOC. Each one of these executives serves as a candidate for the management a SOC will serve under and, thus, its organizational placement. This can occur intentionally through charter, by the demands of the larger mission or business needs, or simply by accident of where a SOC was first formed.

Despite all this, the SOC is distinct and separate from almost any other part of the constituency, even though it may be near IT Operations or NOC, CISO, or security function. Its skills, attitude, mission, and authorities always set it apart. As a result, the following is almost always true:

Regardless of where the SOC is organizationally located, it must have integrated budgetary, logistical, and engineering support in place to serve sustained operations.

2.2.1   Subordinate to the CIO or the CISO

This arrangement makes the most sense for most organizations. It is especially common in large constituencies where IT operations and the NOC also fall under the CIO or CISO. The

CIO or CISO will have an operations slant while maintaining strategic visibility and authority. If this is not the case and the CIO is mostly oriented toward IT policy and compliance, this can be a difficult arrangement for the SOC as it will most likely lack credibility among the larger constituency and will not maintain a cyber operations focus. Sometimes a deputy CISO or deputy CIO position may be created whose sole responsibility is to manage the SOC. A SOC organized under a CIO or CISO who can support a true ops tempo with tactical visibility and connections can work very well. In many cases, where the SOC is organized somewhere other than the CISO, it will have some sort of matrixed or dotted line relationship whereby the CISO or CIO influences SOC actions and focus.

2.2.2 Subordinate to the Chief Operations Officer

This can be a positive arrangement for the SOC, assuming operations functions of the constituency are consolidated under the COO. In such a scenario, the SOC is more directly involved in meaningful conversations about the daily operations and mission of the constituency. If there is a daily ops stand-up, the SOC may have direct representation. It is also more likely that the SOC's needs will be met through adequate policy, budget, and authorities. This arrangement can be looked upon fondly by some SOCs because of its visibility, but the SOC must be careful what issues it brings to the COO's desk, and ensure the topics and concerns are of an appropriate level.

On the downside, it can be a challenging position because the SOC will likely compete for the COO's limited time and money. If the COO does not have direct, meaningful control over constituency operations or the COO's function is seen as overhead, the SOC can inherit this reputation as overhead and be vulnerable to cuts during budget time.

2.2.3 Subordinate to the Chief Security Officer

A SOC almost always leans on security professionals located across a constituency to help establish visibility and support response at remote sites. Alignment under physical and personnel security can help strengthen this. It can also help if these security bodies are able to seamlessly take care of IT compliance and misuse cases. Doing so (as is the case with many ISSOs in government) leaves the SOC to focus on more advanced cyber threats. However, constituents' potentially negative perception of security, overhead, or compliance functions may not help. In addition, there is a likely skill mismatch as CSOs and security personnel are usually not expected to have IT experience. Therefore, it may be a stretch for an organization responsible for physical and personnel protection to take on a large portion of the cybersecurity mission.

The biggest challenges to a SOC organized under a security function are (1) maintaining strategic perspective and partnership with the CIO or CISO while having day-to-day visibility and communication with IT and network ops, (2) separating its function from IT and security compliance, and (3) ensuring the right mix of technical expertise. Again, the SOC must be careful, from a budgetary perspective, because security functions are often seen as overhead.

Also, many organizations do not have a separate security organization apart from the CIO and CISO, ruling this out as a possibility.

2.2.4   Peered with the Network Operations Center Under IT Operations

Another common organizational placement of a SOC is having it collocated logically, physically, or both, next to, or as part of, IT Operations, or NOC. This provides a number of obvious virtues: 24x7 operations are merged, and network response actions can be swiftly adjudicated through a single authority that balances the real-time needs of security and availability. Furthermore, there are many devices such as firewalls, EDR, MDM, that are seen as shared security and IT capabilities. Consolidating both functions onto one watch floor (with distinct staff and tools) may save money, especially for organizations that cannot justify having a separate SOC.

However, network operations and cybersecurity have related missions, but they are not the same. While both organizations manage risk and incidents, the focus of a NOC is typically on availability and service level agreements, while the focus of the SOC is on data protection and cyberspace defense. Keeping network and security operations as distinct peers with separate people, tools, and funding will help avoid sidelining security in the name of network availability. Additionally, the SOC will need to have clearly defined processes for engaging directly with system and data owners that is unique from the responsibilities of IT Operations or the NOC.

In order to support a healthy constituency, IT operations and SOC functions should be viewed equally, rather than one subordinate to the other.

Critically, if the NOC and SOC are both merged and both organized under IT operations, the IT operations leadership chain above the SOC must also have accountability to cybersecurity outcomes. If this is not the case, availability will frequently trump security when issues are escalated, and resourcing choices made. One permutation of the NOC/SOC arrangement that is well-postured to satisfy this outcome is that both entities are physically collocated, but still belong to separate organizations and leadership chains: NOC reporting up through the IT operations division, and SOC reporting up to the CISO (or equivalent). If IT operations, NOC, and/or SOC do not have a physical operations floor, typical virtual work arrangements such as participating in real-time persistent chat, VTC, and other electronic forums of real-time collaboration are vital to cross-collaboration.

If there is a desire to combine the NOC and the SOC there are several ways this could occur. The lowest level of integration includes sharing only some processes and only at specific times (e.g., emergencies only). A greater level of integration includes more processes and some technologies for specific purposes (e.g., detection and response). And an even greater level of integration includes people, processes, and technologies together through shared dashboards and workflows [33]. For example, a SOC will always field some calls from users.

However, directing some of these calls to a nearby help desk could help reduce the call load. Similarly, other parts of IT ops can perform network scanning and patch management, removing that from the SOCs responsibilities while still giving the SOC visibility into the process and data. The next step of integration would combine the NOC and SOC into a single, comprehensive NOSC under the same organizational structure. This provides the most opportunities for efficiencies but also the most risk that one mission will take priority over the other. If any of these options are selected, having the right leadership in place will help facilitate the integration. To avoid bias in favor of the NOC mission or the SOC mission, the decision maker must have direct operational experience on both sides or have a technical advisor with operational expertise on both sides. And the decision maker for the NOSC must have unambiguous authority to direct changes for both missions.

In summary, here are tips for making the NOC/SOC relationship successful:

- Ensure positive resourcing for both teams, clear differentiation of roles, technical specialties, and career paths, both for direct insourced hiring and any outsourcing/ contracting pursued.
- Ensure lines for real-time decision-making are clear, both security and operations missions are given equal voice, and that the decision-maker has experience in both IT operations and security both.
- The management and leadership above the SOC must be held accountable to security outcomes and above the NOC to IT and availability outcomes, equally.
- Implement lines of separation, security and enclaving for security systems, data, cases, and alerts for the SOC commensurate the greater enterprise security posture and risk environment.
- Be sure that no matter how the SOC is organized, it has communication to- and sponsorship by- a cybersecurity executive such as the CISO.

2.2.5 Embedded Inside a Specific Mission or Business Unit

Placing cyber operations within a given business area may limit visibility to only what is within that business unit. This may be a positive arrangement in federated or hierarchical models; conversely if the SOC serves a constituency outside that business unit, its focus may be skewed, and authorities may be strained.

That said, this presents unique opportunities for the SOC to be mission-oriented in how it monitors and responds. If a specific portion of an organization's mission is very sensitive, having a SOC just for that mission can help. However, these efforts must be tied back to an organization-wide visibility and coordination capability (such as a hierarchical SOC model). Alternatively, a distributed SOC model with representatives in each business unit and the main SOC viewed as a headquarters function may work. For small SOCs without additional personnel in other parts of the organization, this is a difficult model to successfully execute.

2.3 Summary – Strategy 2: Give the SOC the Authority to Do Its Job

2.1. Written guidance that grants a SOC the authority to exist, procure resources, and enact change is an important component of building and operating of a SOC.

- Elements of the SOC Charter should include the SOC's function, scope, and authorities along with expectations for partnering with other parts of the constituency.
- The SOC requires support and enablement by a raft of other cybersecurity and IT governance.
- The SOC should take an active role in reviewing other existing constituency policies that are support execution of their functions or developing new policies if needed.

2.2. The SOC draws its authorities, budget, and mission focus from the organization to which it belongs.

- The SOC can be housed in many places within an organization, each with its own pros and cons. The most common placement is under the CIO or CISO.
- Other options include under the COO; under the CSO; under IT operations; or inside a specific business unit.

Strategy 3: Build a SOC Structure to Match Your Organizational Needs

There are thousands of SOCs around the world, each with a shared sense of identity and mission. Yet, due to business needs, risk posture, and other considerations no two SOCs are organized exactly alike. This strategy discusses the drivers for choosing a SOC structure, various SOC organizational models, physical SOC location considerations, and address the questions "Should your SOC go 24x7?" and "Should you outsource your SOC?"

3.1 Drivers for Choosing a SOC Structure

3.1.1 The Mission & Business Served

The SOC must find a structure that allows it to organize its functions in a way that reflects the needs of the constituency and is not prescriptive of a single approach. Elements to consider include:

Business Need
As discussed in "Strategy 1: Know What You Are Protecting and Why," the SOC must first align with the mission or business needs. If the constituency is small and new it may not need a dedicated SOC team to start. Instead, other IT and security professionals can do double duty supporting core SOC functions. Large and mature organizations will likely want a formal SOC organization with dedicated staff given the size of operations. Additionally, when considering the structure of the SOC, the structure of the organization itself is important to know. For example, is the constituency a single operating unit where a centralized SOC would be easy to implement and there are clear lines of responsibility, or are there many independent operating units each with its own distinct needs and management chains in which case situational awareness across the entire constituency is more important than centralized management of all SOC functions?

Risk Posture
The risk posture of the constituency is also important to consider. All constituencies are potential targets, even if it is just because they are running a vulnerable application. However, constituencies working with highly sensitive information such as financial records, healthcare records, intellectual property, or government classified materials likely will want a more formal SOC organization regardless of their size or maturity. They may also want to offer more SOC services for their constituents with a tighter coupling to cyber threat intelligence. Additionally, the risk posture will inform the need for redundancy, or continuity of SOC operations. The

idea of continuity of operations as it related to a SOC's physical location is discussed in more detail later in the strategy.

Constituency

The SOC must maintain a real and ongoing relationship with its customers. For the analyst, this means having ready access to network maps, scan results, and service descriptions. It also means routine engagement with those service and system owners by the SOC or its parent security apparatus. Consequently, a SOC must be small enough to sustain those relationships, and ideally near the people it communicates with most frequently. When a SOC grows too large to sustain relationships with many of its customers, in particular the high-criticality customers, it will struggle to have that context and understanding that often proves so critical. In these cases, creating a hierarchy of SOC organizations may be an alternative.

*The further an analyst is separated from monitored assets,
the less sense of context on those assets they are likely to have.*

3.1.2 Intended Accountability and Responsibility of the SOC

As was discussed in "Strategy 2: Give the SOC the Authority to Do Its Job," the SOC must have a set of authorities from which it derives its ability to function, such as directing various response actions. Some organizational arrangements can help or hinder this. For example, if various security operations functions are strewn in among disparate organizations, with nothing to bind those functions together, they may find themselves at odds and lacking recognition from constituents during key incident response phases. Therefore, the SOC must find a structure that enables it to have a shared identity, set of responsibilities, and authorities that align with the resources available to it.

3.1.3 What Services the SOC Will Perform

As discussed in the Fundamentals chapter, there are a variety of functions that could be performed by a SOC but not all organizations will execute all those functions. If a SOC is only executing core SOC functions such as real-time alert monitoring and triage, they will likely structure themselves with fewer teams than if they are offering vulnerability scanning, pen testing, hunting, or deception operations. It is necessary to understand what SOC services the SOC itself will perform versus what security services are offered by other parts of the constituency. Defense of the constituency is a shared responsibility. In larger constituencies, it may be more likely that some roles are distributed. For example, services such as vulnerability management may be a SOC responsibility, or they may be handled by another part of the constituency. Additionally, sometimes these other security services necessitate additional support from within the SOC. For example, if there is an Insider Threat team or a Supply Chain Risk Management team separate from the SOC, those functions may require additional specialized analysis support from the SOC. The placement of engineering resources is of particular importance and is covered in more detail in Section 3.3.1.

3.1.4 Operational Efficiencies

The balance between supply and demand in the cybersecurity job market is one of the most severe in government and private sector today and will likely continue to be for the foreseeable future [34]. Even the largest and best-resourced companies and governments have a hard time managing retention and keeping key positions filled. Therefore, constituencies will want to consider the potential benefits to consolidating SOC staffing, resourcing, and authorities to achieve a cadre of security experts who can specialize in what they are doing and operate as one unit as compared to having SOC functions distributed across the constituency.

3.1.5 Scope of the Environment the SOC is Responsible for Monitoring and Protecting

Most SOCs will be responsible for monitoring and responding to incidents in the traditional IT environment (on-prem and remote), in the cloud, and for the mobile infrastructure. These different environments will drive the skill sets needed for the SOC personnel and may influence the need for engineering capabilities. For those constituencies that have connected OT systems, there is also a need to monitor the OT environment. Some constituencies may have a separate SOC or set of security services dedicated to OT, however there are advantages to integrating monitoring and response for IT and OT into one SOC. Combining SOC functions can result in faster response times as there likely will be fewer communication breakdowns. This is especially important as the understanding of OT functionality, priorities, and concerns is often very different than the functionality, priorities, and concerns for IT systems. At the same time, there are likely efficiencies to be gained as the same personnel, technology, and facilities can support the different environments even if the personnel require cross training and the technology needs to be adapted to ingest and process OT specific information. Overall, there is more opportunity to get fully integrated visibility into threats resulting in better situational awareness if IT and OT SOC functions are aligned.

3.2 SOC Organizational Models

When thinking about SOC structure, there are two dimensions to consider. First, what overarching model will your SOC employ when it comes to security operations? For example, will SOC-like functions be performed by a small dedicated SOC team, or is the constituency so large that a hierarchical SOC structure makes more sense where SOC functions are distributed across multiple teams? The other dimension is how to structure the personnel inside of the SOC given the model chosen. In other words, how will functions be translated into personnel roles and an organizational structure.

3.2.1 SOC Organizational Model Overview

Building on the idea that there is no single SOC structure that works for all organizations, this section presents different approaches that serve constituencies small to large. Variations on these exemplar models are certainly possible, but Table 2 covers many of the main SOC organization constructs.

Table 2. SOC Organizational Models

Organizational Model	Example Organizations	Remarks
Ad Hoc Security Response	Small Businesses	No standing incident detection or response capability exists. In the event of a computer security incident, resources are gathered (usually from within the constituency) to deal with the problem, reconstitute systems, and then stand down. Results can vary widely as there is no central watch or consistent pool of expertise, and processes for incident handling are usually inadequately defined.
Security as Additional Duty	Small businesses, small colleges, or local governments	No formal SOC organization. However, SOC-like duties are part of other duties. For example, a system administrator that also looks for unusual activity in system logs. Some procedures for incident response may exist.
Distributed SOC	Small to medium-sized businesses, small to medium colleges, and local governments	Formal SOC authorities. Comprised of a decentralized pool of resources housed in various parts of the constituency. Staff may have other duties as well.
Centralized SOC	Wide range of organizations including medium to large-sized businesses, educational institutions (such as a university), or state/province/federal government agencies	Resources for security operations are consolidated under one authority and organization. SOC personnel have dedicated roles in the SOC. This model is the most frequent focus of this book, the most frequent operating model, and the simplest way to think about how most SOCs operate.
Federated SOC	Organizations with distinct operating units that function independently of one another such as businesses that have acquired other businesses but have not integrated them together	A SOC, likely centralized but could also be hierarchical, that shares a parent organization with one or more other SOCs, but generally operates independently. It may have some shared policies and authorities.
Coordinating SOC	Large businesses or government institutions	A SOC responsible for coordinating the activities of other SOCs underneath it. Focuses primarily on SA and overall incident management. Does not direct the day-to-day operations of the SOCs it coordinates.
Hierarchical SOC	Large businesses or government institutions	Similar to the Coordinating SOC structure; however, the parent organization plays a more active role. The parent organization may offer SOC services to lower-level SOCs and has greater responsibility for coordinating a wider range of SOC functions (such as engineering, CTI, malware analysis, etc.)

Organizational Model	Example Organizations	Remarks
National SOC	Country level governments	Responsible for strengthening the cybersecurity posture of an entire nation. Creates opportunities for sharing SA of vulnerabilities, threats, and events across multiple constituencies. May orchestrate activities associated with significant cyber incidents.
Managed Security/ SOC Service Provider	Organizations of all sizes	Provides SOC services to external organizations via a business/fee-for-services type relationship.

3.2.2 Impact of Constituency Size on SOC Organizational Model Selection

The size of the constituency is a key driver in determining the appropriate type of SOC organizational structure. Although there are no hard lines to be drawn, Figure 6 gives a general feel for the relationship between constituency size and the different SOC structures. None of these structures are necessarily better than another. However, they do come with different challenges, opportunities, and expectations which will before discussed in the next few sections.

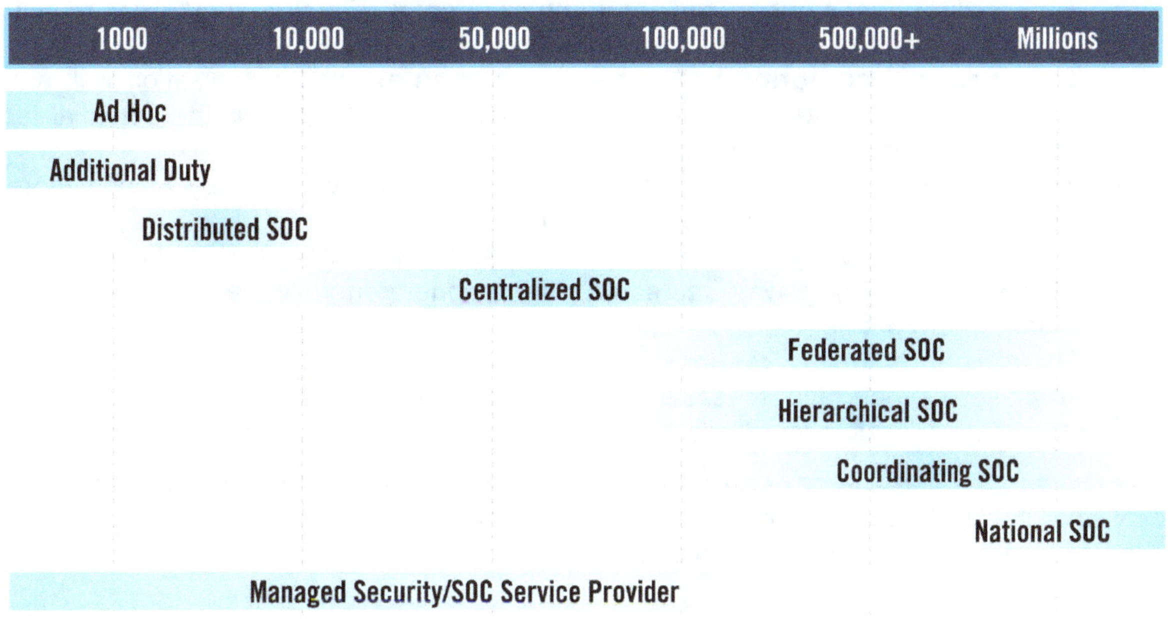

Figure 6. SOC Organizational Model Versus Constituency Size

3.3 Centralized SOC Organizational Structures

Centralized SOC structures are what most people think of when they think of a SOC. This structure is the most obvious way to organize a SOC: consolidate functions of security operations under one organization whose sole mission is executing the security operations mission.

Centralized SOCs come in multiple sizes and can be as varied as the constituencies they support. However, a few notional organizational charts are presented in this section to give readers some ideas on how to structure their own SOC to better support smooth operations, without getting into every permutation of what a SOC might look like.

By placing all SOC services within one centralized organizational structure, the SOC gains many benefits when compared to ad-hoc or distributed organizational models, including:

- **Dedication of resources and focus:** Security operations for the centralized SOC is what they do, and not treated as an additional duty or distraction
- **Ownership and shared identity:** The team comes together with a shared sense of mission and purpose
- **Centralized visibility and management of incidents:** Synchronize elements of security operations so all elements are working in concert toward the same goal, especially during a critical incident
- **Better collaboration and unity of effort and integration among SOC service elements:** There will be fewer organizational barriers to working together
- **Potential for cost savings and economy of force:** A centralized model can reduce duplication of effort and maximize the use of technologies
- **Stronger SOC authority:** Limits the likelihood an external organization will take it upon themselves to perform SOC like functions, which reduces the potential for conflict or disorganization during a response
- **Staff growth:** Allows the SOC to build its own staff over time by having more opportunities for growth and advancement
- **Self-reinforcing progress toward maturity and effectiveness:** With the elements of the SOC working toward the same goal, as one, generally they progress much faster toward greater capability than a distributed or decentralized capability
- **Unambiguous area of responsibility and mission:** The SOC is responsible for a given set of organizations, assets, and networks (the constituency); the lines between who are responsible for what should be clear and not subject to controversy

This is not to say that ad-hoc or distributed SOC functions might not be the right choice for very small constituencies with limited security risks or resources. However, at a certain point, bringing together SOC resources into one organization likely makes the most sense.

3.3.1 The Notional Centralized SOC

As a starting point for centralized SOC organizational chart, this section puts forth a notional SOC structure. This is not an ideal state to be achieved, rather this notional structure is presented to demonstrate how the various functional elements of a SOC might be organized.

Figure 7 takes all the SOC services from Table 1, other than Vulnerability Management, and places them in the context of an organizational structure. The SOC should aim to achieve proficiency in the functions listed above the dotted lines before expanding into functions below.

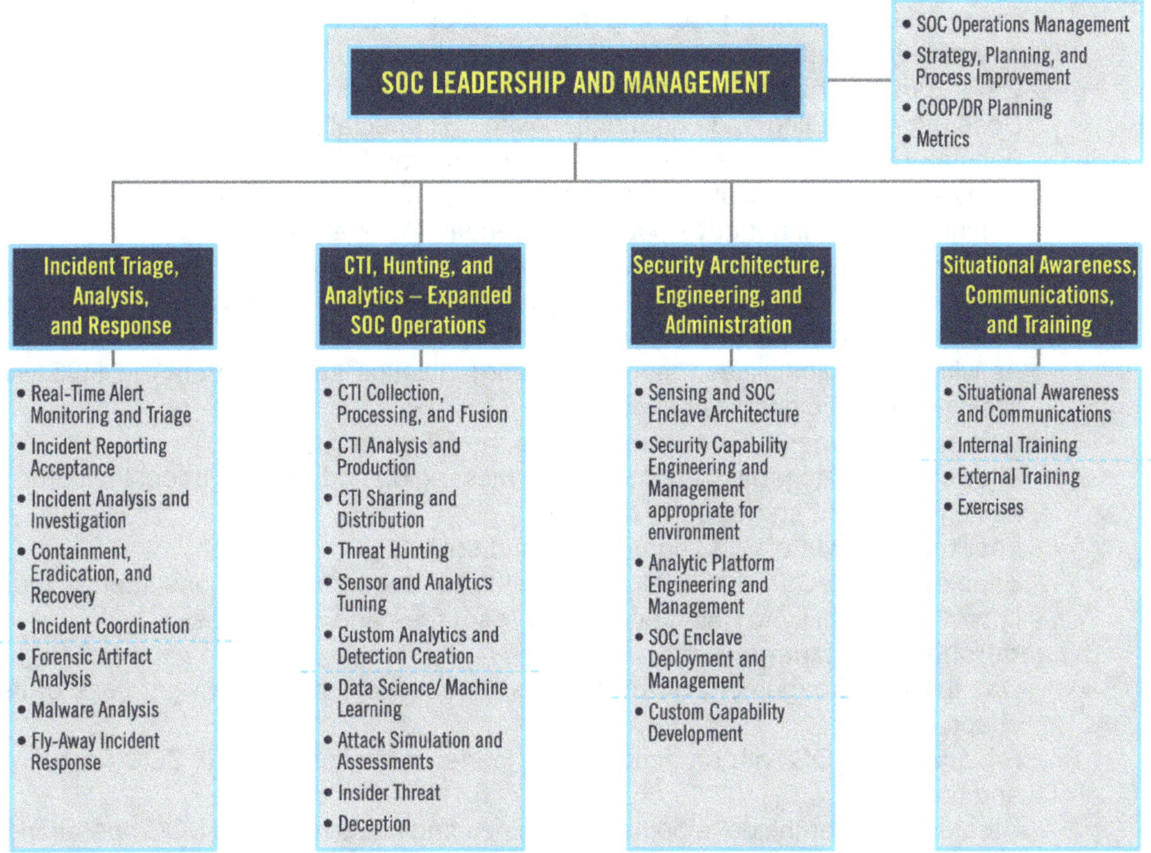

Figure 7. Notional Centralized SOC

In addition to the general benefits of a Centralized SOC, this structure has benefits aligned to each element of the organization chart.

- **Incident Triage, Analysis, and Response:**
  - This is the core function of a SOC. Without this internal, inherent capability, the SOC is no longer a SOC. "Strategy 5: Prioritize Incident Response" discusses this key function.
  - If organized elsewhere, may mean double work because the SOC will have to do its own line of triage anyway.
  - Allows the SOC (particularly incident investigators and responders) to better control the quality of alerts being escalated.
  - Creates a central place to connect humans and systems since some human must triage alerts, emails, phone calls, etc., coming into the SOC, regardless of where they come from, or how much automation, pre-processing, filtering, enrichment, and prioritization is done to that incoming flow.

- **CTI, Hunting, and Analytics – Expanded SOC Operations:**
  - Making this a SOC function helps the SOC orient on the adversary more that it would otherwise.
  - Helps align resourcing against CTI needs of the SOC vice that of other organizations, including focusing the outcomes of CTI collection and curation on operations, rather than CTI "for CTI's sake."
  - Other parts of the SOC, in particular detection authors and responders, require close understanding and coordination with CTI specialists.
  - For many SOCs, the line between analytics creator, CTI analyst, and threat hunter is invisible or non-existent; separating these functions, especially in smaller organizations, will not drive positive or beneficial behavior.
  - This enables the SOC to embrace all methods of identifying and investigating the adversary: hunt, detections, alert triage, and investigation TTPs are synchronized and reinforce one another.
  - As with other capabilities, having a CTI organization separate from the SOC will likely drive inefficiencies in execution.
- **Situational Awareness, Communications, and Training:**
  - Allows the SOC to provide more focused messaging and updates about the current risk environment and value of the SOC.
  - Facilitates growth of SOC capabilities and staff.
  - Enables the SOC to demonstrate its value to leadership and constituents, thus improving its ability to gain and expand resources and authorities.
- **Leadership and Management:**
  - Day to day management of the SOC is a necessary function, regardless of its chape or size.
  - Provides the SOC with an accountable leader responsible for all SOC functions and outcomes.
  - Creates a hub for the development of metrics and plans related to SOC operations.
- **SOC Tools, Architecture, and Engineering:**
  - The organizational alignment of this set of services is often debated and has many considerations so it is further described next.

SOC Engineering

Identifying the right place to house SOC tools, architecture, and engineering resources is a decision that often faces much scrutiny within an organization. When aligned appropriately, these functions enable the SOC to execute its mission with agility, precision, accuracy, and completeness. Poorly aligned, they are a source of frustration for SOC analysts that can result in wasted resourcing, limited SA, inadequate detection, reduced ability to investigate incidents, and insufficient response.

A SOC will support its own tool administration and engineering capability out of necessity, regardless of whether it sanctioned. It is best to enable strong engineering support through correct organization, resourcing, and authority.

The benefits of placing SOC engineering resources directly in the SOC include:

- SOC needs are the priority
- Engineering resources are better aligned to SOC priorities
- Engineers better understand the SOC mission; there is more likely to be an environment of trust between analysts and engineers
- Solutions are delivered in a timely manner and are sized to timescales relevant to the need in question
- Analysts spend less time working around solutions that do not meet their need and engineering their own capabilities to solve problems
- Analysts and engineers are better able to mutually support one another and synthesize healthy professional relationships

The benefits to having engineering resources more closely tied to other engineering efforts include:

- Potential to leverage a broader set of engineering skills sets
- The solutions being developed directly serve customers primarily outside the SOC
- The SOC does not have to pay or budget for its own capabilities

---

SOC analysts must be empowered to translate their operational challenges into rapid innovation they themselves are directly involved in solving.

Typically, when SOC engineering is not in the SOC, a number of very negative outcomes ensue:

- Analysts will struggle to meet this condition, and resort to acting outside organizational boundaries and swim lanes to get the mission done.
- Inefficiencies will form as engineers' priorities diverge from operational need.
- Analysts and engineers will expend a lot of effort around defining and refining requirements and measuring whether they have been met or not.
- Effort expended on- and complexity of- solutions will be frequently misaligned to the mission need.
- An environment of mistrust will form by analysts who do not feel their needs are being met or met in a timely manner.

All SOCs are encouraged to embrace a culture that carries the spirit, if not the execution, found in approaches like DevOps [35], agile development [36], and continuous integration/continuous delivery [37]. That arrangement should adhere to the following best practices:

- Operations and engineering both report directly to the same leader (at some point in their immediate reporting structure—meaning just one or two levels up the organizational structure) who:
  - Exercises authority over tool budget and resource allocation
  - Ensures accountability between ops and engineering

- Engineering is made aware of SOC operations at both the strategic and tactical level, and is included in daily/weekly ops sync
- Ops and engineering understand, influence, and are held to the same configuration management standards
- Routine project update meeting between ops and engineering:
  - Requirements and projects are brought to the table for prioritization
  - Both ops and engineering must check off before projects transition to ops
  - Security, risk acceptance, Office of the Chief Information Security Officer (OCISO)/Office of the Chief Information Officer (OCIO), and other reps are present for SA & support as necessary
- Ops dedicates personnel to tracking requirements and projects
- Both ops and engineering are empowered to contribute to ongoing projects and capability improvement in a harmonious and organized manner—meaning no one has a job relegated to "just maintaining the 'baseline'"
- Tools and capabilities are routinely and iteratively updated over their lifespan
- Engineering resources supporting the SOC have adopted and gained formal agreement supporting risk management and configuration management governance that enables timely delivery of capability without getting caught in "red tape"
- Improvements directly related to ops, such as: detection deployment, sensor tuning, workflow changes and the like follow a process that is lightweight
  - Supports implementation on timescales of minutes, hours, and days
  - Does not get held up by larger scale engineering processes that would otherwise slow down such actions

If for whatever reason the SOC itself cannot "own" the engineering resources and budget, and thus engineering resources must be located elsewhere organizationally, five tips for success should be followed:

- The engineers must have a written agreement with the SOC that they are responsible to the SOC as a customer, and there are process safeguards in place to ensure that accountability
- The SOC has resources dedicated to curating and driving closure on SOC requirements; this role is referred to as a program manager, requirements manager, or product owner
- The SOC still has its own resources for day-to-day upkeep and improvement of SOC systems, in particular "quality of life" type improvements such as:
  - ML models, signatures, SIEM content, analytics, and related detections.
  - Workflows, integrations, and other user-facing customizations to SOAR and case management systems
  - User-facing application upgrades such as those workstations and virtual machines (VMs) sitting behind user thin clients, such as for scripting languages, web browser, and various SOC-specific tools
- The engineering personnel allocated to SOC products:
  - Are dedicated solely to the SOC
  - Meet with SOC analysts routinely

- Ideally are physically located with the SOC if the SOC is has a physical ops floor and/or rotate into SOC roles so they understand challenges operators face and can thus better attend to those pain points.
- There is a budget allocated solely for SOC tool capital improvement, sustainment, and cloud resources that cannot be "robbed" by other competing programs and groups

Regardless of their organizational topology or location, SOC Operations and SOC Engineering should have a relationship of mutual respect, positive reinforcement, and open sharing of ideas and solutions.

The approach of combining all functions into one organization is one of the most prevalent cases historically. However, as more organizations recognize the need for a SOC, there are more permutations and variations where the notational scenario either does not make sense, or simply is not possible. Regardless of the structure, keeping the lines of communication flowing and considering how SOC functions are implemented across the entire ecosystem of the constituency are important considerations.

3.3.2 Small Centralized SOC

The notional centralized SOC in Section 3.3.1 considered all the SOC services other than Vulnerability Management. However, smaller SOCs, in the range of five to 20 people, often find a relatively simple approach to arranging their staff to be more beneficial. This is because with few people, there is comparatively less diversification of roles and there are few positions that do not involve full-time analyst work. A classic Small SOC will often be broken into the following section as seen in Figure 8:

- **Incident Analysis Response:** Which will also include some CTI and sensor tuning functions
- **SOC Tools, Architecture, and Engineering:** With much less distinction between the Sensing infrastructure, SOC Capabilities, and the SOC enclave infrastructure
- **SOC Leadership and Management:** With SOC leadership sometimes also responsible for some level of support in the areas of either incident response and analysis or engineering
- **Vulnerability Management:** If assigned to the SOC

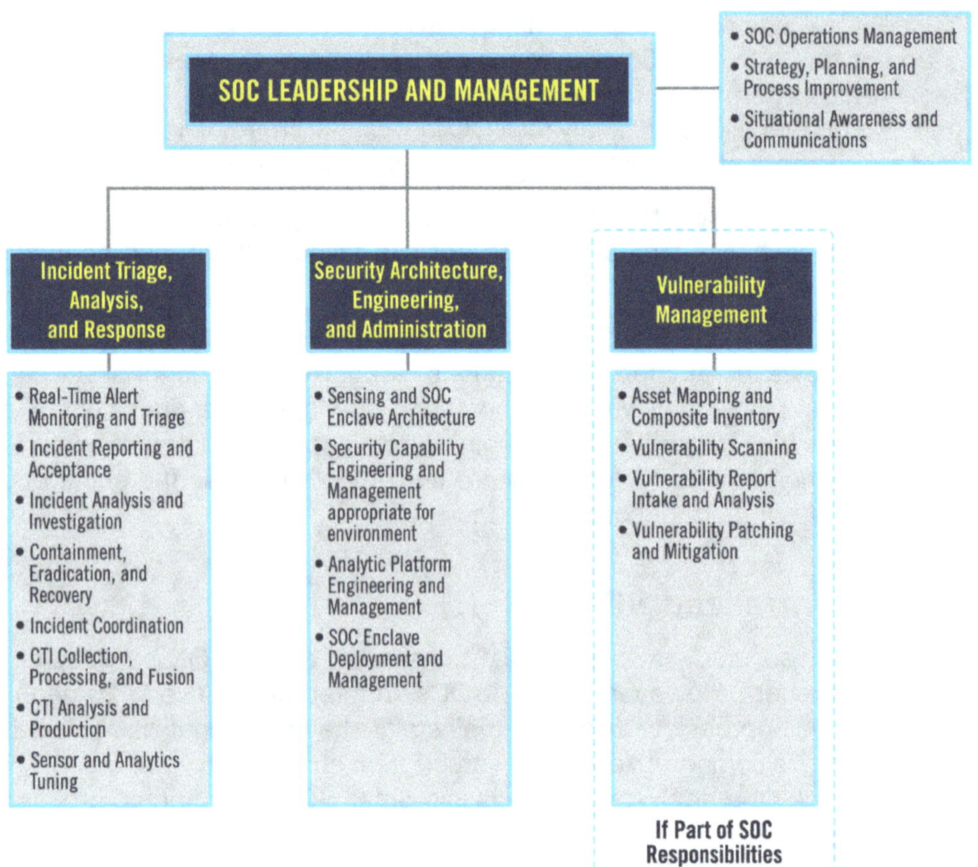

Figure 8. Small SOC

The services listed in Figure 8 can be considered core SOC services, the ones which all SOCs should strive to develop as a starting point for their SOC.

3.3.3 Large Centralized SOC

A large constituency can support a SOC with an advanced set of capabilities and full-fledged division of roles and responsibilities. However, just because a SOC is large does not mean it needs to perform every service or perform every service at the same level of maturity. Consideration will still need to be given to what is required by the constituency the SOC serves. The following changes are ways a large centralized SOC may differ from a small centralized SOC:

- In General, the SOC will likely have more sections and more leadership positions, as well as having more senior technical roles such as a lead architect.
- **Analysis and Response:**
  - Incident triage may be separated from incident response and analysis. The creation of a tiered structure like this is discussed more in Section 3.3.4.

- Incident response and analysis in large SOCs may also include the capability to perform fly-away incident response.
  - Handling of certain common alerts, escalations, and incident types are likely to be subject to substantial regimen and repeatability.
  - There is often a dedicated section for forensics and malware analysis given the unique skills and lab environments required by these functions.
- **CTI, Hunting, Analytics, and Expanded SOC Operations (see * in Figure 9):**
  - These capabilities are most likely to grow in scope and maturity compared to other SOC functions.
  - There is often a dedicated section responsible for cyber threat intelligence.
  - Threat hunting could align with either CTI or expanded SOC operations or be its own section in a very large organization, with its own planning and execution cycles.
  - There is sometimes a dedicated section responsible for expanded SOC operations. This may be more than one section. For example, if insider threat is a significant concern the personnel supporting that effort may be isolated from other activities due to the sensitivity of their work.
  - Pen testing, if present in the SOC, is likely to have its own team and deep specialization both as a team, and for each of the types of computing environments assessed. There is often a strong alignment between Pen testing and Vulnerability Management so the Pen Testing team may be part of that organization instead (see ** in Figure 9).
  - The creation of custom analytics is given more emphasis and the utilization of data science and machine learning methods are expanded.
- **SOC Tools, Architecture, and Engineering:**
  - Sensor support may be separated SOC enclave support.
  - The SOC may choose to develop custom capabilities as well as using FOSS and COTS capabilities.
  - Support to specific tools is likely to be broken down into different areas of specialization, if not entirely separate teams each with their own lead.
- **Situational Awareness, Communications, and Training:**
  - These services may have their own dedicated section rather than being an additional duty of the SOC Lead.
  - There will be additional emphasis on coordination among the SOC elements as well as with external stakeholders and partners.

A potential organizational model for a Large SOC is depicted in Figure 9.

When a SOC gets this large, it is important to ensure there is effective cross-training and cross-pollination so that individual sections see themselves as part of a larger team. See "Strategy 4: Hire AND Grow Quality Staff" for ideas on this topic. Additionally, engineering groups must stay cognizant of the ops group's main challenges as discussed in Section 3.3.1. Moreover, even though this model has multiple layers of management, operators in one section should not hesitate to communicate directly with any other part of the SOC. See "Strategy 9: Communicate Clearly, Collaborate Often, Share Generously" for more on internal SOC communications.

Figure 9. Large SOC

3.3.4 To Tier or Not to Tier

In this book, the constructs of "tier 1" and "tier 2+" are sometimes used to describe analysts who are primarily responsible for front-line alert triage and in-depth investigation/analysis/response, respectively. However, not all SOCs are arranged in this manner. In fact, some readers of this book are probably very turned off by the idea of tiering at all [38]. Some industry experts have outright called tier 1 as "dead" [39]. Once again, every SOC is different, and practitioners can sometimes be divided on the best way to structure operations. SOCs which do not organize in tiers may opt for an organizational structure more based on function.

Many SOCs that have more than a dozen analysts find it necessary and appropriate to tier analysis in response to these goals and operational demands. Others do not and yet still succeed, both in terms of tradecraft maturity and repeatability in operations. Either arrangement can succeed if by observing the following tips that foreshadow a longer conversation about finding and nurturing staff in "Strategy 4: Hire AND Grow Quality Staff."

> *Highly effective SOCs enable their staff to reach outside their assigned duties on a routine basis, regardless of whether they use "tier" to describe their structure.*

Making tiered SOC structure work well:

- **Ensure that service level objectives (SLOs) are not only defined, but understood, agreed to by the analysts themselves, and updated as necessary:** Do not give the analyst a rigid 120 seconds to investigate every single alert *or else*. Give them the opportunity to dive a little deeper so that they have more ownership for the process while not ignoring the incoming queue for hours.
- **Do not allow other staff look down on tier 1:** Most SOC professionals started as junior analysts. This is a growth position, not an underclass. Tackling an alert queue is a transformative experience and will change the perspective of most anyone engaged in this role. It is experience-building and a necessary position just like any other.
- **Promote, advance, and automate, as appropriate:** It is important to prevent role stagnation. Once an analyst has mastered a role, a) evolve their job function and/or b) give them the ability to automate functions of their existing role, allowing them to move on to new tasks.
- **Do not force an "up or out" culture:** If an analyst is doing well in their current position, and they do not want to move on, or should not, do not force them to. If advancement is overdriven, it can cause people to leave prematurely.
- **Enable flexibility in role:** An analyst should not feel limited because of their role title. In a tiered scenario, allowances should be made for analysts that show growth in skills or abilities, even if their title does not change. This could include efforts like building in job rotations for staff.

Making "tierless" SOC structure work well:

- **Nurture new staff in a way that does not jeopardize investigative quality.** If new staff are going to be asked to evaluate an event and determine if it is an incident, they will need assistance. This means they should be provided with all resources such as playbooks to aid in their decision making and they should have appropriate oversight, often by being paired with a senior analyst who can review their work as needed. In this regard, even SOCs without tiering can still share qualities of those that are.
- **Help staff prioritize their work:** If staff members are responsible for both triaging incoming alerts and doing longer incident response, or other functions (like maintaining SOC tools or hunting) they may find it difficult to know what they should work on first. Depending on the analyst, certain responsibilities may get more or less attention. Try to balance staff feeling ownership for how they manage their time with clear written and verbally communicated expectations for what is a priority. Keep an eye out for analysts that always seem to focus on one aspect of the job over another, this may indicate a need to step in to better understand the situation.
- **Manage separation of duties and reduce duplication of effort:** As the SOC gets larger, separation of duties becomes increasingly necessary. Without formal separation of duties, staff may become very fragmented in their time. This does not mean a tierless structure cannot work, it just means that perhaps the role descriptions are slightly narrowed and not everyone does everything.
- **Create a culture where staff work together and help each other out:** Some incidents just take longer to work through than others. If one staff member gets involved in a significant response activity that takes them away from the more routine event triage have other analysts jump in to fill the gap. They should then pay it back to their colleagues by supporting them in the same way.

When considering if the SOC should be tiered or tierless, consider which model will best allow you to do the following:

- **Respond to all alerts:** Ensure that every alert purposefully curated and sent to the SOC for triage is, in fact, triaged.
- **Respond to alerts in a timely manner:** No alert should sit untriaged beyond the triage SLO, which is usually less than an hour. However, be cautious about closing out alerts too quickly and sacrificing quality for quantity. Auditing or reviewing some alerts to ensure proper handling can help address any potential gaming of the metrics.
- **Investigate every alert properly:** Ensure each alert is given due attention by the analyst. Routine alerts are dispositioned with routine handling, and unusual alerts are given the hours of in-depth investigation they deserve, without unduly rushing the analyst to prematurely click alerts to closure.
- **Give everyone the opportunity to work at their level:** People new to security operations do not usually have the practice and knowledge to perform in-depth analysis and incident closure. Those with years of SOC experience may feel this role is behind them. Without compensating controls, pooling a large group with varied

experience can cause errors in analysis. In fact, some analysts may never progress past "front line" alert triage type functions.

- **Give everyone the ability to advance:** In a tiered structure, that may require more deliberate rotations or more formal promotions.
- **Maximize operations quality and repeatability:** A recurring theme is that SOC leadership may only want seasoned responders making the call on response activities. Asking newer analysts to do this alone with no mentoring can be very risky, and frequently will cause disruptive behaviors and outcomes, from system administrators to executives.

Regardless of the approach that is best for each reader, this book sometimes uses the construct of "tier 1" and "tier 2" analysts because: many SOCs choose to take this approach; it is a nomenclature pervasive in IT and thus easy to understand; it's an efficient way to convey meaning; and there must always be someone in the SOC that must analyze a curated feed of alerts. This is not an endorsement of tiered over tierless but is used to convey those different duties are being performed, even if they are being performed by the same person. For more discussion on alert triage and SOC tiering, see [40] and [41].

3.3.5 Hierarchical SOCs

We have introduced the concept of a hierarchical security operations architecture where multiple SOCs operate in a federated or structured manner within a large organization. There are many examples where such an arrangement might be appropriate (e.g., the US Government: within each branch of the Department of Defense (DoD), Department of Treasury, Department of Justice (DoJ), Department of Homeland Security (DHS), large multi-national corporations and conglomerates, and so forth). Although each of these entities has one SOC with purview over the entire department or corporation, there exist several subordinate SOCs beneath each that perform the majority of security operations "heavy lifting" for the constituency. These include regional Network Operations and Security Centers (NOSCs) under the U.S. Army, Financial Management Service under Treasury, Drug Enforcement Administration under DoJ, and Immigration and Customs Enforcement under DHS. In all these cases, there is a department- or branch-level SOC, as well as multiple SOCs beneath each.

Both the parent and subordinate SOCs have a meaningful role to play, even though those roles can be quite different. Going back to a previous point, these organizations must balance the need to maintain strategic SA with the need to be close to mission assets. Most people familiar with security operations are used to operating down in the weeds. This can become a source of conflict in a hierarchical SOC scenario.

In hierarchical models, the central SOC emphasis is on strategic situational awareness and enabling security operations across constituencies, rather than direct instrumentation and alert triage.

So, what is the best way to differentiate these roles? Table 3 describes some ways these two SOCs may interact and share the security operations mission.

Table 3. Differences in Roles for Hierarchical SOCs

Responsibility	Central SOC Role	Subordinate SOC Role
Real-time Alert Monitoring and Triage	Across constituency assets not covered by subordinates, such as Internet gateways or constituency-wide services such as email	Within assigned constituency
Incident Analysis and Response	Cross-constituency coordination, operational direction. Receives summary information and incident reports from subordinates; analysis and retention of data from assets not covered by subordinates, such as Internet gateways. May provide fly-away incident response support during significant incidents.	Intra-constituency response. Analysis and retention of own data, augmented with data from other organizations
Cyber Threat Intelligence	Strategic across enterprise, reporting to subordinates, trending of adversary TTPs	Tactical within constituency, consumer of central threat analysis, focused on supporting SOC detections
Expanded SOC Operations	Maintain a cadre of SOC staff that can support hunting, malware analysis, red-teaming or other expanded operations that are not needed on a day-to-day basis by subordinate SOCs	Maintain a cadre of SOC staff for expanded operations if the subordinate SOC is of sufficient size or has more frequent needs for these functions
Situational Awareness and Communications	Strategic across entire enterprise and with external parties	Tactical within own constituency
Training	Coherent program for all analysts in constituency	Execution of general and specialized training for own SOC
Reports to	Constituency executives, external organizations	Own constituency executives, central SOC
Security Architecture, Engineering, and Administration	Enterprise architecture, enterprise licensing, and lead on tool deployment and refresh	Chooses monitoring placement, specialized capabilities when needed

It is also important to recognize that not all coordinating and subordinate SOCs fall cleanly into these roles. Some SOCs that sit within a large constituency can support better resourcing, more advanced capabilities, and more strategic reach. Larger constituencies can afford more capabilities and, thus, have the potential for greater independence, even though they fall underneath a coordinating SOC. The constructs presented here are only a starting point for establishing roles among hierarchical SOCs.

3.3.6 Unique Value for Coordinating SOCs

For any SOC sitting above other SOCs: the central SOC in a hierarchical model, coordinating SOCs, and national SOCs, they are in a position to offer services others do not. As stated previously, while there may be a tendency to operate "down in the weeds," these organizations can bring other differentiated value.

Instead of focusing on direct reporting of raw event feeds or promulgating detailed operational directives, the coordinating SOC may better achieve its goals by providing a unique set of capabilities that its subordinates usually cannot. These include:

- **Performing strategic analysis on adversary TTPs:** Coordinating SOCs may have access to a larger set of finished incident reporting and therefore are uniquely positioned to focus on observing and trending the activity of key actors in the cyber realm.
- **Providing a clearinghouse of tippers, sensor signatures, ML models, and SIEM analytics that other SOCs can leverage:** A coordinating SOC could harvest indicators from human-readable cyber threat intelligence and provide it back out in both human- and machine-readable form for ingest by subordinates' analysts and SIEM, respectively, such as through Structured Threat Information eXpression (STIX) /Trusted Automated eXchange of cyber threat Intelligence Information (TAXII) [42], [43]. For this to work, however, CTI should be turned around in a timescale and with detail that is beneficial to its recipients. This will likely mean processing and redistributing CTI in timeframes of hours or perhaps a few days, and in so doing preserving as much original detail and adversary knowledge as possible.
- **Providing malware analysis, forensic services, and emergency incident response to constituent SOCs:** These areas either require advanced skills that hard to staff and maintain currency in or are only needed intermittently by any particular subordinate SOC. In this fashion, some coordinating SOCs act in a capacity like an outsourcing MSSP. Malware services can include an automated Web-based malware detonation "drop box" or in-depth human analysis of media or hard-drive images.
- **Aggregating and sharing SOC best practices, process documents, and technical guidance:** This can include enterprise guidance the coordinating SOC develops itself or best practices developed by subordinate SOCs that it helps propagate across the larger organization
- **Providing secure forums for collaboration between subordinate SOCs:** This may include collaboration hubs, persistent chat, message boards, and wikis.
- **Providing enterprise licensing on key SOC technologies:** This can include network and host monitoring tools like EDR, vulnerability scanners, network mapping tools, and SIEM, provided the following two conditions are met: (1) subordinates are not forced to use a specific product, and (2) there is enough demand from subordinates to warrant an enterprise license.
- **Providing SOC training services:**
  - On popular commercial and open-source tools such as SIEM and malware analysis
  - On the incident response process
  - On vulnerability assessment and penetration testing
  - Leveraging a virtual "cyber range" where analysts can take turns running offense and defense on an isolated network built for Red Team/Blue Team operations
  - Running SOC analysts through practice intrusion scenarios, using real tools to analyze realistic intrusion data

These services can be seen as less glamorous than flying big sensor fleets or collecting large amounts of raw data. From the perspective of the constituent SOCs however, they are more valuable. By providing these services, the coordinating SOC is likely to achieve its unique goals better than if it tries to provide the same capabilities as its subordinates.

3.3.7 Data Flows for Coordinating SOCs

Once the roles and responsibilities of the central and subordinate SOCs are clarified, the SOCs can consider how data and information will flow between the organization. See Figure 10 for a potential construct.

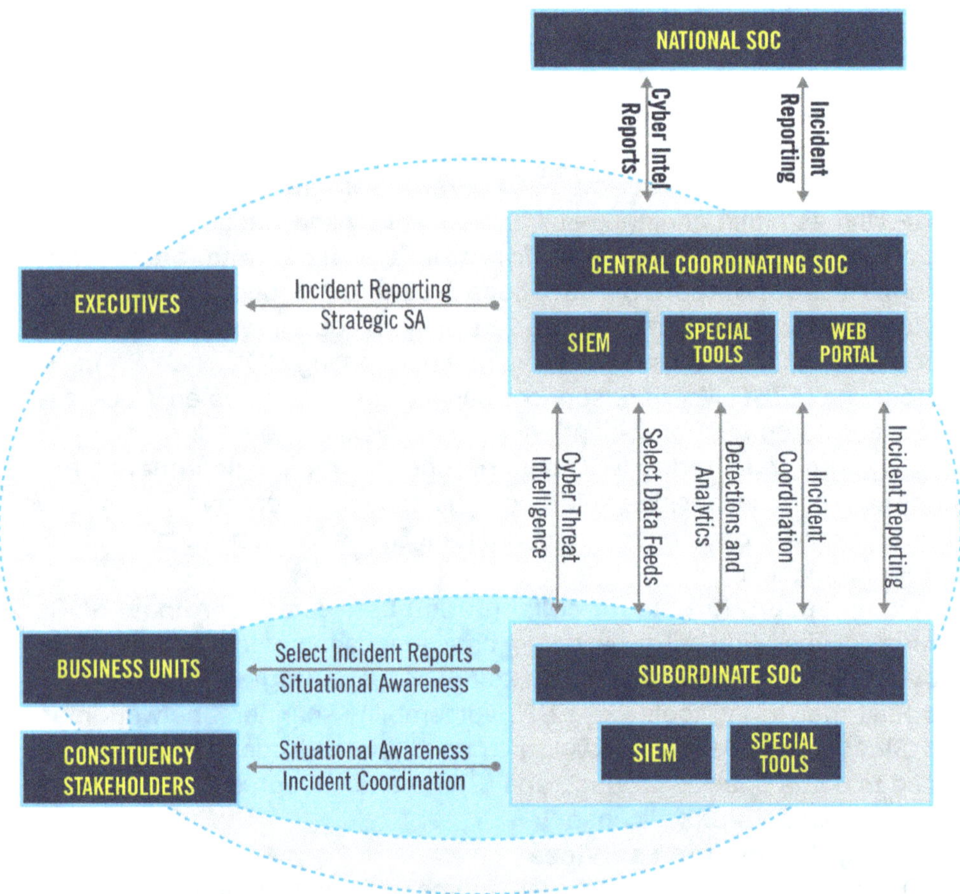

Figure 10. Data Flows Between Central and Subordinate SOCs

There are several themes that emerge here:

Coordinating SOCs are well positioned to handle tasks that scale across the constituency and can be done in one place
For instance, their expertise in advanced tools and adversary TTPs makes them a good place to formulate training programs for the subordinate SOCs. It is also a great place to

perform cyber threat intelligence curation/fusion and TTP analysis because they should have the analysts, the tools, time, and knowledge of constituency networks to make sense of the artifacts handed to them by subordinate SOCs.

Subordinate SOCs perform most of the tactical hands-on monitoring, analysis, and response to incidents

The coordinating SOC is there to make sure its entire constituency is working toward a common goal, and that they have shared SA. While the subordinate SOCs may provide a limited event stream to the coordinating SOC, it is unlikely the coordinating SOC analysts have the context to make sense of that data. Incident reporting and trending from the subordinate SOCs support coherent SA formulated by the coordinating SOC.

Coordinating SOC often have a significant role in big technology purchases, security subscription services, and custom tool development

Requiring subordinate SOCs to use a specific product may be too heavy-handed. Instead, what may work better is to mandate the use of a type of tool and have the coordinating SOC manage the budget for copies of one or two specific brands. As a result, the subordinate SOCs can use the enterprise tool if it fits their needs. Or, they can pay for their own if it does not, as long as it is interoperable with the enterprise tool.

Coordinating SOC must work diligently to maintain relevance and usefulness in the eyes of the subordinate SOCs

The subordinate SOCs are typically sitting on a large data estate while the coordinating SOC themselves may have little of their own sourced data. Rather, the coordination SOC may act as a consumer of feeds form subordinates and external entities. They must also be careful that downward-directed guidance and tasking are perceived as relevant and useful. As discussed in "Strategy 8: Leverage Tools to Support Analyst Workflow," if the disparate SOCs can leverage tools (such as EDRs, SIEMs, or log storage) that support distributed query execution (sometimes known as federated query or cross-cluster join), it can be hugely enabling for collaboration and leveraging data in place.

Coordinating and subordinate SOCs must work in a symbiotic relationship that stems from perceived value and analyst-to-analyst contact, certainly more than mandate and policy. The coordinating SOC may offer substantial help in the form of surge support, such as: in-depth malware reverse engineering or forensics capabilities; incident response fly-away rapid response teams; cyber threat intelligence; and SA to its subordinates, perhaps in exchange for the subordinates' processed incident reports. The subordinates turn data into information; the coordinating SOC turns information into knowledge. This relationship is self-reinforcing over time and, usually, must begin by the coordinating SOC offering something of value to its subordinates that these subordinates cannot get on their own, such as tools and authority.

3.3.8 Federated SOCs

Some very large constituencies land on situations where there is no one coordinating or superior SOC that has authority over the entire constituency. Instead, in such situations, there are multiple SOCs each responsible for their portion of the constituency with no higher-level authority other than perhaps IT executives like the CISO, CIO, and CRO. This arrangement can work well, provided the following guidelines are recognized and adhered to:

- The SOCs should all agree upon how they divide up the constituency through formal governance.
- Because ambiguity is likely to exist regarding IT/OT asset alignment with organizational alignment, the SOCs should have a shared asset tracking capability that assigned clear SOC ownership down to the individual IT/OT asset and/or network. This tracking capability should leverage other asset tracking databases, update routinely and dynamically as organizational shift, and ideally reflect a single source of truth about the entire constituency's IT, OT, and cloud assets.
- The SOCs will need both tactical deconfliction rules around monitoring, scanning, and response activities, along with a plan for how to handle cross-SOC, enterprise-wide, high-criticality incidents.
- As with any other model where many SOCs work in close proximity, the SOCs should have routine forums for information sharing regarding tools, tradecraft, CTI, analytics, detections, and related matters.

3.3.9 National SOCs

At the highest end of constituency size are coordinating SOCs working across constituencies typically at the level of a national/federal government or entire nation. National SOCs serve constituencies that are massive. Unlike coordinating SOCs operating in a hierarchical relationship inside a given government agency or corporation, these national level coordinating SOCs often have constituencies of millions of users and dozens to thousands of constituency SOCs. These coordinating SOCs have a unique mission and role; their goals include:

- Forming a coherent SA picture at the national scale, focusing on constituency vulnerability to threats, and adversary TTPs, usually without access to subordinate SOC telemetry or substantial directive authority.
- Enabling operations among their subordinate SOCs.
- Setting broad best practices and standards for subordinate SOCs.
- Bringing their subordinate SOCs up to a baseline set of capabilities.
- Developing and promulgating educational and training materials intended for a very wide audience.
- Maintaining and promulgating threat situational awareness across many disparate verticals of government, education, nonprofits, and private industry.
- Providing SA and cyber threat intelligence curation/fusion serving leaders at high levels of government.

For a listing of national SOC/CSISRTs, see [44].

3.4 Selecting SOC Functions and Services

There are only a few prerequisites for being a SOC: the ability to (1) accept incident reports from constituents, (2) help constituents with incident response, and (3) communicate information back about those incidents.

This represents only a fraction of the SOC's potential duties. The question is, what other capabilities should it provide? This section explores the possibilities for what capabilities a SOC may offer. The primary objective of the SOC is to support the risk management needs of the constituency. Therefore, the services offered should align to meeting that need. Going along with this, the SOC should: (1) carefully manage expectations of constituency members and executives, (2) enhance trust and credibility with the constituency by handling each incident with care and professionalism, (3) avoid stretching limited SOC resources too thin, and (4) take on additional roles or tasks only when resources, maturity, and mission focus permit.

Before the SOC can decide which capabilities to provide, it must ask some critical questions:

- What is the intended scope of the SOC mission? What does it think it needs and how does this reconcile with the demands of its constituents?
- Is it appropriate for the SOC to engage in direct monitoring and response activities, or is it more productive for the SOC to coordinate and harmonize the activities of other SOCs?
- How does the SOC's organizational placement bias its focus? For instance, is it relied upon to provide SA to constituency executives, or perhaps to implement rapid countermeasures? Can the SOC balance these obligations against other mission priorities?
- What capabilities exist elsewhere that the SOC can call on when needed, malware analysis, vulnerability assessment and penetration testing, artifact analysis and malware analysis, or countermeasure implementation? Are any of these performed by an organizationally superior coordinating SOC? Is outsourcing some of these functions a good decision?
- Does SOC resourcing enable it to reach beyond routine incident response to incorporate additional capabilities such as those described in "Strategy 11: Turn Up the Volume by Expanding SOC Functionality"?
- If the SOC offers a given capability, will it be exercised enough to justify the associated costs?

The SOC will always share control over the scope of its mission with external forces such as edict and policy handed down by constituency executives. Moreover, careful attention must be paid to the perceived expectations of the constituents, versus what capabilities the SOC is in a position to support. Taking another cue from the *Handbook for Computer Security and Incident Response Teams (CSIRTs),* this section emphasizes *quality* of capabilities offered versus *quantity*—do a few things well rather than many poorly [6].

3.4.1 Capability Template

Table 4 illustrates a typical capability offering for several of the SOC organizational models described in Section 3.2.1. It is important to recognize that this table describes the capabilities of each SOC once they have matured into a steady state. In other words, it outlines a target state, not a maturation path; a handful of typical maturation paths are described further in the capability maturation section. Additionally, this table only serves as a starting point—a SOC must always tailor what they take on and how they fulfill organizational needs.

Table 4. Capability Template

	Security As Additional Duty	Distributed SOCs Small/Young Centralized & Federated SOCs	Large/Mature Centralized & Federated SOCs	Hierarchical SOCs	Coordinating & National SOCs
Incident Triage, Analysis, and Response					
Real-Time Alert Monitoring and Triage	b	b	a	a	n
Incident Reporting Acceptance	b	b	a	a	a
Incident Analysis and Investigation	b	b	a	a	a
Containment, Eradication, and Recovery	b	b	a	a	a
Incident Coordination	b	b	a	a	a
Forensic Artifact Analysis	n	o	b	a	a
Malware Analysis	n	o	a	a	a
Fly-Away Incident Response	o	o	b	a	a
Cyber Threat Intelligence, Hunting, and Analytics					
Cyber Threat Intelligence Collection, Processing, and Fusion	o	b	a	a	o
Cyber Threat Intelligence Analysis and Production	n	o	b	a	a
Cyber Threat Intelligence Sharing and Distribution	n	o	b	a	a
Threat Hunting	o	o	a	a	o
Sensor and Analytics Tuning	b	b	a	a	o
Custom Analytics and Detection Creation	o	o	a	a	o
Data Science and Machine Learning	n	o	b	a	o
Expanded SOC Operations					
Attack Simulation and Assessments	n	o	b	a	a
Deception	n	n	o	o	o
Insider Threat	n	n	o	b	o

	Security As Additional Duty	Distributed SOCs Small/Young Centralized & Federated SOCs	Large/Mature Centralized & Federated SOCs	Hierarchical SOCs	Coordinating & National SOCs
Vulnerability Management (if performed by the SOC)					
Asset Mapping and Composite Inventory	b	b	a	a	o
Vulnerability Scanning	b	b	a	o	o
Vulnerability Assessment	n	o	b	a	b
Vulnerability Report Intake and Analysis	b	b	b	a	a
Vulnerability Research, Discovery, and Disclosure	n	n	o	b	a
Vulnerability Patching and Mitigation[4]	b	o	o	n	n
SOC Tools, Architecture, and Engineering					
Sensing and SOC Enclave Architecture	o	b	a	a	o
Network Security Capability Engineering and Management	o	b	a	o	o
Endpoint Security Capability Engineering and Management	b	b	a	o	n
Cloud Security Capability Engineering and Management	o	b	a	a	n
Mobile Security Capability Engineering and Management	o	o	b	o	n
Operational Technology Security Capability Engineering and Management	o	o	o	o	n
Analytic Platform Engineering and Management	o	b	a	a	a
SOC Enclave Engineering and Management	o	b	a	a	a
Custom Capability Development	n	o	b	a	a
Situational Awareness, Communications, and Training					
Situational Awareness and Communications	b	b	a	a	a
Internal Training and Education	o	b	a	a	a
External Training and Education	o	o	o	o	a
Exercises	o	o	b	a	a
Leadership and Management					
SOC Operations Management	b	b	a	a	a
Strategy, Planning, and Process Improvement	o	b	a	a	a
Continuity of Operations	o	b	b	a	a
Metrics	o	b	a	a	a

[4] Even if the SOC has a vulnerability management responsibility, they may not be the organization performing the patching and mitigation. However, it is important that this function is assigned to some group within the organization.

For each SOC organizational model(s), and each potential SOC service, a recommendation is given:

- **Basic (b):** SOCs in this category typically offer this capability/service at a basic level of performance inside the SOC.
- **Advanced (a):** SOCs in this category offer this capability/service at a more advanced, mature level of performance inside the SOC.
- **Optional (o):** SOCs in this category may or may not offer this capability or function. Their choice to do so usually has more to do with their maturity, resourcing, focus, and external requirements than necessarily their organizational model.
- **Not recommended (n):** SOCs in this category are unlikely to offer this capability or function in house. This is usually due to foundational capability and competency not being present, resources being limited, or scoping the focus to what is most appropriate for the organizational model type.

It is important to recognize that with one exception, Table 5 describes the capabilities of each SOC once they have matured into a steady state. In other words, it outlines a target state, not a maturation path; a handful of typical maturation paths are described further in the capability maturation section. Additionally, this table serves as a starting point—a SOC must always tailor what they take on and how they fulfill those needs.

In addition to the categorization provided in the capability template, consider this additional context when developing the list of SOC functions:

Real-Time Alert Monitoring and Triage
This function requires the SOC to first curate a feed of alerts that a set of analysts must triage. New, small, and virtual SOCs may operate in a purely reactionary mode, meaning no such triage is needed.

Real-Time Alert Monitoring and Triage/Incident Analysis and Investigation
This capability is core to virtually all SOCs. However, in very large coordinating SOCs, this will manifest in a slightly different way. Rather than looking at raw data and receiving tips from end users, the SOC will likely receive incident reports and calls for assistance from other SOCs.

Malware Analysis/Forensic Artifact Analysis
These capabilities are hard to staff in many SOCs as there needs to be a steady stream of incidents requiring in-depth postmortem or support of legal action, and the personnel qualified to do this are in high demand. This is a strong candidate to outsource for many SOCs, even well-established ones.

Fly-Away Incident Response
The need for this is a function of whether a SOC's organizational model incorporates elements of the internal distributed SOC model and how geographically dispersed the constituency is. In addition, when there is a series breach or incident, a SOC may call in an MSSP specifically to augment its own internal capability rather than maintaining an internal fly-away team.

Sensor and Analytics Tuning
If a SOC has its own fleet of monitoring equipment (e.g. EDR, SIEM, UEBA, SOAR), tuning must be a sustained, internal activity driven by feedback from analysts. Therefore, this capability is considered a requirement for any SOC that deploys monitoring capabilities. It is also important to recognize that sensor tuning is a continuous process necessary for the correct functioning of the sensor fleet. It is, therefore, an operations function and not an engineering or development function.

Vulnerability Assessment, Attack Simulation and Assessments
For some constituencies, the SOC is a natural place to house and coordinate these activities. It provides a unique basis for enhanced SA and raises visibility of the SOC as a resource to system owners and security staff. For other constituencies it will make more sense to perform these functions outside of the SOC due to the amount of focus on these functions, the size of the team itself, of the desire by the constituency to have these teams provide a perspective independent from the SOC.

Vulnerability Patching and Mitigation
As most SOCs grow in size and maturity, if they had this function in house for anything other than their own assets, it tends to get split off into an IT management function or its own organization.

Vulnerability Assessment and Vulnerability Scanning
These capabilities may be viewed as sustaining IT functions rather than security operations functions. Consequently, the choice to include these in the SOC should occur on a case-by-case basis. Whether or not these are placed elsewhere, the SOC will want to maintain access to the results as up-to-date network maps and vulnerability statistics are key inputs to incident analysis efforts.

Network Security Capability Engineering and Management
As is discussed later in "Strategy 7: Select and Collect the Right Data," network IDS/IPS and firewall functions have merged for many devices and deployments. As a result, this may shift over to IT operations, while the SOC is a recipient of this data and has purview over device and signature tuning.

3.4.2 Capability Maturation

This section provides some potential growth patterns for several popular functional areas, showing how a SOC may expand from one capability to the next. This can also be used to identify dependencies and relationships among capabilities.

Cyber threat intelligence collection and curation + sensor and analytics tuning, with growth to hunting and custom signature creation
Continual exposure to multiple sources of cyber threat intelligence will train analysts to be more discriminating in what they gather, help them recognize how their own defenses can be enhanced, and drive new hypotheses on how to find the adversary "left of hack." Knowledge

of adversary TTPs, constituency environment, and how to write advanced analytics naturally leads analysts to crafting their own custom detections.

Incident analysis, malware analysis, custom signature and analytics creation to cyber threat intelligence creation and distribution
Over time, analysts' experience with individual incidents should grow into a more macroscopic understanding of adversary behavior. This can lead analysts to draw observations and conclusions they may wish to share with constituents or other SOCs, further reinforcing their SA.

Incident analysis to forensic artifact analysis, and/or to malware and implant analysis
As the volume of incidents increases, so too should the SOC's consistency and efficiency in handling those incidents. Analysts' need to establish root cause analysis often leads them in one or both of the following directions. First, as the volume and complexity of incidents caught increase, so too will the number of traffic capture and media artifacts. What may start as ad-hoc artifact analysis will likely turn into a repeatable, rigorous process involving dedicated forensics specialists and tools. Second, the amount of malware caught will likely increase, and thus the need to understand the comparative threat posed from one incident to the next—is this typical malware or a targeted attack? The SOC may evolve proactive means to regularly extract suspect files from network traffic and perform static and/or dynamic malware analysis on those files.

Tool engineering and deployment to tool research and development: As mission demands grow, the SOC will likely run into the limits of COTS and FOSS capabilities, leading the SOC to develop its own tools. This will likely start with projects that "glue" multiple open-source and commercial capabilities together in new or different ways. In more extreme examples, well-resourced coordinating SOCs may put together, from scratch, polished capability packages for their constituents.

For additional examples of capability maturation, consider the following resources see: [5], [45], [46].

3.5 Should We Operate Our Own SOC?

Not every constituency is best served by operating its own in-house security monitoring and incident response team. This approach must be evaluated against several factors, including constituency size and IT budget. In a lot of cases, especially for small- and medium-sized businesses, partial or complete outsourcing may be most appropriate.

The question is not if the constituency requires security monitoring and response capabilities, but how it will procure them.

Every organization requires SOC capability, with varying degrees of sophistication. Some are very clearly ready to make this investment as an operation that is largely achieved natively, in-house. Others may be better off asking or paying someone else to do it. To help put these choices in context, including data on SOC outsourcing trends, consider SOC survey results at the following: [33], [47], and [48]. For more on SOC outsourcing, see [49].

3.5.1 For In-House SOC Capability

An organization that is best served by operating and maintaining an in-house SOC capability will most likely answer "yes" to most or all of the following:

- Has an in-sourced IT staff and operations and recognizes IT as a core competency of the business. This may include having a designated CIO position and having dedicated IT operation staff, and potentially center, such as a virtual or physical NOC.
- Owns and operates many thousands of IT, OT, and networked devices, either on-prem or in the cloud and/or operates IT systems with many thousands of active user accounts (or more).
- Has an incumbent internal investment in security, including:
  - Risk measurement and risk reduction apparatus.
  - A leader with the sole, designated responsibility for cybersecurity, such as a CISO.
  - Other security functions such as governance, compliance, security engineering services, and vulnerability management.
- Have specific needs that it assesses an MSSP cannot address well:
  - Specific monitoring, defense, response requirements that diverge substantially from standard offerings and SLAs or security risks.
  - Desires service-, application-, or mission- specific instrumentation, detection, and response capability.
  - Assesses specific and substantial business or confidentiality risks with procuring security services from an outside entity, such as personnel issues or having a large portion of assets that are high criticality and cannot connect to the cloud or the Internet.
  - Seeks an overall high level of engagement from the security operations team.
- Is willing and able to dedicate resources to fund several dedicated SOC staff and technologies and is interested and willing to build security maturity as an investment over time.

3.5.2 For Outside SOC Capability

An organization that is better served by seeking security operations and response services externally will likely answer in the affirmative to most or all of the following:

- Does not have substantial resourcing and competency in IT
- Owns and operates less than a couple thousand of IT, OT, and networked devices and total active monthly users measured less than several thousand unique accounts
- Does not have an incumbent investment in security, is new to risk management, and does not have strong SA across its organization
- Does not have an IT and security budget to proportionally allocate more than a couple people to security operations

- Is unable or does not wish to operate sensing, monitoring, scanning and analytic equipment, either on-prem or in the cloud (as a service)
- Assesses its security needs are readily service by an MSSP, or if the constituency belongs to a large organization such as a multi-national conglomerate or government, can be served by the SOC belonging to that parent entity
- Assesses that another entity will be able to instrument and respond on/against its assets without onerous risk or integration cost
- Is not seeking daily or weekly engagement from the security team.
- Needs a security monitoring and response apparatus immediately and/or is not strongly interested in building that competency over time

Organizations that self-assess the above questions in a split manner or with an ambiguous outcome may wish to consider a hybrid partial outsourcing arrangement. This might mean that the SOC keeps some functions in-house like monitoring and incident response but outsources other functions such as pen testing or malware analysis.

3.5.3 Succeeding at Outsourcing

In security operations, outsourcing, the practice obtaining services from an outside or foreign supplier, is common, with an increasing set of service providers [50], [51], [52]. Outsourcing in one form or another can go well or poorly, depending on just a few key factors:

- How the outsourcing contractual and business relationship is structured
- The expectations set inside or outside of that formal business relationship
- The human relationship between the outsourced staff, in-sourced security staff, and constituents
- The overall culture and accountability of cybersecurity across the constituency

This leads to an important point:

> *No matter the outsourcing arrangement, constituency system owners and executives are still accountable for security and risk reduction.*

This point is often forgotten, "Oh, we paid Acme Cybersecurity, Inc. to take care of this problem, I don't have to worry about it anymore." Nothing could be further from the truth. Execution may be outsourced, but the authority and accountability to ensure the constituency is not breached still falls on executives (CEO, CIO, CRO, CISO, etc.) and their staff. A breach is still a breach.

> *Expect and demand that any outsourced security staff or function will require routine, daily support from the constituency, because it cannot operate in a vacuum.*

Nevertheless, there are several good reasons to outsource security operations staffing or functions. This may involve using on outsourcing on a temporary basis or planning for outsourcing in the long term. Reasons for outsourcing include:

- There is a huge disparity between need and availability of qualified cybersecurity professionals.
- The constituency simply cannot afford prevailing salary and benefits rates for experienced SOC professionals within the geographic region the constituency wishes to operate a SOC.
- The constituency is of insufficient size to sustain a SOC at all.
- While a SOC may exist, it is of insufficient size or resourcing to fill roles with the frequency (malware analysis) or hours (24x7 alert triage) as needed.
- It anticipated that its staffing needs will ebb and flow, such as for surge incident response support.
- It needs to stand up a SOC capability much faster than ordinary, organic hiring will allow.
- The SOC has lost too many key staff members and needs to fill a talent gap while rebuilding their own internal team.

Some common keys to success across all outsourcing models are as follows:

Establish expectations, usually as part of the contract language
These expectations should include the scope of services to be provided along with details such as:

- Staffing and head counts, including personnel vetting, background/investigation requirements, staff removal
- Staff qualifications (depending on the contract type)
- Hours of operation, on-call procedures, and surge capacity
- Response times and other SLAs
- Intellectual property and data handling requirements

Be realistic about SLAs
Have mechanisms in place to measure SLAs, e.g., outsourcing management of a sensor fleet to an ordinary IT vendor can go poorly if things like signature update cadence is not specified and measured. It is common, for example, to demand certain notification times for detection of a potential serious incident, but mandating closure times for major incidents is generally not possible. Consider that some well-meaning SLAs may drive toxic behaviors, e.g., mean time to close SLAs may force analysts to perform weak alert investigation. See more about SLAs, both positive and negative outcomes, in "Strategy 10: Measure Performance to Improve Performance."

Be clear about the risk being induced by outsourcing
Outsourced cybersecurity by its nature will mean giving an external organization access to the constituency's and the SOC's most sensitive secrets; be sure there are safeguards for both staff vetting and legal resource if the outsourcing company (or its people) are at fault. Work with constituency IT staff to limit the exposure through limiting access where possible, and ensuring staff are vetted, with some monitoring from an oversight perspective.

Evaluate the entire offering

It is necessary to look at the complete picture, particularly if the outsourcing model in question is more than just people. If an external business is being hired to provide a complete solution, not just staff, ensure the efficacy of that solution package, such as sensors and tools, will meet expectations.

Build in flexibility and growth

Watch out for hard-coded language about specific solutions and capacities that cannot be changed without a major update to the contract. Specifying "put this many sensors in location X" will be inaccurate as soon as the contract is signed. In some historical examples, certain contracts would go on for five plus years with static technology purchases, far outlasting meaningful technology cycles; meanwhile the mission could have iterated tremendously in that time. This can be disastrous to the SOC mission and demoralizing to the analysts. By contrast, it may be better to drive an outsourced solution to a certain percentage of coverage, with some guard rails around cost and scale that protect both parties, and language that drives evolution of SOC tools over the entire period of performance. Beware of the old saying "but that's not in the contract"; flexibility is key.

Employee engagement may be a persistent issue

In any outsourcing situation, it is very easy for the contract staff, particularly the individual contributors, to take on the attitude of "this is just a job, I punch a clock, and at the end of the day this is not my problem." There are always exceptions to this rule, but it should be a consideration when outsourcing. To minimize this, some of the managers, and particularly IT staff, might maintain regular contact and relationships with contract staff and strive to build more of an integrated team.

The service provider may have limited understanding and engagement with the constituency mission

By placing SOC functions or staff outside the walls of the constituency, it is very easy for them to lose touch with the constituency, and thus not be tuned into their needs, or specific lines of business. On the other hand, some of the best MSSPs will have monitoring and cyber threat intelligence capabilities tailored to specific industry verticals which could result in a better monitoring and detection capability than the constituency ever could on their own. One thing to look for in an outsourcing provider is evidence of services tailored to the SOC's computing environment, cloud provider, non-traditional IT, and industry. This should take the form of special monitoring tools, tailored cyber threat intelligence, and a cadre of other companies that attest to the quality of these special services.

Ending the contract could means erasing progress

When a managed service or outsourcing contract ends, the institutional knowledge (and sometimes the tools/technology) of the analysts usually leaves too. This can be devastating. Incoming contractors are known to try and hire the outgoing contract staff; sometimes this helps but notoriously the incoming contractor may undercut the incumbent's salary/benefit package, and everyone will leave anyway. When a constituency chooses to outsource, it is making the choice to lose much of the knowledge accumulated by that vendor when its contract ends. A way to counter this knowledge walking out the door at the end is to write into the contract or create SLAs or other agreements for some knowledge transfer and transition to IT staff. Although not optimal, some knowledge transition is better than none. Language should also be included to ensure outgoing contractors are required to provide all data and documentation generated during their contract period. This should include sensor data, playbooks, access information, contact lists, etc. Without this information the SOC's parent organization could find itself trying to build up a SOC from ground zero again.

At the end of the day, digital transformation is causing many organizations to recognize IT and cybersecurity as fundamental portions of their business that must be largely in-sourced. At the same time, even more organizations recognize the need for managed cybersecurity services. As a result, the need for quality security outsourcing should continue for the foreseeable future [53], [54], [55].

The subsections that follow discuss major outsourcing models, including some tips for success for each. It's worth noting that these models are not mutually exclusive and can be used in combination [56], [57], [58].

3.5.4 Staff Augmentation

In this approach, the SOC is executed and "owned" by its parent organization as an indigenous, in-house capability, where an external vendor is hired to provide staffing to the SOC. In such an arrangement, some teams may be mixed with internal employees and outsourced staff, or some teams may be entirely composed by the outsourcing contractor. Most importantly, in this arrangement, the SOC is performed in owning constituency parent facilities, and likely is managed by constituency full-time staff.

This arrangement is particularly prevalent with organizations that can sustain an in-house SOC, and need a way to augment their own personnel sourcing, or want to reduce costs. It is particularly common in the US Government but can also be found in private industry. When planning for staff augmentation consider the following:

- Contractors may find it difficult to find and retain quality staff, just as anyone else, particularly due to cost constraints from the contract itself, profit-seek motives by the service provider, or both. Strong personnel qualifications should be built into the contract to avoid this.
- Underperforming staff can become a problem, especially if the customer has no contractual recourse. This is easily avoidable by giving the customer vetting control over who comes in the door (not just through background checks), and the ability to eject underperforming contract staff without laborious paperwork.
- Poor employee engagement and problem/solution ownership can be a challenge. One of the ways to help mitigate this is to treat contract employees as full-fledged members of the team. "I'm just a contactor" can be a toxic behavior that should be recognized and addressed. Ways to mitigate that include:
  - Strive to include security contract vendor staff in team building events or other group activities whenever legally possible

- ◦ Ensure that the contract and behaviors of the contractor support upward mobility, just as they would for internal staff
- ◦ Drive a set of operational metrics that support positive behaviors, such as problem ownership and innovation
- ◦ Recognize successes by the contractor, as appropriate

3.5.5 Select Functions Outsourced and Surge Support

Some SOCs choose to outsource certain functions that they do not feel they can support internally. This arrangement is very prevalent for centralized and distributed SOCs that are small or young. There are two prominent drivers here: cost (keeping a single seat filled 24x7 is not cheap) and availability of skilled resources (good malware analysts are hard to find).

One of the keys to making this arrangement work well is ensuring clear expectations and good integration of any sustained capability, as noted above. The other is mitigating the challenges that are likely to crop up with SOC functions are broken apart: disparate teams, and thus potentially disparate perceptions, rhythms of business, and priorities. The contract in place needs to protect both parties. For example: when outsourcing alert triage, the contractor must be protected from being flooded with unfiltered alerts, and the contracting SOC must be protected from being flooded with poor quality escalations. Additionally, it is valuable to have a feedback loop between different parts of the SOC; outsourcing select functions can create friction in this feedback loop and is a major drawback of this outsourcing arrangement. However, compared to the alternative of full-blown outsourcing, it may be the best choice for some. Here are some popular functions that SOCs will outsource:

Alert triage outside business hours
In this arrangement, alert triage, traditionally associated with a tier 1-type function, is contracted out. This saves the SOC from a substantial labor cost and may be most appropriate when the SOC needs a 24x7 watch but does not want to sustain that from an operations perspective. Beyond the points made above, the SOC needs to manage its risk around opening access to its tools to the service provider and be careful about what investigation it expects the service provider to perform. Additionally, many SOCs look to alert triage as a "pump" to bring on new and inexperienced staff; SOCs outsourcing alert triage will need to find other methods to train up personnel new to security operations that present managed risk to SOC operations.

Malware analysis and media forensics
With this approach, the SOC outsources one or both duties to the service provider. This is especially appropriate for SOCs that only occasionally need this kind of expertise AND feel as though their needs exceed that of what automated analysis tools like cloud-based file detonation can provide. Moreover, many SOCs may only be able to occupy a malware analysis a quarter of their time, leaving the other three quarters to non-malware duties; for someone aspiring to be a pure-play reverse engineer, this can become boring and may lead to attrition. Outsourcing helps address this issue.

Surge incident response support

During a major incident, such as a breach, the SOC's own ability to process data, investigate, and coordinate response may become overwhelmed. Having a service provider under contract, ready to activate with only hours' notice can reduce risk and damage in a serious incident scenario. In fact, even mature and older SOCs may consider seeking IR surge support in advance, as trying to find a service provider in the heat of an incident can cost precious time. This is doubly important for constituencies with complicated and time-consuming procurement practices.

Penetration testing and red teaming

Here, some or all of this function is contracted out. This can be as a "one time" annual arrangement, to sustained operations. Some security organizations will occasionally seek a third-party penetration test out of due diligence, a desire for an independent party to bring a fresh perspective, regulatory compliance, or a combination thereof. Contracting a third-party penetration test can be especially powerful when internal politics may be a barrier to shedding light on critical weaknesses or systemic issues.

3.5.6 Turnkey Tools as a Form of Outsourcing

As discussed in "Strategy 8: Leverage Tools to Support Analyst Workflow," the SOC will frequently be confronted with the choice of "build versus buy" for various technology investments. Today it is possible, for example, to compose a SIEM out of different technologies that together provide roughly the same features as a commercial-off-the-shelf (COTS) SIEM product; executed well, this can save acquisition and sustainment costs in the $millions and grant the SOC a more flexible solution that it might otherwise have. This *may* be offset by an increase in personnel cost, but just as important, keeping skilled engineers on staff. Considering SOC engineers are in short supply just like SOC analysts, buying turnkey solutions, including those available as a cloud service, may be appealing as there is comparatively low cost, schedule, and performance risk for acquisition.

There may also be a lower cost to maintain. For example: tools like Sysmon, OS Query, and Elasticsearch together provide a superb set of capabilities that displace *some* of what a good EDR product will do [59], [60], [61]. However, they come at a cost not only for maintenance, but for writing and maintaining a huge library of detections and analytics. Buying a best-of-breed off the shelf EDR product would not present nearly the same issue. On the other hand, many commercial products can have extremely high annual maintenance charges-30% or more of the original purchase price per annum.

Increasingly, SOCs today can look upon cloud based SOC tools, and turnkey commercial on-prem solutions to shift their staffing challenges away from SOC tool engineering/sustainment and toward other functions like analysis. However, this does not absolve the SOC of requiring good engineers-someone must keep even cloud-based SaaS offerings running optimally.

3.5.7 Fully Outsourced

Some organizations find it necessary or appropriate to outsource the SOC to an external provider in its entirety. In such an arrangement, the constituency itself has no SOC at all, and an external contractor, such as an MSSP, takes on the entire SOC mission.

This arrangement is usually most appealing to small constituencies-usually smaller than 10,000 nodes or users-that cannot sustain a SOC, but nonetheless need the monitoring and incident response mission executed. Because full outsourcing of SOC services to an MSSP is a well-established industry, the value factors used to differentiate candidate vendors are well known. They include:

- Their overall business and pricing model
- How they propose to instrument the organization and the depth of that instrumentation (See "Strategy 7: Select and Collect the Right Data")
- How to they propose to maintain strong understanding and awareness of constituency hosts and networks (beyond ordinary vulnerability scanning)
- What routine engagement do they provide, such as through reporting
- Under what circumstances will they reach out, meaning what incidents and issues will be escalated out of routine communications
- SLA response times
- A ratio that gives a sense of how stretched their analysts may be; this could be: analyst to sensor, analyst to node count, or customer per analyst
- Their expertise and ability in supporting regulations specific to the constituency such as payment card industry (PCI), Sarbanes-Oxley Act of 2002 (SOX), HIPAA, etc.
- How do they care, train, and rotate their staff, and what is their attrition rate?
- What controls do they put in place around the security of the data collect, and downstream tool and cloud providers

For more information on being successful with fully outsourcing cybersecurity to an MSSP, see [62], [63], [64], [6].

3.6 Should the SOC Go 24x7?

The adversary works 24x7. Should the SOC as well? To answer this question, the SOC must carefully examine the risks to its mission(s) and the daily/weekly patterns of activity across the constituency, the scope of the SOC's mission, and SOCs expected staffing resources.

Considerations for moving to a 24x7 SOC include:

- Is there a specific set of targeted or advanced threats against the constituency that suggests intrusion activity is likely to happen outside of normal business hours?
- What is the size of the constituency and what are its normal business hours? Does its user population have after-hours access to IT resources?
- Does the constituency's mission encompass 24x7 operations that depend on IT, which, if interrupted, would significantly imperil revenue or life?

- Have members of the SOC come in after hours to handle an incident, or have they discovered an incident outside SOC duty hours that could have been prevented if someone were there to catch it? Has this happened several times?
- If an attack occurred during off hours and there was an analyst there to notice it, are there resources outside the SOC that could stage a meaningful response before the following business day?
- How big is the current SOC's staff pool? Are there only four people, or is it funded to support a team of 50 or more analysts? Is there the financial ability to increase the size of the team to meet a 24x7 staffing need?
- Is the host facility open 24x7? If not, can the SOC be granted an exception?
- Is there an organizational stigma associated with not being operational all the time? Some SOCs are considered a legitimate operations center only if they function 24x7.
- Is there a tolerance for a delay in identifying suspicious events? A SOC that is not 24x7 must catch up every morning on the previous night's event data; for those that do not staff on the weekends, this is especially challenging on Monday.

For many SOCs, going 24x7 is not an all-or-nothing decision, there are multiple options for expanded staffing beyond Monday through Friday 9-5:

- Staff only certain portions of the SOC 24x7, such as event monitoring and detection; leave other sections with a designated on-call pager or cell phone.
- Extended hours Monday through Friday (12x5). Expand operations beyond 8x5 so that there are SOC analysts on shift during the bulk of time that constituents are logged in, such as 6 a.m. to 6 p.m., if the constituency resides primarily in one or just a few adjacent time zones.
- Extended hours Monday through Friday plus some time on the weekends (12x5 plus 8x2). Add one shift (8 or 12 hours) of two or three analysts during weekends. This eliminates the problem of clearing a weekend's worth of priority alerts on Monday and provides coverage if the constituency performs business operations on weekends.
- Staff all capabilities 24x7, however have the bulk of the resources present during regular business hours.
- For SOCs that run operations out of multiple locations, stagger shifts between them— as in follow the sun or backup/contingency ops floors in disparate time zones.
- Outsource to another part of the organization. If the SOC has a dedicated triage section, it could, during off hours, hand off those operations to the parent coordinating SOC or sister SOC—assuming they are 24x7, can easily access the SOC's monitoring systems or data feeds, and are able to familiarize themselves with the SOC's constituency mission and networks.
- Another approach is to contract with a SOC managed service provider and hand-off monitoring and detection during off hours or contract with them to provide some or all capabilities 24x7.

If the SOC does move to 24x7 operations, some additional things to consider are:

- **24x7 SOCs will usually want to maintain a minimum staff of two analysts at all times:**
  - Two-person integrity is a best practice in monitoring since having only one person there with access to a lot of sensitive data and systems can present problems, no matter how well-vetted the employees.
  - There are logistical and safety concerns with keeping the floor staffed and secured when someone needs to leave the room.
  - With multiple analysts always on shift, they can cross-check each other's work.
  - Being the sole person on shift can be very lonely and monotonous.
- **Each 24x7 seat requires roughly five full time equivalents (FTEs), including fill-in for vacation and sick leave:**
  - This is very expensive compared to 8x5, 12x5, or even 12x7 staffing.
  - Assuming a minimum of two filled analyst seats, that means roughly 10 FTEs, potentially 11 depending on leave and vacation policies within the organization.
  - Therefore, taking a SOC from 8x5 to 24x7 requires an increase of at least eight FTEs.
- **Analysts on night shifts may feel underinformed, underutilized, and unrecognized:**
  - The night shift does not see the fruits of their labor as much because detailed analysis, response, and feedback usually occur during normal business hours. It is important to have regular deliverables/work output from those on night shift and regular feedback regarding what happened during the day. Because night-shift staff have far fewer interruptions, it may be effective to give them unique tasks that require several hours of focused work, such as in- depth trending or cyber threat intelligence analysis.
  - Casual information sharing is less likely to occur when only a few positions are staffed, and a dedicated effort must be made to ensure collaboration across the entire team regardless of the shift time.
  - One way to offset the issue of staff feeling disconnected is by rotating staff between days and nights every couple of months.
  - Collocating the SOC ops floor with a NOC or IT Operations floor may help, especially with loneliness and management supervision.

During each shift, watch-floor analysts may be encouraged to keep track of important pieces of information in a centralized log (e.g., phone calls, interesting events, visitors to the SOC, or anything else out of the ordinary). This log can be instrumental in reconstructing timelines of events, enforcing accountability, and demonstrating SOC due diligence.

At the end of each shift, the leaving team will perform what is often referred to as shift pass-down or shift handoff, whereby the outgoing team briefs the incoming team on various issues of note seen during their shift. It is a good practice to use both the centralized station log and a pass-down log to formally document this information handoff. Again, the purpose of this process is to enforce continuity of ops, support nonrepudiation and accountability of the floor staff's actions, and serve as a historical record.

In non-24-hour environments, this pass-down log should probably still be maintained, although any sort of person-to-person handoff will need to work differently due to non-continuous staffing. Regardless of the staffing model, here is what can go in the pass-down log:

- Names of SOC operations staff on duty
- Issues passed down from the previous shift, especially those mentioned verbally and not captured in the previous pass-down log
- Case IDs opened and closed during the shift
- Tips and referrals from other parties such as the help desk or users
- Any issues escalated to parties outside the SOC
- Sensor or equipment outages seen
- Any sort of anomalous activity seen, especially if it has not yet spawned a case
- Any anomalous activity that was seen that requires the incoming shift to continue analysis or escalation procedures
- A check-off of duties that the team was required to accomplish
- Details on any tasks assigned to the shift that were not completed or need further attention
- Anything else of significance that was encountered during shift that is not covered in the SOC's SOPs

It is best to have a standard pass down log template that each shift uses, which is usually filled out by the shift lead or team lead on duty. While the log may be captured electronically, it is important to print the log and have all analysts on the outgoing team and the incoming shift lead sign it before the outgoing team leaves for the day. This is key to maintaining accountability for what was done and ensure that nothing is dropped.

3.7 SOC Physical Location and Maintaining Connections Among Distributed Staff

The overall physical environment of the constituency will drive the physical environment of the SOC. If the constituency is located in a single location with everyone working on-site, the SOC will likely mirror that arrangement. If the constituency is geographically distributed with strong remote work options, the SOC may reflect a more distributed approach with analysts working from home.

The SOC can find its physical location a great help or hindrance, depending on several factors. There is a balancing act between SOC size and the need to maintain closeness to the end asset and mission. This may indicate the need to place SOC personnel at multiple sites. This section addresses the following intertwined issues:

- Where the SOC should be physically placed.
- How to arrange SOC resources distributed among several locations.

There are plenty of affordable real-time telecommunication's capabilities—Voice Over Internet Protocol (VoIP), video teleconferencing, real-time chat, and desktop webcams— many find that this is not a replacement for physical presence. Walking across a watch floor to check on a relevant screen can be very powerful. When different sections of the SOC are moved apart, even to different rooms in the same building, collaboration must become even more of a priority. For this reason, the SOC can mix elements of centralized and distributed organizational models, such as leveraging "forward deployed" analysts.

3.7.1 Goals and Drivers

Physical location of the SOC can be a tough choice for some constituencies. Table 5 discusses many of goals and considerations when making this choice:

Table 5. Considerations for SOC Placement

Goal	Discussion
Provide the SOC with a physical space that meets the SOC's mission needs	With the exception of virtual SOC, a physical operations floor will be needed. This usually entails an ops floor, back offices, and potentially a server room, all with secure access. This also means having ample bandwidth to constituency wide area networks (WANs), cloud resources, campuses, and data centers. Existing constituency office facilities and data centers may meet these needs better than other options.
Synchronize operations among the sections of the SOC	The four main functions of security ops should be brought into one organization, the SOC. It is also highly useful to bring them under one roof, supporting regular, healthy, and usually informal collaboration.
Maintain clear lines of separation between SOC functions and IT and cybersecurity functions	Plenty of other people in the constituency believe that some element of security is in their swim lane, fueling conflict with the SOC. Physical presence near these other entities gives the SOC an advantage by supporting close, ongoing contact that can help keep the SOC's interests and impact visible to other stakeholders.
Keep close contact with constituency leadership and other relevant groups	The SOC must leverage support from parent organizations and coordinate with various groups in response to incidents, especially major ones. While this can be done virtually, it is best if they can be brought together physically.
Provide analysts better constituency mission and operations context, speeding analysis and response efforts	Being at a site where computing assets and users reside automatically gives the SOC many advantages: the ability to interact with constituents, perform touch labor on sensors, and execute on-site incident response.
Ensure the SOC's focus on the constituency is not biased to any one organizational or geo-graphic region	Analysts tend to orient on what is going on at the site where they are located. This is both a blessing and a curse. If analysts are biased toward their site, what are they missing at the others? In more extreme cases such as large, heterogeneous enterprises, remote sites will perform their own security monitoring and incident response functions without coordinating with the SOC, in large part because they feel the SOC is out of touch with their mission.
Better position the SOC for staff hours that mirror when the constituency is open for business	The SOC's business hours should encompass those of the constituency. It helps to place the SOC in the same time zone as a plurality of the constituency. If the constituency's users and digital assets are spread all over the world, the SOC may have more options to maintain 24x7 operations while keeping analysts employed only during the daytime.

Goal	Discussion
Ensure continuity of operations of the SOC mission through geo- graphic diversity of SOC assets	The SOC should ideally be considered integral to the constituency's mission. This may compel SOC management and constituency executives to create one or more additional redundant or "load balanced" operations floors, giving the SOC some geographic diversity and resiliency.

3.7.2 Where to Physically Locate the SOC

In theory, the SOC can operate from any location that has office space, connectivity to its tools, and connectivity to its constituency. For some SOCs, this includes work from home and virtual SOC arrangements. If the constituency has consolidated its IT into one or a few data centers or offices, the SOC could operate there, providing on-site response for a large proportion of incidents. Doing so would also allow the SOC to orient toward mission systems, enabling them to focus more on what is going on with the computing environment and less on routine politics. In practice, this is not always the best strategy.

Practically speaking, most non-coordinating centralized SOCs, meaning SOCs that are members of their constituency, coordinate with three entities daily: cybersecurity apparatus/ office of CISO, constituency IT staff, and constituent users themselves. Furthermore, they rarely have absolute authority in incident prevention or response; in this regard, their most important contact(s) are those from whom the SOC derives power, such as the CIO or CISO. The SOC must maintain routine contact with constituency executives to stay relevant.

There are constant changes to policy, monitoring architecture, threat, and incidents, all of which require regular coordination. If there is insufficient power, space, and cooling for on-prem SOC servers or no suitable place for a SOC operations floor in the headquarters building, it may be better for the SOC to find space at a nearby office building.

3.7.3 Succeeding with Virtual SOCs and Work from Home

Some SOCs find it necessary to locate elsewhere or virtually for one or more of the following reasons:

- Lack of physical space at/near the headquarters
- Lack of available security talent at/near the headquarters
- A predominate virtual workplace, flexible workplace, or work from home culture
- Widespread health or geopolitical event, such as the COVID19 pandemic

The global phenomenon of COVID19 response forced virtually all SOCs to shift to a partial or total work from home conditions. Regardless of the reasons, the SOC should observe the following tips and tools for making virtual and work from home cultures successful:

- Computing and tool infrastructure that supports remote work.
  - Leverage remote access virtual private network (VPN), virtual console like Integrated Lights-Out (iLO)/Integrated Dell Remote Access Controller (iDRAC), and cloud-based technologies

- Workstations and access that support remote work, such as virtual thin clients and low-trust/zero-trust architectures
- Robust collaboration tools, including video, persistent chat, and VoIP, supporting:
  - Team synchronization, such as with daily ops standups and staff meetings
  - 1:1 analyst collaboration
  - Virtual "presence" with key stakeholders, business owners, executives, and partners
  - Focus on specific incidents, adversary hunt, and targeted operations
  - Movement to fully paperless technology, including digital document repositories and all digital forms, approvals, and workflow
- Plan for reduced or no access to physical secure storage safes:
  - Use of managed, potentially cloud-based password, key, and certificate management for SOC credentials
  - Reduction or elimination of physical media collection, forensics, and storage, shifting focus to network-based media collection, and storage area network (SAN) or cloud-storage of digital media artifacts
- Ensuring quality work product and attendance by all analysis, by utilizing more metrics and automation to support consistent delivery
- Greater focus on a common operating picture, metrics, and KPIs that are portable and reusable across different parties, discussed further in "Strategy 10: Measure Performance to Improve Performance"

One of the most difficult situations for a virtual SOC is one where the constituency served is predominantly located in offices and does not embrace work from home culture. The SOC will likely need a physical presence in these situations to avoid being sidelined during normal operations, and close coordination during a major disruptive incident.

3.7.4 Centralized SOCs

In centralized SOC models, it is best for the SOC have a presence at remote sites for purposes of incident response, equipment touch labor, and engagement with local constituents. This is crucial when the constituency headquarters is far from major elements of constituency IT/OT operations. Here are some compensating strategies for a centralized SOC model with a geographically dispersed constituency:

- Have at least two designated POCs at each major location where the constituency operates. These POCs:
  - Are usually sysadmins, security officers, or security champions: people whose day job is other parts of IT but who have been designated by their organization as a focal point for security matters for their organization.
  - Watch over security-relevant issues at the site, such as new system installs and changes to network architecture.
  - Hold the keys to SOC racks or rack cages and are the only people who can physically touch SOC systems.
  - Are the default contacts for on-site incident response.

- Are customers of the SOC's audit collection/distribution capability if one exists.
  - Serve as champions for SOC interests at the site.
- Make contact with site security POCs at least quarterly to ensure they are still in the same position and that their contact information is still current. Having multiple security POCs at a site will help ensure that if one person leaves, the alternate person can still engage with the SOC until a suitable replacement is found.
- Have SOC representatives participate in IT CM/engineering boards for digital assets that operate at remote sites.
- Send SOC representatives to quarterly or annual collaboration forums run by IT people at sites where they discuss major initiatives in site IT.
- Keep up-to-date rack diagrams for all SOC equipment, local or remote.
- Have access to updated network diagrams of site networks and enclaves.

Consequently, the line between centralized and distributed SOC models may appear to blur. The main distinction here is that the site security POCs do not work for the SOC as their main job. Therefore, the SOC cannot heavily task them outside the scope of incident response and sensor touch labor. In hybrid and distributed models, this is not the case, as described in the next section.

3.7.5 Democratizing Security Operations

Taking the model of security POCs one step further, some SOCs with critical and high-needs customers can draw them further into security operations. Specifically, those security POCs can be "deputized" by the SOC as holding a part time role, with SOC tool access. If done right, this can be a huge win for everyone involved:

- Better tailored detections written against constituency systems and services
- Better integration of constituency networks and systems into alert triage processes
- Faster and more accurate incident response

There are a handful of key items that should be executed when doing this:

- The expectations of both parties need to be made clear, meaning for example what is the "deputy's" steady-state time commitment toward supporting the SOC.
- The deputy analyst cannot share their tool and data access further.
- The deputy analyst must not escalate malicious or anomalous activity outside the normal SOC reporting chain.
- The deputy analyst must not perform response actions except as directed by SOC incident response coordinators or leads.
- Routine syncs with the SOC and the deputy's management, as appropriate.

This arrangement can sometimes be enhanced further by adapting some of the points in the next section- incorporating remote analysts.

3.7.6 Incorporating Remote Analysts

In addition to, or instead of, deputizing constituents into the SOC, the SOC can deploy personnel to remote sites, thereby augmenting resources at the central SOC operations floor. While these individuals report to the SOC, the lion's share of the SOC's main analysis still occurs centrally, and incident calls are routed to main operations. However, the SOC now has people who perform all the roles of the deputized personnel, above, making security ops part of their day job, and are accountable to SOC leadership.

Keeping members of the SOC working in concert while spread across multiple sites will certainly be a challenge. Here are some tips on how to keep the whole SOC in sync:

- **Develop processes that integrate across sites:**
  - The SOC CONOPS and escalation SOPs need to support site escalation and response coordination with SOC operations leads. The SOC does not want anyone at the site taking response actions without the knowledge of SOC leadership.
  - Consider hiring analysts at remote sites who previously held IT security-related jobs at that site, thereby leveraging their familiarity with local operations and IT culture.
- **Build a culture that supports analysts at different sites:**
  - Ensure that analysts at remote sites go through the same personnel vetting and indoctrination process as all other SOC analysts.
  - Bring remote workers back to the SOC as budget allows, for team cross-pollination and refresher training.
  - Consider having a "virtual ops floor" where all floor analysts and site analysts join an open chat room, video session, or VoIP session while on duty.
  - Call attention to successes by site analysts to the rest of the SOC team.
  - Schedule regular visits and telecons by SOC leadership to analysts at remote sites, giving them "face time" and keeping leadership abreast of site activity.
  - For sites that host more than a few analysts, consider consolidating their desks to adjacent cubicles or offices where site SOC personnel can interact.
  - Consider keeping site analysts on the job during their site's business hours.
  - Site analysts may demonstrate skills worthy of promotion beyond their current job function. Give them appropriate room to further tailor SIEM content, detections, and other analytics to use cases specific to the site.
- **Leverage technology across sites:**
  - Ensure all SOC data feeds and sensors are integrated into one unified architecture. While the site may have its own specific source of log data and monitoring systems, this should be part of one unified, coherent architecture, with analytics tailored to that site or region.
  - Consider approaches for extending the SOC enclave to the remote site for use by the analysts there, perhaps leveraging one of the following approaches:
  - Connect SOC workstations back to the SOC through a strongly authenticated VPN and ensure that sensitive SOC material is under close physical control.
  - Use a remote thin-client capability with strong authentication if remote site SOC materials cannot be cordoned off from other users.

3.7.7 Continuity of Operations

A continuity of operations plan (COOP) is an effort within an organization to ensure that primary functions continue to be performed during a wide range of emergencies, including localized acts of nature, accidents and technological or attack-related emergencies.[5] For the SOC this means being able to continue the essential function of monitoring and defending the constituency. There are many aspects to a COOP plan, one being the consideration of continuity facilities. Depending on the geographic footprint of the SOC, constituency executives and SOC management may decide that some level of physical redundancy is necessary to support their COOP plans.

When building the physical aspect of a COOP capability for the SOC, there is often an impetus to implement a full-blown, top-of-the-line "hot/hot" capability whereby a complete duplicate of the SOC's systems (including a second ops floor) is stood up at a location distant from the primary SOC ops floor. This can be expensive and is not always necessary. Before rushing into a decision to create a COOP site, the SOC should consider the following:

- **Mission needs:**
  - In a COOP scenario, how long can the SOC be down? How quickly must an alternate capability be brought fully online? Will a partial duplicate suffice?
  - If activation of the SOC's COOP capabilities were called for and there were any impediments in the process of executing the COOP, would the security mission actually be a priority in the eyes of constituency executives?
  - Does the creation of a secondary COOP site, even if only partial, outweigh other competing resource needs such as more sensors or more personnel?
- **Likely COOP scenarios:**
  - What contingencies is the COOP plan designed to address? How realistic are they, and how often are they likely to occur?
  - In the scenarios considered, if the SOC was taken out along with the rest of the site where it is located, what constituency systems are left to defend?
  - If the main site constituency systems or SOC enclave were compromised, are the COOP SOC systems designed to be insulated from that compromise?
- **Technical and personnel considerations:**
  - Is it already a physically distributed SOC, where the sister site can take over full capability? Can IT solutions supplement COOP, such as using replication across sites?
  - Is the SOC already partially or fully virtualized such that either personnel or systems are less likely to be impacted by an event in a single location and therefore could easily pick up responsibilities?
  - Would a secondary COOP site need to be regularly staffed? If so, should it be for the same hours as the main SOC (such as 24x7) or will regular business hours suffice (8x5 or 12x5)?

---

[5] Derived from the definition of COOP contained in National Security Presidential Directive-51/Homeland Security Presidential Directive-20 (NSPD-51/HSPD-20).

○ For the COOP site(s) under consideration, does their functionality (such as WAN and Internet connectivity) depend on infrastructure at the SOC's main site? If so, it may not be an ideal choice.

Even for SOCs that have a hot/hot COOP capability with servers and analysts at separate sites, it is somewhat rarer that every section of the SOC resides at both locations. Sometimes the SOC has redundant systems such as the SIEM and sensor management servers, local networks sensors, triage analysts, and, perhaps, a couple of sysadmins at the COOP site. In this scenario, it can be easier to coordinate operations between sites than if the SOC also spread in-depth analysis, response, hunting, CTI fusion, sensor management, engineering, and all SOC capabilities between two places. It should be noted that cloud based SOC tools may benefit from the COOP capabilities of its cloud provider.

Regardless of what functions reside at the secondary site, the SOC CONOPS should carefully integrate compensating controls to keep both sites in sync. It also helps to have a lead for the secondary site to coordinate operations with the main site leads and to provide care and feeding for the local analysts. One strategy that may work for SOCs with a hot COOP site in a different time zone is to either match or stagger shifts. By staggering shift changes for the two sites, there is always someone watching the console. For instance, if the main site is an 8x5 operation, the working hours for the secondary site two time zones away could be shifted by an additional two hours, giving four hours of overlap. By doing this, each site is up for eight hours, but together they provide 12x5 coverage.

If a SOC wishes to have a secondary COOP "lukewarm" site that it does not staff every week but is up to date enough to support emergency operations, it may consider the following strategy:

- Choose an existing constituency office building or data center with at least a few spare racks and cubicles.
- Deploy a redundant instance of key SOC systems such as SIEM, sensor management systems, thereby providing failover capability.
- Find a good spot to place some SOC workstations, perhaps near the TA's office or cubicle.
- Ensure all security data feeds are directed to both sites or mirrored from the primary to the secondary, at all times.
- If the primary site goes offline, having the log data immediately available at a secondary location could be invaluable.
- When performing COOP, the amount of time to bring the secondary site online should be minimized. If monitoring systems at that site are online and up to date even when not actively being not being used, transition is that much quicker.
- Regularly check (perhaps on a monthly basis) to ensure COOP servers and systems are functional and up to date with patches and configuration changes.

- Schedule semiannual practices of the SOC COOP to ensure processes and procedures are up to date and work as expected.

There are many more aspects to COOP beyond the physical location considerations, for more on this topic see: [65].

3.7.8 Follow the Sun

In the "follow the sun" model, the SOC has two or three ops teams, each separated by many time zones. Each ops floor is on the watch during local business hours (e.g., 9 a.m. to 5 p.m.). In a three ops floor arrangement, at roughly 5 p.m. local time, one ops floor roll to the next ops floor, where it is 9 a.m. This pattern continues every eight hours, giving 24x7 coverage but without making people come to work in the middle of the night. A similar pattern ensures for two ops floors working 12 hours each.

This approach is very common for IT help desks that serve wide geographic regions (e.g., with major IT vendors and very large corporations). There are several advantages to follow the sun, including:

- Far fewer analysts, particularly those in triage roles, are routinely required to work at night.
- Analysts on shift are more likely to share the language and culture of those calling during their shift.
- In terms of labor costs, it also may be more affordable than a single ops floor staffed 24x7 because:
  - Paying people during normal business hours may be less expensive than paying them to come in at night.
  - Some of the ops centers may be located in geographic regions with lower median income for security analysts.
- In the case of a high-criticality incident, in contrast to a single site asymmetrically-staffed SOC, it may be easier to keep more staff in the office and working the issue 24x7 until resolution.

That said, additional care should be taken when considering this approach to staffing and COOP.

First, disparate SOCs need to be in sync regarding myriad issues- operations, expectations, technical solutions, TTPs, etc. In particular, the following practices should be implemented:

- Routine travel between centers for leads and managers
- Routine travel for some individual contributors and analysts, particularly those that show strong career progression
- Sharing responsibilities in virtual teams that cross geographic regions; conversely, having an entire function like SOC tool engineering carried out in a single location can cause both a business continuity risk, and a risk of isolation or siloing
- Routine pass down, usually daily

Second, each SOC response analyst will work several threads for several hours or days. Handing off an incident from one analyst to the next every eight hours may cause issues.

For some incident handling, particularly non-emergency cases, it may be wise to have one analyst "own" an incident cradle to grave.

Third, SOC management will need to ensure attention is paid to ensuring strong diversity and inclusion behaviors are followed by SOC staff, not just in hiring but in routine interactions [66], [67].

Follow the sun can be a strong approach to staffing and COOP for SOCs supporting worldwide, distributed organizations and/or that must maintain 24x7 operations. Bottom line, SOC management and the executives above them must ensure distributed staff are working harmoniously.

3.8 Summary – Strategy 3: Build a SOC Structure to Match Your Organizational Needs

3.3. Dimensions of a SOC organizational model include both the internal SOC structure (mapping of functions to roles) and the overarching model of how the SOC is placed within the constituency and its overall objectives. The primary overarching SOC models include:
  - Ad Hoc Security Response
  - Security as an Additional Duty
  - Distributed SOC
  - Centralized SOC
  - Federated SOC
  - Coordinating SOC
  - Hierarchical SOC
  - National SOC
  - Managed Security/SOC Service Provider

3.4. It is very important that these elements of security operations are brought together into one organization for any given constituency, as they are self-reinforcing and do not lend well to being broken apart. Pursuing a DevOps culture of continued engagement in improving SOC tools and processes is essential.
  - The primary internal elements of a notional SOC include:
    - Incident triage, analysis, and response
    - Cyber threat intelligence, hunting and analytics
    - SOC tools, architecture, and engineering
  - The choice to using "tiering" within the internal SOC model should be based on operational needs and organizational personnel management philosophies. Both tiered and tierless models can be effective.

3.5. There are many functions and services a SOC could offer. The choice of services should support the constituency's risk posture and the defined SOC mission. Maturity, funding, what the constituency requires, operating model play major factors here. These are also likely to vary as the SOC matures over time. No one SOC will offer every function and service possible. Some functions like vulnerability management and penetration testing should be performed somewhere; the choice whether they go in the SOC or elsewhere is often enterprise specific.

3.6. Cybersecurity and other stakeholders in the constituency should weigh various options including resourcing when deciding to insource or outsource the SOC entirely or certain SOC functions. Outsourcing tends to be more common with smaller constituencies and tasks or functions that are carried out more rarely.

3.7. The SOC will need to decide if it wants to go 24x7. The decision should be based on criteria such as considering the constituency's mission and hours of operation. The decision for the SOC to go 24x7 is not an all or nothing choice. There are alternative models such as working extended (but not 24x7) hours, staffing only certain functions 24x7, or outsourcing some or all the functions during some hours.

3.8. The SOC physical location will probably mirror its constituency. If the constituency is primarily remote workers, the SOC may be as well. If the constituency primarily works from a central location, the SOC will probably be located there as well. The SOC should be able to maintain continuity of operations in the face of natural disasters or other outage scenarios.

Strategy 4: Hire AND Grow Quality Staff

People are the most important aspect of operating a world-class SOC. However, many surveys indicate that finding staff is the number one concern for those working in the cybersecurity field [68] and that the skills shortage has a significant impact on their organizations [69]. This strategy discusses the importance of hiring the right people when possible and emphasizes the need to grow and develop team members in-house to meet staffing needs. Additionally, it discusses creating an environment that encourages staff to want to stay while also planning for the eventual turn over that will happen. Finally, it addresses the frequently asked question, "How many analysts do I need?" and provides guidelines for sizing SOC staff.

4.1 Whom Should I Hire?

This section discusses the traits of SOC hires, recognizing that there are many different backgrounds and skill sets that can make up a successful SOC team. It examines candidate qualities in relation to a candidate's mind-set and soft skills as well as their background and skill set. It distinguishes between hiring staff that are new to cyber operations versus those with experience since, as will be covered in Section 4.2, growing staff is just as important as trying to hire experienced staff.

4.1.1 Mind-Set and Soft Skills

Perhaps the number one quality to look for in any potential hires to the SOC is passion for cybersecurity, regardless of the position or level of experience. Intrusion monitoring and response is not just a job where people put in an eight- or 12-hour workday (or shift), collect a paycheck, and then leave. When it comes to cyber, look for enthusiasm, curiosity, and a thirst for knowledge. In fact, the gamification of cyber, wherein the cybersecurity specialists see themselves in natural competition with the adversary will, if channeled properly, drive their achievement. This passion and competitive spirit are what will keep them coming back to the job, day after day, despite the stress and challenges inherent in operations. Additional foundational skills to look for include:

- Good intuition and ability to think "outside the box"
- Attention to detail while seeing the bigger picture
- Ability to pick up new concepts quickly
- Critical and creative thinking
- Ability to thrive in high ops tempo, high-stress environments
- Solid sense of integrity and identification with the mission

- A strong desire to "win" against an adversary by preventing intrusions or discovering and stopping them when incidents do happen

Arguably as important as the above set of skills is the need for every person in the SOC, without exception, to be a strong team player. This includes:

- Good communication skills
- A preference, and demonstrated ability to put the team before oneself
- The ability to take ownership over not only wins but losses
- Growth mindset and the ability to accept and embrace feedback
- Emotional intelligence and stress management
- Ability to influence without authority
- Self-initiative and strong time management
- Conflict management skills
- Ability to drive clarity and remove ambiguity in any kind of interaction

When hiring more experienced personnel, also be on the lookout for staff that demonstrate:

- An understanding of how to clearly communicate their findings and ideas across the SOC staff, with system owners, and up to leadership
- The capacity to establish and grow relationships with members of the SOC and partner organizations, sharing best practices, tools, and techniques
- An ability to put themselves in the mindset of the adversary, look at the structure of the constituency and their mission, and assess where there is cause for concern
- A willingness to look for new solutions and a desire to improve efficiency and effectiveness.
- A desire to mentor and grow other staff around them and the ability to provide on-the-job training and knowledge sharing to other analysts

This last point is particularly important and can make the difference between SOC that is good for a time and a SOC that thrives over time.

A SOC team member that can help build a pipeline of talent multiplies their own effectiveness and sets a SOC up for long-term success.

One way to get great candidates is to encourage high-performing SOC team members to refer friends interested in working in cybersecurity operations, even if their experience in IT is not security focused. Talent attracts talent and friends working together help build a sense of community within the SOC.

4.1.2 Background and Skill Sets

With the growing number of undergraduate- and graduate-level programs specifically in cybersecurity, forensics, and malware analysis, there are a rising number of applicants who deliberately tailor their formal education to a career in cybersecurity. Factors that make a candidate stand out include participation in events such as capture-the-flag, contribution to open-source projects related to cybersecurity, previous general IT and cloud experience, and internships in the cyber operations field.

That said, making a formal degree or five years of experience in IT a universal requirement for incoming analysts is not necessary. Candidates who have not been in IT very long, but demonstrate solid problem-solving skills, are prime candidates to grow into more advanced roles as their skills develop. In fact, there are some who argue that the shortage of cybersecurity professionals is, at least in part, self-inflicted [70].

For people making the switch into cyber operations there are several previous positions that candidates may come from: IT help desk, other areas of cybersecurity, software and systems development, and system administration. Candidates with a background in either system administration or penetration testing can usually pick up the SOC specifics in a matter of weeks and months. In any of these cases, it is important to assess candidates' breadth and depth of technical knowledge, ability to assimilate and use new information, and appreciation for the realities of IT operations. Candidates from outside these more traditional careers who are looking to switch career tracks should also be considered for entry-level positions. If a candidate has the passion for the role and good analytical skills, they may still be a strong candidate to make the transition if the SOC can provide robust on-the-job training.

Experienced candidates should be able to demonstrate a general "literacy" of IT and cybersecurity. They should also possess deep knowledge in at least one or two areas related to cybersecurity operations—a concept known as the "T-shaped person" [71]. Additionally, they should show sound performance in soft skills and teamwork. For the first two, this knowledge is usually gained through a combination of the following four things:

- Formal training in IT, computer science (CS), electrical or computer engineering, cybersecurity, or a related field
- On-the-job experience in IT operations, cloud-based services and systems, system/ network administration, or software development, incident response, forensic/malware analysis, threat hunting, cyber threat intelligence, and vulnerability assessment/ penetration testing
- Previous experience in the systems and mission area specific to the SOC's constituency- retail, finance, healthcare, energy, manufacturing, government, etc.
- Self-study in any of the above areas, perhaps achieved in candidates' spare time

No one of these experiences is necessarily better than another, each brings with it certain expectations. Consider the whole of a candidate's experience when deciding on a hire. In fact, it is very common in any given SOC to find plenty of staff with either no college degree, or a degree in a field that has little to do with IT and cybersecurity. The authors of this book have encountered successful SOC specialists with backgrounds and degrees in liberal arts, fine arts, criminal justice, hard sciences, "soft" sciences, management, education, and most any other field.

When diving into the technical skills of an experienced candidate, recognize that cybersecurity has become such a broad field that no one candidate will be an expert in everything. Be clear about what skills are most important to the SOC organization and focus on bringing in a highly diverse team with complementary skills rather than a single person who is expected to know everything. Job requisitions that ask for 10+ years of experience with knowledge of every possible skill set and tool used in a SOC are much more likely to sit unfilled.

Requisitions should also focus on describing the job to be performed, and the expected outcomes to draw in creative thinkers. As long as the candidate has the core skills for the position and also demonstrates aptitude and interest in learning the other skills needed you have probably found a good candidate. While it is easy for SOC to focus on hiring against rote technical knowledge, maybe even specific tools, this paints an incomplete, perhaps lop-sided view. Rather, a balanced approach is recommended. Hiring managers should ensure that they are also evaluating candidates' behavior skills and past successes, such as in accordance with behavioral interviewing techniques [72], [73].

> *Ensure interviews balance technical evaluation with growth mindset,*
> *soft skills, past performance, and strong team player behaviors.*

Keeping in mind the different roles being filled in the SOC, here is a menu of possibilities:

- Knowledge of OS internals (such as Windows, macOS, Linux, Android, iOS).
- Working knowledge of Open Systems Interconnection (OSI) network protocol stack, including major protocols such as IPv4, IPv6, Transmission Control Protocol (TCP), User Datagram Protocol (UDP), Simple Mail Transfer Protocol (SMTP), Hypertext Transfer Protocol (HTTP), HTTP/2, HTTPS, Server Message Block (SMB)/Common Internet File System (CIFS), Secure Sockets Layer/Transport Layer Security (SSL)/ (TLS), Remote Desktop Protocol (RDP).
- Understanding of how web applications, email applications, and other common internet-facing applications work.
- Understanding of modern, distributed authentication and access control mechanisms and architectures like SSAML, OAUTH, and ADFS.
- Working knowledge of cryptography basics, protocols, and concepts such as Public Key Infrastructure (PKI), asymmetric key exchange, and symmetric cryptography.
- Knowledge of common networking services such as network address translation (NAT), DHCP, DNS, and network time protocol (NTP).
- Hands-on understanding of using major cloud service providers such as Amazon AWS, Microsoft Azure, or Google Cloud as well as an understanding of how to configure, secure, and monitor systems and data in these environments.
- Familiarity with Mobile architectures, protocols, and security.
- For those doing cyber threat intelligence activities a knowledge of common threat models such as the Diamond Model of intrusion analysis [74] Cyber Kill Chain [75] and MITRE ATT&CK; also, an understanding of public and private data sources to support cyber threat analysis along with strong report writing skills.
- For malware analysts, the ability to tear apart a suspect file and formulate a working understanding of its attack vector and likely purpose, and ability to break down executables using disassemblers, debuggers, and hex dump. This could include

Portable Executable (PE) and Executable and Linkable Format (ELF) binaries along with Jscript, java, and python bytecode or Microsoft Office macros.

- Understanding of mission-specific systems in the constituency, be they building automation, satellites, connected vehicles, health care devices, or SCADA systems.

For more details on relevant skill sets for SOC personnel, refer to NIST's NICE framework [76] and materials from Carnegie Mellon University Software Engineering Institute (SEI) CERT/CC [77].

4.1.3 SOC Leaders

Finding strong leadership for the SOC often presents a somewhat unique challenge given the mix of skills required. A SOC greatly benefits from having leaders with a background in security operations as it takes strong technical understanding of the problem space to be able to make informed risk decisions. The most effective SOC leaders will have a mix of:

- Qualities common to any leader such as people management skills
- Technical and cyber literacy
- Orientation toward high-ops tempo, high-stress positions requiring "guts" to make tough decisions and negotiate difficult decisions with leaders at all levels of the food chain

As the size and scope of management responsibilities increase, other business and communications skills become increasingly important. Consequently, hiring for SOC leadership positions is different from individual contributors. Look for someone who is good at handling sensitive issues, can coach others through a performance improvement plan, has strong administrative skills, and has a vision for the SOC's future. They also need to be able to 'manage up and out' with the ability to anticipate and answer executive questions as well as to work with line of business owners. They must also be open to taking feedback and engage in continual learning as much as the SOC team members do, given how quickly the cybersecurity field evolves.

4.1.4 Final Thoughts on Hiring

When thinking about building the SOC team, make sure to work with human resources professionals to build and execute a process that will support identifying and selecting a diverse set of candidates. This includes how to source resumes and candidates, how candidates are selected for interviews, who serves as an interviewer, how interviews are conducted, how final hiring decisions are made, and how salaries are determined.

In addition, once the SOC grows beyond a couple of people, management will need to budget for continual recruiting and hiring. The pace at which they do that may change based on organizational size and rate of turnover, but do not never expect the SOC will look exactly the same next year as it does today. Either staff move on to other jobs, or the skills sets SOC staff need evolve and change. This leads to the next key topics, which is the need to grow SOC staff.

4.2 Grow Your Own SOC Staff

Even with an increase in programs aimed at developing cybersecurity skills in high school and college students, military personnel, and career switching professionals, there just are not enough qualified staff to go around. So, what is a SOC to do? The answer: Grow your own talent!

Rather than just investing in recruiting and hiring senior staff, investing in on-the-job training for a specific role, cross training to build skills beyond a single role, exercising the SOC so management knows where to spend time improving processes and staff knowledge, and providing external training opportunities to augment in-house knowledge will pay deep dividends. The rapid pace of change in cybersecurity means there is always something new on the horizon, so no one is ever 'done' with their learning, so look to develop plans that support both early career staff as well as more experience staff.

> *Growing the SOC team requires a consistent investment of time and resources but leads to long term success.*

When growing SOC expertise from within, the team can choose what subject areas to focus on and what technologies or topics they would like to be an expert. An advantage of growing this in house is that different team members can choose to come up to speed in different areas based on what experts are already there. If there is already an expert in a particular area, they can mentor or answer questions. If there are expertise gaps are in the SOC, that is a prime area for the SOC to invest in training even if the SOC also wants to hire externally for those skills as some expertise may be better than none. As a basic list to get started in understanding what skills and expertise may be needed, here are some suggested topics:

- Network forensics and intrusion analysis
- EDR or SIEM deployment, tuning, and maintenance
- Media and file forensics
- Secure coding and vulnerability exploitation
- Malware analysis and reverse engineering
- Vulnerability assessment and penetration testing
- Operating system, database, and network management
- Cloud-based services and administration.

Ideally, SOC analysts will balance breadth and depth in these areas. They may dive deeply into EDR or SIEM deployment, tuning, and maintenance but know just enough about media and file forensics to understand what the person skilled in that area does and what types of information are valuable to them. Almost all incident responders however are well served to be expert in whatever operating systems are the most prevalent in the constituency. If Windows is the primary operating system used, understanding the aspects of Windows Registry, and how to use Sysinternals utilities and tools is essential. If the constituency primarily uses Linux, understanding file system, processes, and knowing how to analyze the kernel is useful.

4.2.1   On the Job Training – Building A New Professionals Pipeline

One approach that is showing success is building a pipeline of new professionals with potential, vice only bringing in experienced staff. A robust training program that brings new hires up to speed on the TTPs the SOC uses to execute its mission and on the specifics of executing a given role is a critical component of this approach. In-house training programs may include both formal and informal components.

Formal training plans may consist of:

- Technical qualification or "check-ride" process that each new hire must pass within a certain time period after hire; this ensures all employees can operate with both a base level of technical capabilities and consistent adherence to relevant SOC processes, depending on their job function.
- Enrichment activities such as computer-based training or slide decks.
- Training scenarios where analysts must pick out activity from a real or synthetic set of log data, using the same tools that the SOC has in operation.

Informal on-the-job training in tools and techniques, through formal supervisor, apprenticeship-style coworker, and mentorship relationships, lunch and learns, and workshops, such as:

- Focus on a particular actor's TTPs
- Deep dive on constituency IT, OT, and cloud architecture and mission areas
- Advanced tool use
- Interesting cyber news and CTI items

Using deliberate apprenticeship-style pairings of seasoned and talented personnel is especially powerful for expanding teams. Not only are techniques learned by less experienced staff, the added benefit of depth and continuity of techniques benefits the SOC as a whole. The trick is working with senior staff to encourage this, which means creating a culture that values apprenticeships. In both formal and informal cases, consider archiving presentations and demos for review and later use.

4.2.2   Cross Training and Rotations

Cross training helps to build resiliency across the SOC, ensuring that if one team member leaves there is less of a gap to fill. As staff know more about other job functions, they can better see how their own primary responsibilities fit in to the bigger picture. Additionally, staff may discover that they have an aptitude for another part of the SOC mission that they might not have previously considered. It also helps reduce monotony and keep people interested and excited about the mission.

Cross training and rotations can take a variety of different forms. One option is to rotate staff through certain positions; as discussed elsewhere in this strategy, this is particularly helpful in SOCs that are small or are tierless:

- Tradecraft improvement such as signature tuning, cyber threat intelligence fusion, scripting, and analytics development

- Performance tuning and content management, ensuring SIEM and other analytics platforms operate in a satisfactory manner for all parts of the SOC
- Triage "analyst of the day/week," such as in a SOC of a dozen people, each person takes turn fielding alerts

In any of these cases, this allows members of the SOC to be more aware of the "pain" felt from suboptimal system configuration or misbehaving alerting and be routinely involved in their remediation. In larger SOCs with larger pools of more differentiation, it is very easy to lose sight of different experiences and operational "pain" felt by others; even in these cases, cross training and staff rotation is still necessary, not because it is operationally necessary but because the SOC might otherwise lose cohesion.

4.2.3 External Training

External training opportunities give staff a way to learn about new techniques and technologies beyond their current expertise. They also provide a way for SOC staff members to become part of the larger cybersecurity community. However, keep in mind that external training opportunities can help round out an in-house training program, but they should not be a substitute for it.

External training and enrichment opportunities include:

- OpenSecurityTraining2 [78]
- SANS Global Information Assurance Certification (GIAC) [79]
- Offensive Security [80]
- Others cataloged by US-CERT [81]
- Product-specific training or certification classes
- Product-centered conferences for current and prospective customers hosted by vendors such as FireEye, HPE, McAfee, Cisco, Splunk, Microsoft, Amazon, and many others

Professional conferences include:

- BlackHat—held at several locations, most notably Las Vegas, Nevada [82]
- Defcon—held right after BlackHat, Las Vegas, Nevada [83]
- RSA Conference, San Francisco, California [84]
- Shmoocon, Washington, DC [85]
- Security B-Sides, various locations [86]
- Layer one, Los Angeles, California [87]
- Flocon, multiple locations [88]
- PhreakNIC, Nashville, Tennessee [89]
- Hackers On Planet Earth (HOPE), Queens, New York [90]
- Hacker Halted, various locations [91]
- THOTCON, Chicago [92]
- SANS Summits and conferences, various locations [93]
- IEEE symposium on Security and Privacy [94]

- FIRST.org, various locations [95]
- USENIX, various locations [96]

4.3 Create an Environment that Encourages Staff to Stay

Keeping good people is one of the biggest problems SOC leaders face. SOC staff members cite many reasons why they like working in their organization and choose not to move elsewhere. Three of the most common reasons for staying are:

- They feel like a cohesive, tightly knit team of highly qualified, motivated professionals.
- They experience new and interesting challenges every day and are given the freedom to solve them.
- They believe in the mission of cyber defense—both its importance and its uniqueness.

This section looks at additional factors that help maximizing staff retention.

4.3.1 Pay Fair Market Value

Sufficiently talented and self-motivated employees will quickly pick up a variety of highly marketable IT skills in just 12–18 months working in a SOC. For example, it is possible for a fresh college graduate with drive and a computer science degree to spend a year or two in a SOC and then jump ship for a significant annual pay increase.

All team members are subject to leave for higher paying positions if they do not feel they are being adequately compensated in their current roles. The key with pay is not that the SOC must be the highest paying organization, but that they should pay fair market value.

However, be cautious about muddying the issue between pay and promotion. Good senior technical staff should be financially recognized for their contributions and should not feel the need to move into leadership roles just for the pay.

The bottom line is that a SOC must be able to adequately compensate its employees. This can be especially challenging in government environments, small businesses, non-profit organizations, or with contracted employment. SOC management should ensure that team members receive adequate compensation and that the SOC is granted different or higher pay bands separate from positions in general IT. Also, recognize that pay is only one part of the equation, once a fair pay threshold is reached, management must consider other factors that influence staff retention.

4.3.2 Support Career Progression and Staff Growth

One of the most desirable traits of an analyst—passion for the job—goes hand in hand with the desire to take on new and different challenges. The key is to recognize what drives each member of the team. Some staff members will be very focused on traditional upward career progression looking to move in to expanded leadership roles. Others will be more excited about learning a new skill or capability and growing laterally [97]. As Figure 11 shows both upward and outward growth should be considered viable career paths.

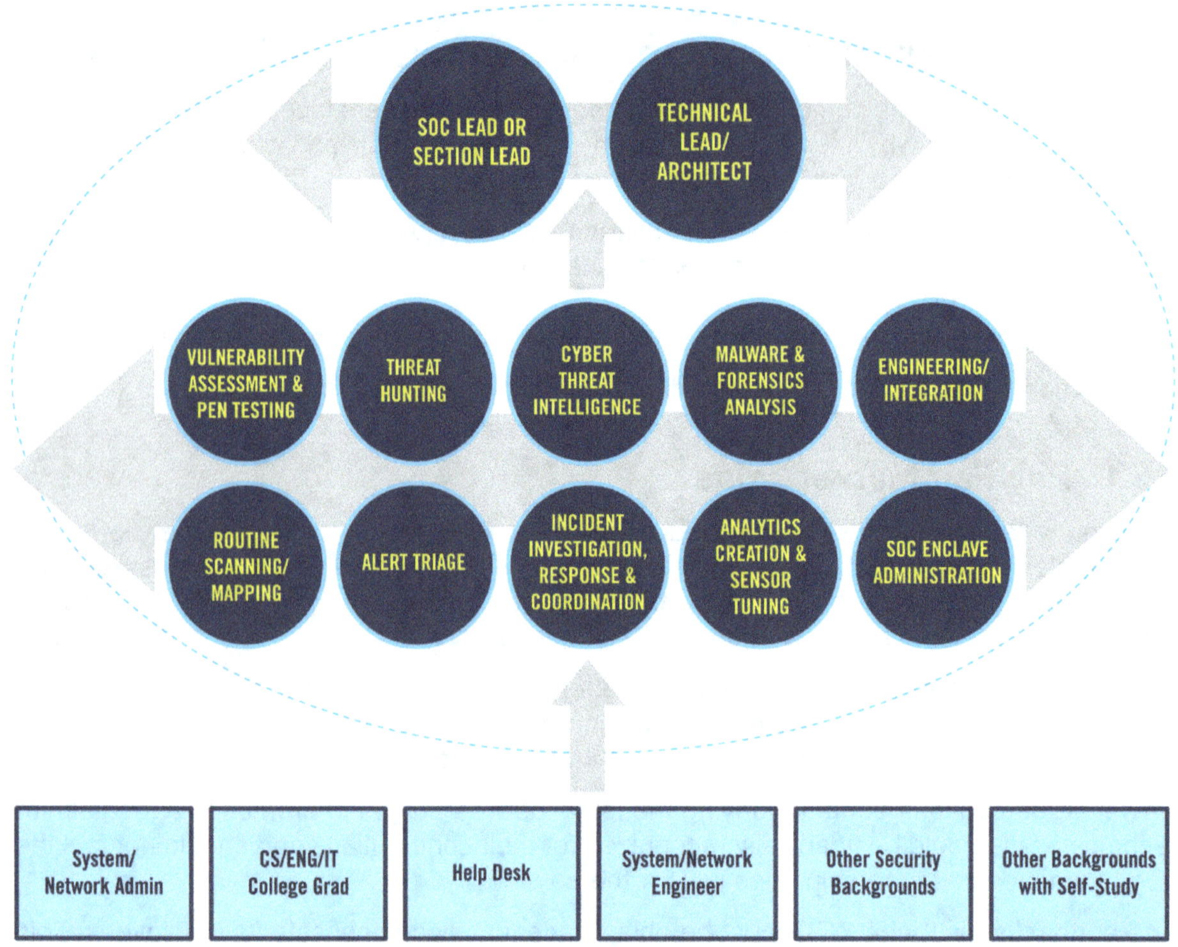

Figure 11. SOC Career Paths

Given the many different ways a SOC professional can grow their career, it can sometimes be difficult to figure out what should be next on a learning path. One of the first questions to ask is if the staff want to expand the knowledge in the area they are already working, so moving from a more junior practitioner to a more senior practitioner in the same area, or if they want to expand into a different area. Once the goals are clearer a roadmap can be developed to help the staff achieve their goals. It can be helpful to plan out a training roadmap for several years, even if that plan may evolve over time, as many skill areas require sustained learning to gain proficiency. When developing a training plan, consider looking at roadmaps from academic institutions, free online training programs, and commercial training courses to get a feel for what order of learning may be useful. Even if the learning comes from on-the-job training rather than external training, having a roadmap can make the experience more focused. For ideas about roadmaps look at those published by SANS [98] and Open Security Training [99].

4.3.3   Encourage Automation and Progression of Capabilities

It is easy to fall into the daily grind where every analyst comes in every day and looks at the same data in the same way, without any change. Hiring and growing staff who do not have a thirst for knowledge and an enjoyment of new challenges will lead to stagnation and staffing flight risks. While some analysts continue to be fulfilled by performing their assigned duties, others are looking for the opportunity to grow into different roles. One way to help people grow is to allocate time in people's schedules for learning new skills which they can apply to driving improvement to SOC tradecraft, such as automation and analytics. Then, as routine activities become automated, even more time is made available for learning new skills to improve SOC capabilities.

High levels of automation are achieved by maintaining an up-to-date, robust, strongly integrated tool set. Talented analysts—the ones who can think creatively—expect access to a robust set of tools that match the current threat landscape and give them results in what they consider a reasonable amount of time. For instance, running basic queries against a day of log data should be doable in a matter of seconds or minutes, not hours. Having old and broken tools is a quick way to lose talent. Things to consider include:

- Creating structure, repeatability and automation around incident analysis and investigation tasks, such as via SOAR and interactive analyst notebook capabilities to automate routine tasks.
- Leverage automated prevention capabilities where it is cost efficient and appropriate to do so, such as with a best of breed EDR, thereby minimizing the routine activities that would otherwise soak up SOC resources.
- Drive improvements to the constituency cybersecurity program as much as possible, so that strategic issues recognized by the SOC are addressed at the right level. If SOC staff feel and see that they continue to suffer the same attacks without any chance to help improve constituency cyber hygiene and posture, then they can become demoralized and leave.
- Make handling of routine incidents more automated, repeatable, and formulaic, as appropriate, so they take up minimal amounts of time and can be handled by more junior staff.
- Refer handling of some routine incident types, such as inappropriate website surfing, to other constituency organizations, as appropriate.
- Foster the opportunities and grant the freedoms for analysts to help the SOC advance its capabilities. Those may include:
  - ML techniques to augment analysis
  - Threat hunting
  - Threat analysis and threat research
- Schedule and provide capacity around activities that drive skill and capability growth, like hunting and purple teaming.

4.3.4 Communicate, Communicate, Communicate

One oft-cited reason analysts enjoy working in a SOC is the sense of teamwork and the mission. And one of the best ways to encourage this feeling is to provide regular feedback to SOC staff on how their contributions are making a difference as well as making the staff feel like part of a community.

Probably the most straightforward way to accomplish this is through regular meetings where SOC personnel discuss tactical and operational issues, as well as less monthly or quarterly meetings to discuss bigger picture issues. The desire to bring the SOC together must be tempered with mission demands. For instance, with a larger SOC, weekly or daily ops "stand-ups" may involve only SOC section leads.

The second way to accomplish this is to provide feedback (known by some as "hot washes") to the entire SOC on the results of recent incidents. If done properly, this supports several goals:

- Provides evidence that individuals' efforts are having an impact, both their daily work and their ideas
- Recognizes individuals' and team's accomplishments and contributions to the mission
- Brings to light techniques that can be used across the SOC, provides a venue for sharing and discussing new ideas
- Keeps SOC members informed of nuances regarding incident handling procedures
- Drives improvement to processes and technologies that are having the most success
- Provides artifacts that can be rolled up into records of accomplishments the SOC can use to justify expanded resourcing and authorities

The third way is to maintain regular analyst-to-analyst sharing among SOCs, a topic touched on many times in this book. By regularly sharing past experiences, team members gain a sense of community and belonging. Continual discussions also improve knowledge sharing and can lead to the 'aha' moments where two analysts realize that their combined knowledge can solve a problem. Many SOCs are incorporating collaboration hub capabilities such as Slack, Google Hangouts, or Microsoft Teams to support team discussions and information sharing.

4.3.5 Support a Diverse and Inclusive Work Environment

Increasing diversity in hiring not only opens the talent pool for staffing, but evidence also continues to weigh strongly in favor of improved outcomes from more diverse teams for many types of work, especially knowledge work. Increasing the effectiveness of the SOC is not only good for the organization, but high-performing organizations are ones where staff are more likely to stick around. For an overview of the topic consider the book *The Diversity bonus: How Great Teams Pay Off in the Knowledge Economy* [100]. Additionally, supporting inclusiveness, the cultural and environmental feeling of belonging [101], [67], means staff feel valued, respected, accepted, and encouraged to fully participate in the organization. This leads to increased engagement, retention, and staff satisfaction. Other ways to help build an inclusive culture include looking for opportunities to help staff get to know each other beyond their daily work activities, building in training to help SOC staff become aware of unconscious bias [102], and recognizing and when possible, scheduling around different cultural holidays. Work with human resource professionals to develop even more ideas for building and maintaining a diverse and inclusive work environment.

4.4 Pre-Plan for Staff Turn-Over

The day that a star analyst hands in their resignation is not the day to start planning for what to do after they are gone [103]. High staff turnover is a reality for the cybersecurity industry and management of every SOC need to be prepared for it.

The good news is that if SOC leadership are building and growing the team continually, and cross training them across the SOC, they are already preparing for staff turnover since continual growth and knowledge sharing is built into the staffing model. But what else should they be doing to ensure continuity during staffing disruptions? Answer: many of the same things that ensure consistent high-quality practices in daily operation anyway. In particular, maintaining appropriate documentation and institutional knowledge is one of the best ways to ensure the SOC is prepared for staff departures. This includes keeping a frequently updated set of SOPs that describe each of the duties, procedures, and skills for each function. This also includes maintaining as much institutional and technical knowledge in lead and management positions as possible, in addition to its existing at the individual level, compensating for gaps between departures and new hires.

4.4.1 Find the Right Level of Process and Documentation

It was mentioned before, but it bears repeating—the SOC must find the right balance between structure and freedom in governing the ops tempo and daily routine of its analysts. With too little process, there is no consistency in what the SOC does, how it finds, or how it escalates and responds to incidents. With too much process or bureaucracy, the analysts do not have the time or freedom to pursue the most important leads that "just don't look right" or to rise to the challenge when called upon. If the SOC slips to either end of this spectrum, staff will leave.

Good candidates for a formalized process include:

- Overall CONOPS that articulate to constituents, inside and outside, the major inputs and outputs of the SOC escalation process, demonstrating rigor and repeatability in terms of overall cyber incident handling
- Daily, weekly, and monthly routines that must be followed by each section of the SOC:
  - Consoles and feeds that must be examined
  - Preventive maintenance and health and welfare checks
  - Datasets and dashboards that must be curated and made available to external parties for SA
  - Websites that must be checked for updated cyber threat intelligence and news
- SOPs that describe in detail what is done in response to certain events:

- ◦ Escalation procedures for well-defined, routine, noncritical incident types such as data leakages, viruses, and inappropriate Web surfing
  - ◦ Escalation procedures for less structured, more unusual, or critical incident types such as root compromise and widespread malware infections
- Downed sensors, data feeds, or systems
- Facility and personnel emergencies (e.g., those initiating a COOP event) such as inclement weather, fire drills, or personnel out on sick leave

In addition to SOPs, it is beneficial for the SOC to record its institutional knowledge in various knowledge management repositories, especially those that capture information around specific customer systems and networks, and those that maintain information about adversaries.

This list does not include how the analyst evaluates security-relevant data for signs of malicious or anomalous behavior. This is the most critical element of the incident response life cycle and requires a combination of process mixed with analytical judgement. For more on this topic refer to "Strategy 6: Illuminate Adversaries with Cyber Threat Intelligence." It should be noted that SOC processes (especially SOPs) are more focused on junior members of the staff such as junior sysadmins and those doing alert triage. This is natural since those newer team members require more structure and routine in their job.

4.5 How Many Analysts Do I Need?

This is one of the most frequently asked questions when shaping the SOC, both from SOC managers and those new to cybersecurity operations. Unfortunately, it is one of the hardest to answer, because there are so many issues at play. This section breaks down the factors that impact overall SOC staffing by leveraging the Large SOC model from Section 3.3.3 to look at how to staff many parts of the SOC.

> *A SOC's capacity to perform its entire mission is usually influenced more by its skill level, maturity, and automation that the number of analysts.*

4.5.1 General Considerations

The easiest way to reason about the size of a SOC is to talk about number of monitored/defended users or digital assets, as is the case in this book [104]. This is a rule of thumb, as industry surveys show a high standard deviation in analyst to organization size [33, p. 11]. In the past, some parties suggested ratios, like one analyst for every 50 to 75 devices [105], or number of alerts per analyst [106]. Other numbers indicate there is an average of 3% of IT employee headcount for IT staff and 3% of IT staff headcount for security [33]. Hands-on experience proves there are more factors to consider:

- SOC mission and offered capabilities (for example do they perform their own malware analysis)

- Size, geographic distribution, and heterogeneity of assets being defended (such as due to disparate identity planes or presence of OT)
- Number of incidents (detected or otherwise) on constituency systems
- SOC organizational model
- Size, coverage, and diversity of SOC monitoring and analytics systems
- Intended and existent SOC staff skill set
- Business/coverage hours offered by each SOC section (8x5, 12x5, 24x7, etc.)
- Level of automation built into SOC monitoring, correlation, and analytics
- SOC funding for staff resources

Consider a few key points. It is expected that one skilled analyst with force multiplier technologies such as SIEM can be as effective as multiple newer analysts with inadequate tools. From "Strategy 3: Build a SOC Structure to Match Your Organizational Needs," it was presented that SOCs can come in vastly different shapes and sizes and that an 8x5 position takes one FTE, whereas a 24x7 position requires 4.8 FTEs.

Together, these factors show that SOC staffing needs require a more complicated equation than a simple ratio—there are many independent variables, each one having a potentially profound effect on the answer. For new SOCs, few of these factors may be set in stone when initial budgets and staffing numbers are decided. For many SOCs, the answer evolves over time, as they grow and mature. SOC managers typically seize one of four different opportunities to grow or shape their staff:

- In the wake of a major incident that has constituency executives' attention
- When the SOC's organizational placement is changing
- At the early stages of annual budget planning
- In the wake of a major inspection or assessment

The following sections examine the primary factors influencing staffing for each section. It is also important to keep in mind that the hours each section may keep can vary. Some analysts may be required to staff 12x5 or 24x7, whereas the rest of the SOC may maintain a more limited 8x5 schedule.

The sections following address staffing as if each role is distinct. However, as discussed in "Strategy 3: Build a SOC Structure to Match Your Organizational Needs," many SOCs also combine those roles or at least have a more fluid transition of staff responsibilities between them. For clarity, this section discusses them as unique roles; however, the SOC organizational structure selected may shift how many of each type of analyst or engineer needed.

4.5.2 Real-Time Alert Monitoring and Triage

Out of any of the SOC's sections, the staffing model for alert triage and initial investigation is the most predictable; in SOCs that are tiered, this refers to "tier 1." Consider the average number of minutes it should take an analyst to evaluate the disposition of an alert and the number of alerts worthy of their attention in a given shift. This will provide a rough guideline for the number of analysts needed. Sort of. One of the most important lessons learned when it comes to event monitoring is:

With too many events appearing in their dashboards, analysts have two options: (1) furiously acknowledge or skip many alerts without fully analyzing them or (2) hunt and peck for random alerts out of their feeds. Instead, analysts should be presented with discrete views into the data that can be fully evaluated over the course of their shift. The number of alerts the analysts must deal with in a shift is highly dependent upon many factors, most notably the quality and quantity of data feeds flowing into their tools, the automation, filtering, and enrichment applied to them. And it is hoped that they are presented with views into the data that are something other than just scrolling alerts. Modern SIEM, SOAR and big data platforms provide all sorts of data visualization tools. As a result, preparing any such mathematical formula to predict analyst staffing requires additional questions.

How well tuned is the SIEM? The sensors? The data feeds? Are all the alerts unified into one or more SIEM dashboards, or are they split among half a dozen disparate tools? That said, deploying a new sensor technology was deployed or add a new dashboard to SIEM does not necessarily equate to needing eq to hire more analysts. In some cases, the addition of new tools might be able to completely automate certain monitoring use cases if the SOC is using an advanced SOAR capability and could result in a reduction of the number of needed staff. Clearly, there is a lot of gray area here.

Also, consider other tasks thrown at analysts performing alert triage. In a small SOC, these analysts may also do routine cyber threat intelligence collection or vulnerability scanning. Or staff may be dedicated to other routine tasks like monitoring constituents' Web-surfing habits or IT compliance-related activities. This certainly adds to their load. The SOC escalation CONOPS will come into play here because many analysts have multiple responsibilities beyond just triaging events.

The size of the constituency is a very important factor in determining the number of staff needed. Large centralized SOCs may have as many as two and six analysts on each shift performing alert triage. A small SOC may only have two analysts working a single 8x5 shift. When looking at a SOC watch floor, this can be quite deceiving to outsiders, because there are many other parts of the SOC residing in back offices, which are just as important to the mission.

Finally, it is worth noting that no matter how much automation, correlation, ML, and artificial intelligence a SOC has, there must be a human that evaluates the output of all that machinery. Some vendors and security professionals may claim, "we have technology X, we don't triage analysts anymore!" That might be a bit of an overstatement. Instead, it might be fairer to say, "in presence of technology X, our staffing needs for triage analysts are less than they would be otherwise."

4.5.3 Incident Analysis and Investigation; Incident Coordination; Containment, Eradication, and Recovery

The number of staff needed to analyze, investigate, respond to, and coordinate incidents is most directly related to two factors:

- The frequency and number of cases that move from an alert to an incident, or the number of cases passed over from parts of the SOC (e.g., cyber threat intelligence analysis and trending or forensics).
- The ability of each analyst to turn over cases, which is influenced by their skill and tools as well as by how much time is allowed for in depth analysis of the incident including development of event reconstruction and lessons learned.

For instance, some SOCs can be stuck in the response cycle: find intrusion, pull box off network, reimage box and their staffing needs will reflect that limited scope. On the other end of the spectrum, there are SOCs that have an advanced adversary engagement, tradecraft analysis, and reverse engineering capability. Furthermore, the SOC may have these capabilities, but they may be split out into different advanced capabilities, cyber threat intelligence, or forensics section. Some SOCs will actually host an entire malware catalog and analysis framework, further increasing staffing needs (but also effectiveness).

Staffing for this section is heavily influenced by the SOC's ability to find (and pay for) staff capable of carrying out advanced analysis, as well as the overall vulnerability and threat profile of the "Strategy 3: Build a SOC Structure to Match Your Organizational Needs." If the constituency is suffering frequent intrusions from low-skilled adversaries, clearly there will be a greater demand for incident responders than if major incidents are primarily perpetrated by top tier APTs. In addition, the SOC may feel compelled to spend cycles chasing down IT misuse cases such as users caught downloading elicit materials on constituency systems. It is easy for the SOC to spend resources on these cases because it has the right tools to investigate them, even though such cases should probably be moved to another organization or simply receive less focus and "policing." Staffing will also be influenced by the constituency's willingness to invest in hunting for potentially malicious activity rather than just responding to existing alerts. The number of staff needed for this activity is driven by the risk tolerance of the constituency.

Depending on all these factors, SOCs may have alert triage to advanced analysis ratios anywhere from 1:1 to 1:3. This means that for every two alert analysts on a day shift, there could be between two to six incident investigators and responders, depending on how operations and escalation are structured. In terms of actual personnel counts, this may be more like 5:1 or 3:1, because the alert triage positions are more likely to be staffed 24x7 (one 24x7 "seat" filled costs roughly 4.8 FTE).

4.5.4 Cyber Threat Intelligence; Threat Hunting

This section of the SOC has the most open-ended portion of the SOC mission. Staff are asked to consume as much CTI and sensor data as possible, in a never-ending quest to uncover anomalous activity. Staffing also in this section will be driven by the SOC's access to, and ability to process and correlate, CTI data. If it has inaccurate or outdated CTI, and is doing everything manually, increasing the number of analysts will not necessarily improve a SOCs ability to provide CTI to the constituency. If, on the other hand, the SOC has strong relations with partner SOCs, comprehensive monitoring coverage, and advanced analytics, the opportunities are almost endless.

Due to limited resourcing smaller SOCs will likely have a small cyber threat intelligence and hunting section. Perhaps equally or more often, this is an additional duty for other analysts such as incident investigators. Perhaps most importantly, capacity reserved for "peacetime" hunting activities will frequently and justifiably be pivoted toward incident investigation and response during high critical incidents and major breaches.

Hybrid tiered, coordination, or national SOCs will likely have a very large cyber threat intelligence and trending section because their focus is largely shifted more toward watching the adversary instead of watching enterprise assets and networks. In the largest examples, national-level SOCs may designate a number of sub-teams, each focused on a specific brand of adversary or geographic region. In all cases, the number of personnel in this category will be driven by the risk posture of the constituency and how much the constituency can afford to invest in these capabilities to help them reduce risk.

4.5.5 Vulnerability Scanning

Staffing a vulnerability or continuous monitoring capability within the SOC is straightforward, in that it is dependent on only a few factors. Some SOCs do not have this capability at all (usually because the function is performed by another organization). For those that do incorporate vulnerability scanning into their SOC, consider the number of systems being scanned, the complexity and efficiency of tools that perform data scanning and scan data aggregation, and whether the scanning targets are broken up into disparate networks. A good scanning tool should both excel at distributing, scheduling, and staggering scans, and collating scan results. SOCs leveraging tools without these benefits will spend proportionally more time with less output.

After a scan is performed there also needs to be staff that can track and follow up on findings and work with system owners to remediate any vulnerabilities that were found. This should be automated as much as possible, meaning system owners should be able to help themselves to vulnerability data pertaining their systems, thus relieving the SOC from manual reporting tasks.

SOCs that perform network or vulnerability scanning and compliance management in house will often have a team of two to five people—possibly more if they have a very large constituency; possibly only one person if their scanning tasks are limited in nature.

4.5.6 Vulnerability Assessment; Attack Simulation and Assessments

As with Vulnerability Scanning, this function may be operated from within another part of the constituency. However, if it is operated out of the SOC, then making staffing choices is in some ways similar to cyber threat intelligence analysis and trending. Vulnerability assessment

(VA), pen testing, purple teaming, and related activities can be very open-ended for many constituencies. That is, there will always be more work to perform. As a result, the SOC can most likely assign as many people as funding permits, with the following caveats.

First, SOC management is advised not to build up a huge VA/PT section while starving other sections like advanced analysis or trending. Second, the SOC must carefully manage its workload based on the authorities and rules of engagement granted by constituency executives. In environments where the VA/PT team has more freedom of action, it may be able to set the agenda for its operational activities. Third, this team may matrix in personnel from other sections. To bulk up teams on large "jobs," this SOC should cross-train staff on defensive and offensive techniques and share knowledge of constituency systems and networks. Caution should be exercised here, as rotating staff out of analyst positions means any cases they were working must either be put on hold or handed off to another staff member. In addition, staff must work on VA/PT engagements with some regularity for their skills to stay current. Fourth, the work the VA/PT team does might create additional work, including alerts and investigations ordinarily destined for routine handling.

This section's capacity can also be directly correlated to staffing. Assume that a given SOC observes that an average assessment requires a team of three, plus one lead, and it takes an average of three weeks to perform a "job," start to finish. That works out to 12 staff weeks per assessment. Four assessments work out to 48 staff weeks of effort, which works out to four assessments per year for every FTE, including training and time off. Capacity and staffing projections can thus be made. Granted, it is not always this simple (some jobs are bigger than others), but at least this is a starting point. The SOC can also use this sort of calculation to predict how often it is able to revisit a given network, site, or program (depending on how its assessments are bounded).

Finally, as discussed in Section 11.4, breach as a service enables the pen test team to achieve substantially greater "revisit" rates and capacity through automation than they would otherwise. Specifically, in cases where a full pen test is not required, or a low-key test of routine protective and detective capability is sufficient, a BAS may turn a two-week effort into a day's engagement with a team. While this may not necessarily enable the pen test team to shrink in size, it will enable better coverage and focus than the alternative.

4.5.7 Security Capability Engineering and Management: Network, Endpoint, Cloud, Mobile, OT, and Analytic Platform

SOC tool engineering and management staff requirements are largely a function of five factors:

- The number of sensors and covered nodes per sensing technology
- The number of different sensing technologies being used
- The degree of automation and reliability of that platform
- Whether those different platforms can be managed centrally, or are on disparate networks or identity planes, requiring the administrator to "swivel chair" between them

- The engineering processes/life cycle requirements of its parent organization and constituency

An important distinction should be drawn here: there are resources needed to maintain the platform, and then there are the resources needed to customize and leverage the platform (such as creating and tuning SIEM content like correlation rules). Cloud-managed host and network sensing, SIEM, case management, and SOAR technologies do not require nearly as much labor to run the *backend*, but they still require someone to do things like sensor deployment, custom detection creation, and use case management. Consequently:

To get started, these bullets provide some very rough estimates, according to experience:

- **Small SOC:** Approximately half time of a person per sensor type of small sensor deployment (for less than 50 network sensors, 2,000 server sensors, or 10,000 desktop host sensors)
- **Medium SOC:** one person per each moderately sized sensor deployment (50–200 network sensors, 2000–10,000 server sensors, or 10,000–50,000 desktop host sensors)
- **Large SOC:** two or more people per each large-sized sensor deployment (for greater than 200 network sensors, 10,000 server sensors, or 50,000 desktop host sensors)

In very small deployments, sensor and signature/heuristic policy management is relatively straightforward. As the number of sensors grows, however, management becomes more challenging, as the variety and number of different deployment scenarios and diversity in rule sets require more management overhead. Adjustments must also be made for sensor platforms that entail greater integration challenges, require constant care and feeding, are operating at the edge of their performance envelope, or need an extraordinary amount of custom rule set creating or tuning. In larger SOCs, it is not unusual to have a team of three or four people devoted just to keeping a single EDR suite functioning properly. This is due, in part, to its tight integration with the server and workstation environment.

This is not the whole story. Many organizations have a shared management model that allows for joint operation of some security tools. In this case, both the SOC and the IT department share responsibility. One example is a corporate "next generation" firewall that provides both packet filtering as well as deep packet inspection, web proxy, and threat prevention. The firewall itself may be managed by the IT group while the SOC manages the firewall policy and threat signatures. Another example is that an EDR solution, including endpoint agents and the server may be rolled out and maintained by IT, however the detection policies might be maintained by the SOC.

Additionally, there are a number of other jobs that the sensor and system administration shop must carry out every day. With the rising complexity of analytic frameworks, many SOCs also feel compelled to commit staff to maintaining, enhancing, and integrate these systems. Such jobs could involve maintaining the SOC SIEM or SOAR, big data platform, CTI repositories, or the malware repositories. Part of these jobs will invariably entail specialized platform or database administration work (e.g., data warehouse tuning and optimization). Most SOCs that make a serious commitment to their SIEM implementation will designate one or more people as SIEM or SOAR content managers; their job is to manage and tune the plethora of custom correlation rules, filters, dashboards, and heuristics built into SIEM. For many of these platforms, one of the major time sinks is ensuring daily data feed health and driving saturation of coverage across the constituency. Consider a SOC that is supporting an audit collection architecture that serves many sysadmins and security personnel in a large constituency—this size user base will require dedicated help.

4.5.8 SOC Enclave Engineering and Management

Staffing the engineering section of the SOC is influenced by these six factors:

- The number and complexity of SOC systems in operation
- How often new capabilities are rotated into operation
- Whether the SOC has any homegrown or custom capabilities to which it must devote development cycles
- Where the line is drawn between system administration and engineering functions, if there is a distinction, and if any of those functions are performed outside of the SOC
- The extent to which the SOC owns its own underlying infrastructure, and whether it is able to leverage cloud solutions or virtualization platforms

Depending on how hardened and isolated the SOC enclave is from the rest of the constituency, the SOC will likely need to also allocate staffing to the following:

- Maintaining SOC analyst workstations, domain controllers, AD objects, and group policy objects (GPOs)
- Patching SOC/analysis enclave infrastructure and sensor systems
- Maintaining internal incident tracking database (if it is separate from the SOC's SOAR or SIEM)
- Maintaining SOC network switch, router, and firewall infrastructure (if on-prem)
- Updating SOC internal or constituency-facing website
- Maintaining SOC on-prem network area storage (NAS) or SAN resources
- Maintaining and securing SOC cloud-based services

Inherent in all these functions is not only the hands-on operations and maintenance (O&M) of systems but also CM, patching, and upgrades. In fact, some larger SOCs may designate someone separate from the sysadmin lead to preside over document management and configuration tracking.

For many SOCs, the team that performs system O&M is the same team that engineer's new capabilities; this arrangement usually drives the best sense of problem ownership, staff engagement, and rapid iteration in capability. In such cases, the SOC can simply take the staffing requirements for system administration, inflate them by some multiplier, perhaps 1.5, and calculate the total number of individuals needed for system administration and engineering together. However, this masks the efficiencies gained by integrating engineering into operations.

By contrast, a SOC without an integrated engineering function usually receives new or upgraded capabilities more slowly and they may not match operators' requirements as well. As a result, operations must devote additional resources to applying bandages and duct tape to problems (i.e., making the tools work as intended).

The point is that by having engineering integrated into ops, the additional staffing requirements to engineer new systems are usually more than made up for by the efficiencies gained, to say nothing of the improvement to the mission. If a SOC pursues this approach, it must be sure to maintain appropriate levels of CM and documentation of its deployed baseline capabilities.

4.5.9 SOC Leadership and Management

Once the SOC team grows beyond a couple of people consider adding in SOC-specific leadership roles. This could range from team leads of various functions to an overall SOC manager. In smaller organizations the team lead or manager may split their time between technical and managerial responsibilities. In medium and large SOCs, the SOC manager may focus almost entirely on day-to-day management, driving SOC performance; communicating the value of the security operations to others; evolving the SOC through new technologies and processes; and recruiting, training, and assessing the staff. Consider the SOC's own organizational standards and practices to identify how many leads are needed given the number of individual contributors. If the SOC is 24/7 operation, it is often helpful to designate a lead per shift- either a manager, technical lead, or senior analyst.

4.6   Summary – Strategy 4: Hire AND Grow Quality Staff

4.1. Staffing is one of the biggest challenges for a SOC, yet it is also one of the most important factors in the success of the SOC mission. When hiring, passion for the role is key indicator of success. Ensuring hiring accounts for mind-set, soft skills, background, and technical skillset.
   - Experienced candidates should be able to demonstrate a general literacy of IT and cybersecurity along with having deep knowledge in one or more areas, a concept known as the "T-shaped" person.
   - For more experienced practitioners, look for candidates that are invested in sharing their knowledge with others.
   - SOC leaders benefit from having a strong foundation in security operations. However, other business and communications skills are just as important, especially in larger organizations.

4.2. There are not enough cybersecurity professionals available, each SOC must also grow talent internally, and support career progression. This can and should be done through not just building a hiring pipeline but cross training, role rotations, and external training.

4.3. Equally as important as hiring the right staff is creating an environment that encourages staff to stay. Factors to maximize staff retention include paying fair market value, supporting career progression and staff growth, adding in automation, and advancing new capabilities, creating a culture of communication and sharing among the SOC team, and supporting a diverse and inclusive work environment.

4.4. Staff turnover is a reality for most SOCs, pre-planning for departures by formally capturing institutional knowledge addresses this issue and supports overall SOC process execution. At the same time, the SOC should balance repeatability and consistency in formal process with enabling analysts to think for themselves and do what is right.

4.5. The number of cybersecurity professionals needed by a SOC will be driven by factors such as the SOC mission and offered capabilities, size of the constituency, level of automation, and funding available. Key factors that drive the number of staff needed for each capability area include:

- **Real-time Alert Monitoring and Triage:** Number of alerts coming in and the level of automation that supports alert triage.
- **Incident Analysis and Investigation;** Incident Coordination; Containment, Eradication and Recovery: the frequency of the number of cases that move from alert to an incident and the ability of each analyst to turn over cases which is influenced by both their skills and the tools available to them.
- **Cyber Threat Intelligence Collection, Processing and Fusion; Threat Hunting:** the most open ended of the SOC capabilities. Impacted by the risk posture of the constituency and the amount the constituency can afford to invest in these capabilities to help them reduce the risk.
- **Vulnerability Scanning:** number of systems being scanned or monitored; complexity and efficiency of the tools being used; the complexity of the environment being scanned.
- **Vulnerability assessment; Attack Simulation and Assessments:** risk posture of the constituency which helps determine scope and frequency.
- **Security Capability Engineering and Management:** Network, Endpoint, Cloud, Mobile, OT, and Analytic Platform: The number of sensors and covered nodes per sensing technology; the number of different sensing technologies being used; the degree of automation and reliability of that platform; whether those different platforms can be managed centrally, or are on disparate networks or identity planes, requiring the administrator to "swivel chair" between them; and the engineering processes/life cycle requirements of its parent organization.
- **SOC Enclave Engineering and Management:** The number and complexity of SOC systems in operation; how often new capabilities are rotated into operation; whether the SOC has any homegrown or custom capabilities to which it must devote development cycles; where (and if) a line is drawn between system administration and engineering functions, and if any of those functions are performed outside of the SOC.
- **SOC Leadership and Management:** size of the SOC; organizational standards and practices.

Strategy 5: Prioritize Incident Response

Responding to incidents is at the core of the SOC, and arguably its most important function. Prioritizing incident response activities to ensure the SOC is prepared and staffed goes a long way to building a successful SOC. As long as there have been computers and networks, there have been cyber incidents. Seminal organizations such as the CERT Coordination Center (CERT/CC) have asserted for decades there is no SOC (or previous incarnations such as CSIRT) without incident handling at its core [5], [107]. Incident responders are the front lines of determining what is currently happening, where, and how security events are unfolding and determining the actions to react.

This strategy addresses the important aspects of responding to incidents in general and touches on specific recommendations for incident response in cloud and mobile environments. Incident response is one of the most discussed SOC functions and there are many great books, articles, and presentations available on the topic. Our goal with this strategy is to emphasize the importance of prioritizing incident response as a core function of the SOC.

5.1   What is Incident Handling?

First, as mentioned in the Fundamentals section, NIST defines a cyber incident as, "Actions taken through the use of an information system or network that result in an actual or potentially adverse effect on an information system, network, and/or the information residing therein" [14]. In other words, anything the incident responders take action to determine, find, or analyze falls into the scope of incident handling. Depending on the constituency, these can include the following (this list is not exhaustive):

- Phishing Attacks
- Malware infections
- Web Attacks (e.g., SQL Injection, Cross-Site Scripting)
- Denial of Service (DoS)
- Successful unauthorized access (compromise)
- Unauthorized information disclosure or data loss
- Device theft, loss, or destruction, such as those related to ransomware and physical loss
- Compliance or hygiene-related issues

Incident handling involves receiving, sorting, categorizing, and prioritizing of incoming incident reports or other requests and responding to requests and reports, and analyzing incidents and events. This includes the full lifecycle of an incident from receiving the first indication there is an incident all the way through closing out the incident. Building on the CERT/CC list of activities included in incident handling, these activities may include [5]:

- Conducting log review and forensics to identify scope, depth, and source of intruder activity
- Working with system, network, device and cloud resource owners and users to assess alerts and logs in context
- Altering security controls to contain or eradicate an adversary
- Rebuilding systems, networks, and cloud resources
- Filtering web, network, or other incident-related traffic
- Resetting passwords, certificates, and other service principles
- Applying relevant advisory and alert solutions

Generally, the terms incident handling and incident response are inconsistently used throughout SOC communities. In some circles, incident handling is considered a broader term than incident response, suggesting it encompasses tracking and reporting, while incident response is specific to responding to the incident itself; although, many SOCs call the function "Incident Response" and include tracking and report writing in the function. For the purposes here, incident response and incident handling are interchangeable.

5.1.1 Models, Frameworks, Methodologies, and Resources

Computer incident response has been around "officially" since about 1988, so many frameworks, models, books, and methodologies have emerged and evolved over the years [1]. Many of them have similarities that include planning, responding, and evolving, or applying lessons learned to improve. They can all be used to inform incident response planning, and the references provide details on each of the steps for each model. Some of the more popular include the following foundational models and resources:

- Computer Security Handling Guide (National Institute of Standards (NIST) Special Publication 800-61) [108] and Guide for Cybersecurity Event Recovery (NIST Special Publication 800-184) [109, p. Figure 16] as shown in Figure 12.

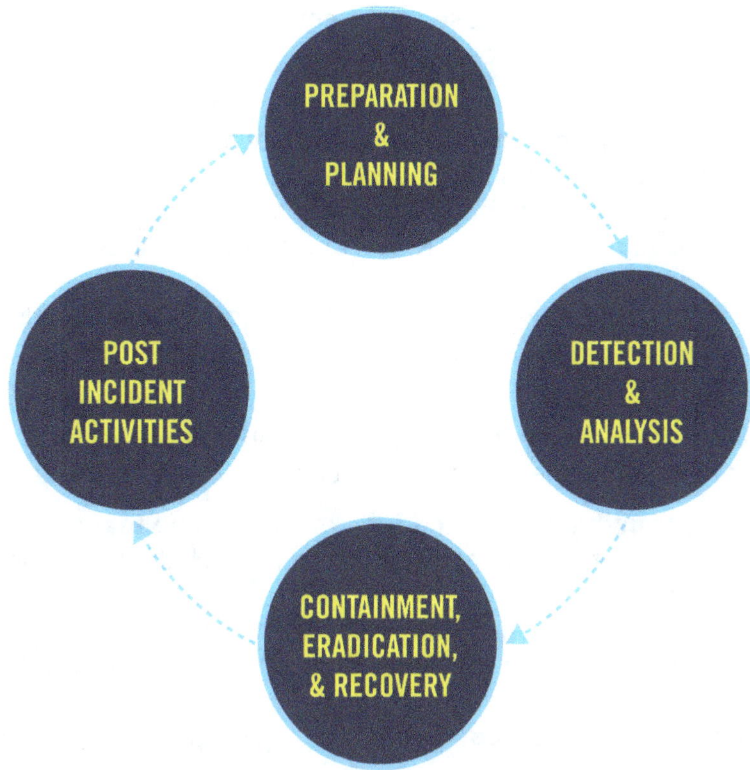

Figure 12. NIST Incident Response Lifecycle, [108], [110]

- VERIS framework: a "common language for describing security incidents in a structured and repeatable manner" [111]
- ENISA Reference Incident Classification Taxonomy: as the title infers, ENISA's approach to categorizing and mapping incident types, including prominent stakeholders such as CIRCL and CERT.LV [112]
- MISP mapping of incident and threat taxonomies: a sort of "Rosetta stone" that connects to other frameworks such as ENISA's and VERIS's [113]
- OODA Loop: applying Observe, Orient, Decide, Act (OODA) to cyber incident response [114]

Using these frameworks and other references are helpful in planning incident response. The OODA loop was described in Strategy 0: Fundamentals, Section 1.1, and is useful in planning for IR to ensure the SOC frames the response procedures around situational awareness as investigations progress.

5.2 Planning

When there is a major incident, all eyes are on the SOC. Planning can ensure the SOC's incident response is efficient, effective, relevant, and complete. Security professionals must render a response that is appropriate given the criticality of the situation. Most incident handling should be routine and not cause for an emergency. The SOC will be prepared for most response efforts if it has the following in place:

- A workforce with strong technical, analytic, and communication skills
- CONOPS, SOPs, and escalation procedures that guide the SOC's actions
- Means to coordinate analysis and response activity among members of the SOC
- Established POCs with whom to coordinate response actions
- Established and ad hoc artifact collection and analysis tools sufficient to establish the facts about incidents
- The authorities to enact swift and decisive response actions when called for and passive observation or incident de-escalation when they are not

The CONOPS, SOPs, and escalation procedures should contain both core topics as well tailored content. Building on some of Cisco's suggestions on what to include in an incident response plan, the following content should be included by most organizations [115]:

- **Roles and responsibilities for the incident response team members:** Who is in charge of the incident, who is performing the IR (including outsourcing information, if that is relevant).
- **Communications, coordination, and contacts:** Who needs to be informed, coordinated with, and when and how, both internally and externally. This should include legal and law enforcement.
- **References and procedures:** SOPs, reporting guidelines, policies, and other documentation.
- **A summary of the tools, technologies, and physical resources:** Including what the SOC has and how to access them.
- **A list of critical network and data recovery processes:** Including step by step plans.
- **Reference to the SOC charter:** This could include other documents that give the SOC approval/authority to operate.

In addition to the above, the SOC can tailor the plan to address specific strategies. For example, if the SOC constituency is large and geographically distributed, listing the different groups and stakeholders may be a necessary part, in addition to the roles and responsibilities. Once some of the basics are in the plan, the SOC can then anticipate the different incidents that might hit the constituency.

5.2.1 Prioritizing Incident Categories

Identifying the types of attacks the SOC might see is an important first step in planning as it helps the SOC prioritize actions. This should include identifying both common attack vectors that the SOC expects to see routinely as well as those vectors that may be less common but would be of high risk for the constituency. Planning for events that do not happen often is as important, or more so, than planning routine incident response; the SOC team will not have as much experience handling those types of incidents and therefore will not have the muscle memory of knowing exactly what to do in those instances.

Capturing a list of common attacks on the constituency that inform incident handling is a good way to start. One method to do this is to map out a table with the type of incident, the priority, and recommended action, specific to the SOC, as suggested in ATT's guide for incident response [116]. Populating the table can be completed using MITRE ATT&CK's matrices, specific to the environment. As a caution, it is possible organizations will have different priorities for different types of incidents, so a direct copy of others' priorities is not recommended. For example, AT&T suggests that a distributed denial of service (DDoS) attack is a higher priority than unauthorized access, which could be true for an organization relying on providing network and cloud services but may not be true for a research organization depending on intellectual property for survival. The research organization with intellectual property may care more about confidentiality and accuracy, in which case incidents and events characterized by unauthorized access might be the top priority. A power utility providing electricity may prioritize denial of service of control systems as ultra-high.

Table 6 shows and example of combining AT&T's prioritization table format with MITRE ATT&CK's tactics and techniques content.

Table 6. Sample Incident Prioritization Planning [116], [117]

Incident/Event	Priority Level	Response or Action
Most port Scanning activity (pre-incident)	Low	Ignore most of these. Block or incorporate into detection if scans are tied to other reconnaissance, a known bad reputation, or there are multiple events from the source.
Malware infection	Medium	Remediate any malware infections as quickly as possible before they progress. Scan the rest of the constituency or enclave for associated indicators (e.g., SHA256 hashes).
Denial of service	Low-Medium, depending on duration	Configure affected externally facing services/systems e.g., web servers) to protect against DoS requests (e.g., HTTP and/or synchronized (SYN) flood). Coordinate with Internet Service Provider (ISP) to block/reroute the activity.
Unauthorized access	High	Detect, monitor, and investigate all unauthorized access attempts; prioritize mission-critical or sensitive data. Remediate through rebuilding accounts, systems, etc. as determined.
Insider threat	High	Identify associated privileged accounts for all domains, servers, apps, and critical devices. Ensure monitoring is enabled. Shut down access and/or coordinate with authorities where appropriate.
Web attacks (XSS, SQL ingestion, Cross-Site Request Forgery (CSRF), etc.)	High	Follow unauthorized access and/or malware response, depending on circumstance. Check web services, application, and database logs for extent of incident.
Phishing	Medium	Follow malware infection response or action. Check e-mail and other indicators for other recipients and attacks.

Incident/Event	Priority Level	Response or Action
Lost or stolen device, ransomware	Low-High	Severity depends on sensitivity of data stored on- or accessed-from device, as well as whether the device used encryption at rest, whether the data can be recovered, and whether the device can be recovered or restored. Follow up with user and IT operations to assess.
Poor security practices observed	Low-Medium	May be handled as routine compliance, depending on relative severity, either by the SOC or a partner team such as a green team, if one exists.

The table shows that this is not an exact science, but more an exercise to understand what matters in the SOC's organization. For example, phishing is a delivery for a malware exploitation, and insider is another form of unauthorized access. In some SOCs it may make sense to pull those items out into their own categories due to their relative importance to the constituency or how the constituency wants to be able to track incidents. In other SOCs it may make sense to incorporate those types of incidents into the broader category.

This type of a table can help the SOC make decisions. Using the content here as an example, the SOC that produced this table would know that it is comparatively better to be off the Internet (as from a denial-of-service perspective) than it is to have unauthorized access or compromise of intellectual property, in the face of an imminent, high confidence targeted attack. This would allow them to make more rapid response decisions than having to identify those priorities in the heat of the moment. To reiterate however, this table is offered as the beginning of a set of guidelines for prioritizing incident handling. Not all unauthorized access cases are higher in priority than poor security practices, such as an unpatched vulnerability exploited in the wild, for example. Every SOC is different and most mature SOCs will have one or more SOPs for each of these incident types.

5.2.2 SOPs and Playbooks

Large and mature SOCs build up sizeable incident handling guidance codified as a "living" set of SOPs or playbooks staff are expected to know and routinely access. The purpose of these is to establish a) clear expectations for staff and b) repeatability in handling of most incidents. As a rule of thumb, if an incident type is handled by the SOC, on average, once a month or more, there should be an approved SOP on file for that incident. These playbooks are equally important for bringing new staff up to speed as they are for ensuring existing staff can focus on new challenges vs having to reestablish the right activities to take with every new incident. Well defined SOPs and playbooks are also a key factor in being able to incorporate automation activities into the SOC. Without clear guidance on what actions need to be taken in what scenario automation tools cannot be programmed to take the appropriate actions. In some cases, playbooks will already be incorporated into security automation and orchestration tools and will just need to be tailored for the SOCs environment.

A playbook is likely to include the following:

- Title
- Intent
- Scope/who it applies to
- Stimulating conditions, meaning under what circumstances it should be used
- Procedures, steps, and expectations to be followed
- Various metadata: who approved it & revision history (modified by whom, reviewed by whom, when, revision notes)

Even in cases where the SOC does not follow a given incident response scenario often, it may still be a good idea to codify that specific incident type in an SOP. This saves the next person the time and effort of re-learning something from scratch. This is particularly true for incident scenarios that are both rare but high impact/high visibility.

There is some art to writing effective playbooks, especially in balancing the amount of time it takes to capture detail and specificity with the understanding that some details will change with every incident. Capture those things that are consistent from a given technology type and incident of priority to the constituency. Some details and specifics are necessary to be effective. If a playbook is too high level, the experienced cyber responders will fill in with their own knowledge, which can be great, and can also lead to inconsistency and others left in the dark. Also, if they leave the organization, the understanding of what they did may go with them. A balance of capturing enough detail so that less experienced cyber responders will understand how to respond to an incident and more experienced responders are consistent in approach. Too much and specific detail, and the playbook becomes obsolete quickly, and may stifle analysts' ability to act on their own intuition and adapt to the incident at hand.

A cornerstone of effective playbooks are checklists. Checklists can be adopted and assimilated from various places and are useful for all responders, from inexperienced to expert. Even experts need to remember to notify the right people at the right times, for example.

Playbooks and SOPs are specific to the mission, business, and organization, as with all SOC functions. However, this does not mean the SOC needs to start from scratch. A good place to start developing checklists and ultimately SOPs and playbooks is to examine various types of playbooks available, and tailor and adapt good practices from them. Here are some resources and examples to get playbook, SOP, and checklist ideas:

- The MS-ISAC Ransomware Guide [118]
- Incident Response Consortium products and community [119]
- CREST Cyber Security Incident Response Guide [120]
- EPA Incident Action Checklist [121]
- SANS Critical Log Review Checklist for Security Incidents [122]
- SANS SCORE Security Checklist for Incident Handling [123]
- Microsoft Cloud Incident Response Playbooks [124]
- AWS Security Incident Response Playbook Templates [125]

Finally, it is useful for the SOC to organize all its playbooks in one knowledge management framework or system. Critically, official SOPs and playbooks should be marked, organized, or otherwise segregated from other casual information capture. This is important because when a cyber defender is looking up "ransomware" they do not want to be troubled with stray, nonauthoritative hunting notes from 3 years ago; it is best for them to index directly what they need, fast. For more on the standing documentation that assists the SOC in incident handling, see Appendix C.

5.3   Detection and Analysis

In examining incident handling, the activities of receiving, sorting, categorizing and prioritizing incidents, or triage, can be grouped, analyzing incidents and events or investigation analysis is another group, and responding and reporting is a third group for discussion.

5.3.1   Incident Triage: Receiving, Sorting, Categorizing, and Prioritizing

Initial incident reports can come into the SOC from e-mail, phone calls, IT service desk ticketing, partner organizations, and alerts from the SOC's own tools, among others. Cyber defenders then triage these incoming signals or make an initial determination on what the next steps are. Incidents usually do not start out as incidents, but anomalous or unexplained activity, which can make categorization challenging.

Triage helps coordinators or analysts separate out potential emergencies from routine response needs.

Of course, these communications need to be monitored, and sometimes these are combined through agreement with IT help desks. For example, the help desk and SOC may form an incident escalation agreement that requires the help desk to directly route all security issues it fields, such as by transferring phone calls and tickets immediately to the SOC. If the SOC and help desk do not have the ability to share and route tickets, usually a simple escalation service level agreement (SLA), measured in minutes or hours, will suffice.

As noted in 5.2.1, organizations have different ways of grouping or categorizing incidents and there is not a universally agreed-upon list of top incidents or attacks. Some SOCs break down incidents into attack vectors and information impact, which could be useful for instrumentation purposes for SOAR or EDR (See "Strategy 8: Leverage Tools to Support Analyst Workflow") but may not be as helpful for the incident responder to understand the attack holistically. One scenario might be an attacker conducts low and slow user account enumeration, then successfully uses phishing to enter initially, pivots to exploit vulnerabilities in other systems to move laterally, for example. Some tips for triaging incidents include:

- **Choose categories of incidents based on response:** For example, successful cross-site scripting (XSS) and Structured Query Language (SQL) injection attack responses likely differ from a malware infection clean-up which is different from a DoS attack. And some responders may be equipped to deal with complex compromises, while others may be better at addressing a virus eradication.

- **Consider categories across different organizational groups:** For example, phishing attempts might be coordinated with a group in charge of e-mail, whereas a web-attack needs to be coordinated with a website administrator.
- **Determine the types of categories before an incident occurs:** This helps with planning. New types can always be added as they arise.
- **Establish guidelines for triage:** Ensure analysts do not spend "too long" analyzing one incoming alert, while others grow stale. Oftentimes, a SOC will apply time boxes for initial triage, to ensure all incoming alerts are handled in a timely manner. One way to help support this is to have a pooled set of users responsible for triage, with backups in place to share load.

5.3.2 Types of investigation analysis

In piecing together what happened, several types of investigation analysis can be applied, and choosing will depend on the data available, the type of incident, and the timeframe (not to mention how long an analyst has to give some answers). As investigators move forward, it is important to assess what is speculation from what are facts. For example, *generally*, an incident cannot be confirmed using TCP/IP traffic captured at a network device alone; so, if the only data analyzed is network traffic, and someone indicates there is a compromise, the investigator can ask more questions, and gather relevant data (from the host) to confirm. For incident analysis, the following techniques can aid in ensuring the right conclusions are derived from the data: [126]

Check the assumptions
Ensure anyone participating in the analysis is aware and careful of making assumptions. Be wary of filling in gaps that satisfy a hypothesis, for example, without the data that confirms the activity.

Seek more data
If there are apparent gaps in the activity that present barriers to conclusions, seek data that will augment the analysis. This may include identifying data sources that might be outside of the SOC purview.

Analyze indicators
The most common technique of SOCs is to examine all observable pieces of information, including malware hashes, hardcoded/reused adversary passwords, IP addresses (with caution), web traffic, NetFlow, etc. to piece together the TTPs of what happened, when dealing with human adversaries.

Create timelines of events
A sequence of dates and time when adversary activity occur provide insight into how long an adversary stays in one account/system/network, what occurred first, and can give investigators ideas on when an intruder moved from one network to another. It can assist with data pulls, as well as establish patterns, such as did the intruder "go dark" for a while and hide, or the data did not show the trail of where they went next.

Compare good and known to suspicious activity

For example, most reputable application, service, and operating system providers have either libraries of "known" files with hashes available or known, trusted signing certificates. These known "good" hashes (or other characteristic) can then be used to compare to suspicious files; this is frequently done for operating system registry or kernel files.

Do not rely solely on IOCs

Discrete indicators of compromise, such as IP addresses or file hashes/checksums are easy for an adversary to change. IOCs are important and useful in an initial incident investigation to provide hints and determine where and what to search for but it is important to look beyond them. TTPs that are not tied to specific indicators are better for identifying known adversaries because they provide correlated patterns that are more durable. Effective investigations strive to create accurate TTPs for intrusion sets.

Scenario or hypothesis analysis

Forming hypothesis based on what is currently known and determining what the options for the activity might be can assist investigators on determining what data to pull next. This assists investigators in finding where else and adversary is resident, and what else was compromised, as well as finding other possible answers.

Be aware of bias

Investigators have built in and often unconscious bias. Once investigators have some experience with incidents, one incident might look very much like another in the data. It is important to treat each incident as a new one, and correlate them with other incidents only with hard, indisputable facts.

Each attacker-based incident is unique, it is important to ensure the response fits the incident.

5.3.3 Forensics for Investigations

Forensics are used for several different purposes in cybersecurity; digital forensics is "the application of computer science and investigative procedures involving the examination of digital evidence – following proper search authority, chain of custody, validation with mathematics, use of validated tools, repeatability, reporting, and possibly expert testimony." [127] There is a difference between conducting forensics that ultimately lead to offenders going to jail, diving deep into reverse engineering malware, and simply determining what happened to address an incident. In "Strategy 11: Turn Up the Volume by Expanding SOC Functionality," some of the more advanced forensics topics are addressed, while this strategy focuses on determining what happened and how to respond.

In incidents, forensics are used to discover suspicious/malicious activity possibly linked to activity being investigated. This involves activities such as pulling EDR data to investigate and respond, for example. The data needed to investigate will depend on the affected system(s) and specific attack, and could include network, system, or other data. Detailed resources on how to conduct incident investigations for specific operating systems are available on the Internet, either directly from product vendors or through third party sites. Here are some examples:

- Applied Incident Response (variety of references including Windows Event Log analysis, Volatility, PowerShell for IR, and others) [128]
- SANS Cyber Security Blog: Cheat Sheets, Handler's Diary, and others (focus area: digital forensics and incident response [129]
- SANS Cyber Security Tools (focus area: digital forensics and incident response) [130]
- Microsoft Security Incident Response Blog [131]
- Redhat Enterprise Linux Security Guide [132]

In general, the tools and operating system interfaces have changed over the years, but the investigation techniques have not. Most incidents are not detected at the initial entry point. In fact, initial entry of an attacker into a system may be through legitimate username and password gained from other systems or enterprises, such as an organization in the same business or a member of the same Windows domain.

Most SOCs find attacks "right of hack," or after the successful initial entry. Example incident activities to look for include upgrading or escalating privileges, moving laterally across user accounts, or downloading exploits or tools. The best data for confirming incidents originate from end hosts or systems and other user account data. This can mean EDR data which is discussed further in "Strategy 8: Leverage Tools to Support Analyst Workflow." While network data is commonly used and is a great starting point and useful to assist in pinpointing likely hosts, confirming malicious activity and finding details are better coming from affected host data. Of course, ensuring host originated data is collected and not stored on the host assists in ensuring audit logs and other user data is not altered in an attack. In multi-system, multi-account activity, host data can help with backtracking to the earliest activity to ascertain the initial entry into the constituency. Identifying "patient zero" enables the investigator to determine where else the attack reached (accounts, systems, networks) and to determine response actions.

Whether looking for activity in traditional systems or EDR, some examples of (operating system agnostic) considerations for investigation include:

- **File system and files:** Unexpected files the user did not create or new hidden directories, mysterious encrypted directories (which offer a challenge)
- **Running processes:** Unexpected or hidden processes or those do not perform as usual (e.g., exceptionally memory intense, or strangely not present when it should be)

- **Scripts, executables:** Mysterious or unaccounted for programs and apps
- **File checksums and signing certificates:** System file hashes that do not match expected hash or signing certificate that do not match the correct certificate authority
- **System logs:** Unexpected user accounts, privilege changes, trusted hosts, important file changes
- **Network, VPN activity:** Unusual network (current and recent), VPN, tunneling, and other communication connections, including RDP, SSH, and other connections that could signify a backdoor

5.3.4 True and False Positives and Negatives

Tools do not always alert when something bad happens, and just because they throw an alarm does not necessarily mean it is time to isolate a host or call in the team to respond on a weekend. Just because a next-generation firewall alerted that an SQL injection attack occurred does not mean that the database was actually hacked. Part of the analysis process is determining the nature of the alert while understanding reasons the alerts may not be telling the full story. Each alert that detection tools generate falls into one of four categories described in Table 7, depending on whether alert fired and whether something bad happened.

Table 7. The Four Categories of Activity

	Bad behavior occurred	Bad behavior did not occur
Alert fired	**True positive** Something bad happened, and the system caught it.	**False positive** The system alerts, but the activity was not actually malicious.
Alert did not fire	**False negative** Something bad happened, but the system did not catch it.	**True negative** The activity is benign, and no alert has been generated.

The most prominent challenge for any monitoring system to achieve a high true positive rate. In both academia and commerce, creators and salespeople strive to prove that their system never misses a successful hack (i.e., it never has false negatives). However, this often comes at the price of a tidal wave of false positives [133]. Too many false positives compel analysts to spend inordinate amounts of time wading through data or, worse, ignoring a tool entirely because the good signal is lost in the noise.

On 20 April 2010, a Transocean oil rig off the coast of Louisiana exploded, killing several workers, and spilling hundreds of millions of gallons of oil into the Gulf of Mexico. The disaster and demise of the rig, Deepwater Horizon, cost millions of dollars in direct business and economic damage, not to mention the environmental disaster and cleanup that followed for months and years [134].

Several factors precipitated this event; one of them was the rig's safety alarm systems. The alarms for one of the safety systems on the rig were disabled because they went off too often. To avoid wakening maintenance personnel in the middle of the night, the alarms were essentially deactivated for a year prior to the explosion [135]. One of the main themes of the Transocean incident was that ongoing maintenance problems prevented the safety systems from operating and alerting correctly. The Transocean incident shows that an analyst becoming numb to even the most serious of alarms can have disastrous consequences. While the Transocean disaster happened for a number of reasons, ignoring safety alarms clearly did not help.

Using terms such as "true positive" and "false positive" make assumptions about the *intent* of the detection author. Detection is not just used for malicious activity. Dismissing a non-malicious event as a "false alarm" assumes that the goal is to detect some sort of malicious activity; sometimes this is not the detection goal. SOCs often choose to activate various detections and accept log feeds because they will leverage the data for contextual or retrospective analysis, not because they serve as tip-offs. A fleet of EDR sensors may generate hundreds of "good" tip-offs a day, while millions of audit logs are collected to back them up.

It is worth nothing that some SOCs use a fifth category of alerting, often referred to as "benign positive." The intent of this term is to describe situations where a detection engine worked properly (otherwise labeled as a true positive), but the activity in question is correctly understood to be non-malicious. Labeling an alert as benign positive says to the detection author "this detection worked properly, but the activity turned out to be non-malicious."

5.3.5   Tips for effective and accurate analysis and investigation

Before acting, incident responders investigate the situation to determine about what action is truly required. Routine incidents are usually straightforward, requiring little analysis, whereas a potential breach that starts with a subtle anomaly, such as an unexpected log in time, requires more. This section discusses some tips to unravel more subtle, yet potentially more serious incidents.

Stay calm

When police, firefighters, or paramedics arrive on the scene of a 911 call, they are cool, calm, and collected. They can assess and stabilize the situation and direct response accordingly. Doing so engenders trust on the part of the complainant or the victim. The SOC should follow the same practice. For those users not familiar with SOC operations, an incident is cause for great excitement and emotion. This can lead to reactions that amplify damage. The SOC will gain the trust of those involved if it provides measured response, no matter what circumstances it encounters.

Establish relationships with law enforcement and legal counsel in advance
This is so the SOC knows who to contact when an incident is happening, and consulting lawyers for legal advice is important. They can advise the SOC on legal considerations such as forensic evidence collection, how to handle incident issues related to personnel, and the constituency's option in criminal or civil action.

Talk to Users, Service Owners, and System Admins
Users are familiar with their account activity. One of the best sources for identifying potential breach incidents, which are often subtle, is the user community. Users are familiar with their file structures, with their profiles, and their activity or non-activity, and may notice when something is not as they left it.

Consider matrixing work to people not normally in the SOC
Not every SOC has all the skills and knowledge in-house to handle every intrusion. For example, if reverse engineering is needed, yet the SOC does not have reverse engineering expertise, consider bringing in a third party, such as another SOC for part of the analysis. And, sometimes an incident investigation can get very complex and involve more data than the SOC can analyze even with expertise.

Put it in context
Determine what the observed activity means in the context of the service and any other events. What are the factors around the activity? Are there other activities and is there a sequence or timeline that can be formed? Is the activity seemingly normal or consistent with service activities or abnormal or not as expected? Where is the activity (virtually and/or physically) and what data might augment understanding?

Avoid premature conclusions and assumptions
It is easy to assume activity is a root compromise and decide where it is from by an IP address. For example, if a SOAR tool fires a correlated network-based rule and sends an alert that has not triggered before, it may seem initially it is confirmation of compromise. But the investigator would want to confirm the activity (for example, using EDR data directly from the system) before deciding. It takes a skilled analyst to correctly interpret what a set of security logs, network data, or media artifacts convey. It is better to err on the side of conservative judgment, pending more data, then to jump to a conclusion that is incorrect, which can cost valuable time, money, and political capital.

Differentiate facts from speculation
Establish techniques in advance to ensure burden of evidence and for distinguishing between correlation and causation.

Do not focus on attribution, but do consider adversary association
It can be helpful to associate current activity with previous activity or adversarial groups as this may give defenders additional context for understanding and investigating an incident. However, creating connections too quickly and without sufficient evidence can lead to incorrect hypothesis that harm the investigative process. At the same time, the SOC may not require full attribution to confidently act on likely related information. For more on attribution vs. association see Section 6.5.1.

5.4 Containment, Eradication and Recovery

Incident responders spend most of the time on routine security events that are definable and have repeatable processes to respond. Cleaning up malware and following up on phishing attempts are examples. Although more infrequent, investigations light the way to determine what response to take in more serious incidents. Response actions that are incorrect, ineffective, or too blunt often result in important cyber resources wasted or legitimate services rendered unavailable on insignificant routine incidents.

Ensure the entire SOC is working toward the same goal

In the heat of the moment, it is easy for members of the SOC to step beyond what they are authorized to do, considering their limited perspective on what needs to happen next. Telling a system owner to disconnect a system or shut off access could be disastrous, even if it seems like the right thing to do at the time. Coordination is not only between the SOC and external parties, but it also starts internally, through both peer-to-peer collaboration and a clear command structure.

Establish Clear Command

When firefighters show up at the scene of a fire, the onsite incident commander is in charge. The district fire chief and the city mayor generally do not show up because there is no need. For most SOCs, these clear boundaries of trust and communication are not as well established as for doctors or emergency responders. It is important to clearly designate an incident leader for that specific incident.

Avoid falling victim to the "fog of war"

Before responding, ensure the investigation details and facts are accurate and complete. For example, if the response to a complex, multi-system compromise is to block the associated IP addresses and re-image the systems, there is a great chance the sophisticated attackers will be back in the constituency (assuming they were removed at all).

Understand the "so what?"

When the SOC explains an incident to stakeholders and upper management, the bottom line is not about bits and bytes, it is about mission, dollars, and, sometimes, lives. The SOC must translate technical jargon into business language. There are four questions that should be answered: (1) what (and/or who) was targeted, (2) was the adversary successful, (3) who the adversary is and what is the motivation, and (4) how to continue the mission.

Report when ready

When a hospital patient goes in for surgery, family members sit for hours in a waiting room, anxiously awaiting news of their loved one's fate. While it would be great to hear frequent updates on their loved one's procedure, doing so would impede the surgeon's ability to complete the operation correctly and in a timely manner. The SOC can maximize reporting effectiveness by negotiating reporting timeframes and recipients. During a serious incident, the SOC may consider two separate regular meetings every day or two. The first is for direct players in the incident who can talk bits and bytes, and usually occurs informally on the SOC ops floor or multi-lateral VoIP/VTC. The second is a more formal SA update to upper management. This ensures everyone is on the same page and allows SOC personnel to focus on operations. Also, the SOC can avoid leaks of details to those not involved by avoiding reporting outside of negotiated parameters.

5.4.1 Deciding When to Respond

The SOC must ensure responders have enough information and data to take informed steps to handle the incident. It is important to recognize that a SOC may not always deploy countermeasures at the first sign of an intrusion. Reasons for this include:

The SOC wants to be sure that it is not blocking legitimate activity
A response action could impact a constituency's mission services more than the incident itself. The SOC may need to first coordinate with other entities, such as system owners, before acting. These other constituents can provide the necessary context to help determine if the activity is authorized or not.

To determine the scope of the attack
Sometimes taking immediate action would jeopardize the SOC's ability to find the understand the full extent and severity of the incident. Not identifying all the malicious activity associated with an incident before eradicating results in adversaries persisting, and even changing behavior to hide more stealthily.

Sometimes, forensic evidence must be preserved, collected, analyzed, and stored in a legally sound manner
In such cases, the SOC must observe greater rigor and traceability in its procedures than would otherwise be necessary. The SOC wants to take no action that would affect the efficacy of any images or other data. This is especially needed for potential criminal cases that will be defended in a court of law. Requirements for potential crimes should be coordinated with the organization's legal department, and law enforcement.

Determining when to respond (to stop the exfiltration or damage being inflicted) vs. gathering more intelligence (to understand what the adversary is interested in) is a judgment of trade-offs. This trade-off of complete situational awareness of how they are moving through the constituency vs. stopping damage presents a natural and continuing tension through the incident response process. Each incident is different, and the right balance requires judgement.

> *There is a natural tension between watching adversary activities to understand what is next and taking action to stop them from further damage.*

5.4.2 Response Scope

Assessing and determining how to respond is critical to being effective in incident response. Incident response moves very fast, sometimes too fast. Responders need to ensure they are taking the right actions at the right times. When incidents become chaotic, some responders will start trying to react without thinking through the outcomes needed to address the incident; there is a phenomenon of acting because it seems like something must be done immediately. To avoid this, keep the end in mind understand what effect or goal is desired to end the incident permanently. Some typical sample responses to attacker-based compromises and coordinated effects are identified in Table 8. The goal of illustrating this table is to convey the need to think about the actions planned and consider the effect it will have not only on the adversary, but on the affected users and systems.

Table 8. Sample Response Action with Possible Effects

Incomplete Response Action	Possible Effects
Taking systems, cloud applications, cloud resources, or accounts off-line	Results in temporary unavailability to attacker and legitimate users
Rebuilding/reimaging systems or VMs	Knocks adversary out of system (if authentication is rebuilt); destroys forensic artifacts if not otherwise preserved
Changing account information such as username and password	Knocks adversary out of account (could be temporary if adversary has username/passwords for multiple accounts)
Modifying files, systems, processes (such as modifying or deleting registry keys in Windows), and cloud resource configurations	Removes/changes offending processes, also may interrupt service availability if implemented incorrectly
Ending or killing a process	Temporarily ends current process—is not necessarily permanent
Removing files	Renders files unavailable, but may be temporary

5.4.3 Action Tracking and Incident Case Management

Understanding the actions that will be taken needs to be coordinated. When there are multiple people involved and multiple systems, coordinating becomes critical. Centralized tracking is best (see Section 8.4 for more on incident case management), to ensure responders can reference the work done, and access data across activities. Above all, SOCs should avoid a system where responders maintain their own spreadsheets of indicators, actions, and artifacts. Spreadsheets limit the ability to coordinate, inhibit incident reporting, and can lead to misfires and misunderstandings. A tracking system should minimally capture the following:

- Incident summary and details
- Timeline
- Incident responder lead and contact information
- Actions completed and in process
- Status

If the SOC deems it necessary to break down specific findings, indicators, and adversary actions into a spreadsheet or similar tool, it is critical that all team members, including supporting IT personnel, snap to a consistent set of data capture and vetting, and all have the access when they need it. Above all, responding to incidents is most effective when following plans and procedures. Stay true to incident plans and standard operating procedures throughout response, and update with lessons learned when the heat of response simmers down.

5.5 Post Incident Activities

Arguably, if the SOC took the time after a major incident (or even minor, daily response efforts) and worked through the collective lessons learned and incorporated them into planning, it would evolve more rapidly over time. Post incident response or post incident review (PIR), also known as a hot-wash or after-action review (AAR), involves capturing and reviewing the lessons learned and action analysis for future incidents. Spending quality time on PIRs improves the SOCs response capabilities, as well as other areas such as detection. Topics to discuss during this review include:

- What worked very well?
- What did not work at all?
- What slowed down the response?
- Was there any confusion? When and where, and what would avoid it?
- Did the right people have access to the data needed?
- Were there any tools missing, not available to enough people, or otherwise not useful?
- What do the group members wish happened or wish was available?
- Who should be involved that was not?
- What skills were needed, and which were used, which were missing?
- Did notification and reporting work well?

Once these and other questions are answered, the outcomes can be translated back into adjusted plans, future investments, and hiring to fill any skills gaps. In addition, it is typical for the SOC to collect observations about suboptimal security practices from services and constituents during major incidents. These should be captured in a repeatable and structured means and assigned back to respective constituents for action. If the constituency's cybersecurity apparatus has a "green" team, a group charged with enhancing security efficiency, this coordination may be best accomplished in partnership with them [136]. To make tracking actions and resolutions more manageable, the PIR should capture findings and action item metadata to include:

- Item title
- Finding/gap
- Desired resolution or success criteria
- Date filed, and by whom
- Status (open, closed/will not fix, closed/fixed, blocked, etc.)
- Tracking number/ID
- Tags or status
- Associated/linked incident(s)

The last item is particularly important. It is rare for any one major incident to not have findings or gaps that were not seen in other incidents. Highlighting these repeating issues across major incidents will help the SOC take a data-driven, repeatable approach to shoring up the constituency's defenses. It will also help reduce frustration around the "we've seen this problem a million times" phenomenon that can sap analyst morale.

It is helpful for the SOC to attach its post incident response process to one used to drive resolution to routine and systemic security issues, such as security bug tracking, security hygiene metrics, and/or green team remediation activities. Often, the SOC itself does not have the time, resources, or attention span to drive this kind of resolution. The SOC, working with the larger cybersecurity apparatus, is strongly encouraged to make this process as repeatable and routine as possible.

5.6 Incident Response in the Cloud

Responding to incidents in cloud environments require a shift in thinking before an incident happens, and the actual response will vary depending on both the type of cloud and the provider. For example, an approach from IBM suggests adopting redundancy, automation, and collaboration techniques into the response [137]. In addition, considering the goals of the cloud service and the constituency expectations of the cloud experience will aid in better response. Building on the Cloud Security Alliance for effective response in cloud, there are several key areas of importance to consider when responding and planning for incidents in constituencies that include cloud services [138]:

Shared responsibilities across cloud provider, SOC, and user or stakeholders
Cloud providers share IR responsibilities with the SOC and others, and the SOC needs to understand the line of where the SOC is responsible to respond, and where the cloud provider responds.

Cloud SLAs
If the goal of having cloud services is high availability or performance described in an SLA, for example, the SOC would want to ensure to address denial of service and other incidents and outages that degrade performance are covered (and parties responsible specified). Also, the response would include fast communication mechanisms, such as contact with live SOC personnel, rather than using e-mail. Finally, the SLA may include IR between cloud provider and customer agreements.

Visibility into cloud environment
Monitoring of cloud-based systems such as SaaS solutions, where the constituent owns neither the host nor the network, typically require both different data collection and different detection mechanisms than on-prem and IaaS systems. In these cases, the SOC needs to adapt its strategy for each resource type and cloud provider used.

Location of data
For constituencies with a hybrid mix of cloud, on-prem, and data centers, data can be stored in various physical locations, with different providers. And, within cloud providers, data is often distributed, and requires a different set of response actions than data all co-located in one place, on one system.

Forensic data capture

The SOC will need to enumerate and practice methods of forensic data capture for the cloud resource types being used. As intimated above, when using SaaS services, PaaS services, or considering the cloud control plane itself, capturing hard drives or memory images will be either impossible or meaningless. The SOC should be prepared in advance, either by collecting logs steady state or be ready to pull them as necessary. For IaaS, the SOC should investigate in advance how VM images can be paused, cloned, and/or recovered using the cloud provider's native automation, or ideally, security tooling.

Response expectations

The set of tools the SOC will have in cloud environments may vary substantially than for on-prem. Due to how cloud services are often architected, performing response at the network layer may be cumbersome or not scale across thousands of disparate cloud resources. Often, the SOC will need to become very familiar in advance with the cloud identity planes and on-prem identity federation services in advance. For example, shutting down an app or service principal may be far more effective than placing block on an IP address.

5.6.1   Tailoring IR for Specific Cloud Environments

Major cloud providers often provide incident response guides that address both general cloud strategies as well as those specific to their cloud environments, and how their tools might be used to address incidents in their cloud environments. Some of them include (in alphabetical order):

- AWS Security Incident Response Guide [139]; Building a Cloud-Specific Incident Response Plan [140]
- Google Cloud Platform Data Incident Response Process [141]
- IBM General Cloud Incident Management Architecture: Overview - IBM Cloud Architecture Center [137]; IBM Resilient Incident Response Platform On Cloud [142]
- Microsoft Azure Security Control: Incident Response [143]
- MITRE ATT&CK Cloud Matrix (TTP descriptions & mitigations) [144]
- Oracle Cloud Infrastructure Blog: Guidance for Setting Up a Cloud Security Operations Center [145]; Oracle Cloud Infrastructure and the GDPR [146]
- VMware Cloud on Dell EMC: Operations Management [147]

For each provider and consumer, the tools, techniques, and specifics will vary, yet there are considerations that generally apply for IR.

- **Automate some of the IR:** For example, AWS provides services such as Amazon Inspector and Detective to detect, track and gather incident information [148], [149]. Microsoft provides automation rules for automating playbooks with responses in Azure Sentinel [150].
- **Ensure logging is enabled and cloud-specific alerts are available to the SOC:** Especially consider servers that are in the cloud and other important devices. To extent possible feed into SOC monitoring (See Sections 7.5 and 8.2 for automation in SIEM/SOAR/EDR).

- **Look for shadow cloud use:** Not all users will notify IT they are using cloud storage for organizational data and processing. The SOC can use various known edge logging to identify possible "rogue" or shadow use of cloud [151].
- **Monitor malicious user behavior with cloud tools:** Cloud user-based analytics exist for several cloud providers [151]. These tools can aid the SOC in better insight within the cloud environment and how users work with the technologies.
- **Mind the technical differences between cloud and on-prem IR:** Data is stored and available in different forms on cloud than traditional systems, and so it is easy to break processes, access, and to miss information if not aware of the cloud environment differences.

5.6.2 IR Planning in the Cloud

Because multiple organizations are often responsible for security, planning for IR has special considerations in addition to those for traditional on-prem IR. To avoid miscommunications and misunderstandings, plans focus on the edges between the provider and consumer. Cloud IR plans should include:

- List of major cloud providers used by the constituency, including contact info for their respective security teams
- Links to asset/resource repositories defining the scope of cloud assets in each cloud
- The scope of security responsibilities for the cloud provider vs the SOC
- Links to s148
- ecurity monitoring capabilities, dashboards, logs, and other situational awareness either:
- furnished by the cloud provider as a general customer-facing service or
- engineered, deployed, and consumed by the constituency or SOC (such as cloud-based protection capabilities, cloud monitoring/log collection and cloud-based SIEM)
- IR plans specific to high criticality and high-risk cloud services (such as key storage and databases)
- Shared response planning, to include roles and responsibilities of providers, customers, and stakeholders as well as physical and logical boundaries for response
- Cloud forensics tools and capabilities, especially those specific to each cloud environment and how to access them

Some SOCs prefer using standards to assist in planning activities, and many cloud specific standards address incident response. The Cloud Security Alliance published a guide, Cloud Incident Response (CIR)Framework, which publishes a table of standards for each step of an incident response lifecycle [138].

5.7 Incident Response with Mobile Devices

With an emphasis on working from home, many SOCs have adjusted to accommodating more constituency-owned laptops remotely accessing resources through VPNs. Incident response on these is similar to other enterprise hosts for incident response. However, mobile devices, specifically small form factor cellular-connected computing devices running operating systems such as iOS and Android, present their own challenges. Bring Your Own Device (BYOD) policies, ability to access e-mail from anywhere on constituency-owned devices, and other evolving mobility enable authentication through traditional organizational perimeters, such as firewalls, into internal resources. Mobile devices are used somewhat differently than traditional hosts in how they access applications, how they are sandboxed on a system, which companies build and deliver applications, and even the access of users and organizations—carriers and manufacturers can have administrative access. All these factors contribute to the need to approach addressing incidents differently, and training responders in these differences and how to adjust techniques. When attackers exploit users' devices, incident response moves beyond the networks and to the devices themselves for investigating, adding to the difficulty. In comparison to traditional cyber incident response in on-prem constituencies, responders with constituencies that include connectivity through mobile devices additionally consider the following elements in responding and planning for incidents.

Geo-location
With Mobile IR, responders must locate the devices, which means tracking down the users' physical locations, often geographically diverse.

Connectivity
Mobile devices are often always-on and always-connected and move from network to network (cellular and wireless networks). This means devices have more exposure to unknown networks and devices. Also, responders need to factor in the adversary's access and how and when to power down a device to address an incident. This connectivity exacerbates the tension between gathering cyber threat intelligence and stopping the malicious activity when dealing with a live adversary. The adversary will likely be tipped off when the device is powered down to rebuild.

Minimal security data
Unless the constituency owns the device and installs security logging and other functionality or stipulates users must install an MDM to access company resources, most users do not install or configure security management functions on their devices; those who do may differ from user to user, so the data needed in incident response may not be captured or available or consistent. As mentioned previously, the best way to confirm an incident is through proper use of EDR bundled with- or installed by- an MDM solution (See Section 7.3.2).

Use in embedded devices
There is a robust set of security management capabilities delivered through MDM software for iOS and Android devices. However, Android may be used in in places the SOC did not expect; a very good example is Video Teleconferencing Capability (VTC), Voice over Internet Protocol (VoIP) and telephony devices. These devices show many of the same risks as ordinary mobile computing but may be patched less often and may be incompatible with MDM solutions.

5.7.1 Mobile Device Incidents

Mobile devices are susceptible to all the types of incidents described previously in traditional organizational environments. In addition, the exposure of mobile devices to different types of networks and third-party apps introduces mobile-specific attacks. The MITRE ATT&CK for mobile matrices provide some specific examples and descriptions of attacks along with mitigations specific to mobile devices, and Table 9 [152] summarizes some of these more common attacks.

Table 9. Sample Mobile Attacks [152]

Mobile Incident	Description
SMiShing	This is similar to phishing in e-mail, only with Short Message Service (SMS) or text messaging [153]. Attackers are attempting to gain personal information or access.
Baseband attacks	By exploiting vulnerabilities in baseband (firmware level software that enables mobile devices to connect to cellular networks), attackers can monitor all the device's communications, as well as place calls, texts, or initiate large data transfers without the user's knowledge
RF attacks or jamming	RF jamming has the same effect on mobile devices as a denial of service on traditional networks. Essentially the RF signal is distorted by an attacker, so that the receiver does not receive the signal
Jailbreaking (iOS), Android rooting as attacks	This is a variation on privilege escalation. Jailbreaking refers to enabling root access to the system and is used legitimately by users to install apps and extensions. Used by an attacker, it enables replacement of manufacturer-installed operating systems with custom kernels, for example.
Rogue/malicious access points	An adversary could set up unauthorized Wi-Fi access points or compromise existing access points and, if the device connects to them, carry out network-based attacks such as eavesdropping on or modifying network communication [154].
Malicious and fake apps	Apps installed by the user that have unwanted or undesired functionality such as: remote monitoring and device bugging, user location tracking, or use of the device in for-profit schemes like cryptomining.

5.7.2 Mobile Investigations

Piecing together what happened in an incident on a mobile device is challenging and requires creativity from responders. And there are a lot of mobile devices, cellular providers, operating systems, and applications, so investigations will vary based on the device and providers. Using more common devices, some tips on where to look can be extrapolated to other types of devices and providers and include:

- **Back-ups:** iTunes, iCloud (iOS) and Android Debug Bridge (ADsB) command results or third-party apps are used to back up systems onto computers and may be used to understand what data and damage might be involved for a stolen device.

- **Wireless access points:** Network traffic from where the device was connected can provide hints to activity. This can provide the protocols used, and possibly identify unexpected or suspicious traffic from apps.
- **App stores:** When examining possibility of malware, or illegitimate files posing as legitimate, the app stores used to download apps in question can be used to compare and determine if apps are legitimate.
- **Mobile device specific data to include:**
  - **Global Positioning System (GPS) data:** Can assist in locating phone logs and text, in addition to understanding location (if relevant to the incident).
  - **App data:** Specific apps may store information relevant to the incident, especially if accessing other apps such as video or photos or GPS or other data. Sometimes the purpose of the app is to collect data, so this is valuable as well.
  - **SMS:** If a live adversary is involved, SMS might provide useful information such a sender and receiver phone numbers, and relevant dates and times. e
- **Local files:** Sometimes files can provide clues to active adversary behavior, and what has been accessed and downloaded from other constituency resources.
- **Call detail records (CDRs):** Usually maintained by service providers, and can inform start and end dates and times, the device that made and received the calls, and originating and terminating cell towers.

Tools and techniques for investigating will vary, depending on the mobile and wireless policies of the constituency. Some of the open source, free, or widely available tools are the following:

- **Santoku:** Open-source tools available and specific for mobile forensics, malware, and security; the toolkit enables investigators to image and analyze devices as well as decompile and disassemble malware and binaries [155].
- **Mobile device management (MDM):** Software installed on clients which support central management and implement security features specific to the type of device. Several commercial providers provide MDM software for clients and can integrate with central management, including Unified Enterprise Management (UEM); for example, Microsoft provides MDM natively, providing ability for clients to be enrolled, and a management server and client MDM protocol [156].
- **Mobile threat defense (MTD):** Software that can actively block enrolled devices from affecting constituency resources when malware and other threats are detected. Several commercial solutions exist and can be integrated into the constituency. For example, Microsoft provides Intune, which can integrate other MTD commercial solutions to actively block mobile devices considered compromised (including with malware) [157].

5.8 Incident Response and OT

Responding to incidents involving OT is slowly becoming more commonplace, although many time OT is managed by an entirely different group. SOCs might first get involved because an incident involves an OT web-based interface with data traversing traditional constituency systems. There are numerous stories about SOCs detecting beaconing activity, which turn out to be connected to the vending machines phoning out to the vendor to indicate a beverage choice needs to be replenished. These "incidents" are a relief to responders, but they do raise the question about who is handling incidents and events for OT, if the SOC is not familiar with where the OT is located. Section 7.7 discusses monitoring OT, and the protocols and use of tools that are compatible across OT and IT systems. Incident response investigations of OT related systems and data has many commonalities with IR for traditional constituency systems. The main difference comes in when the OT device itself has been successfully compromised or rendered unusable. The following are considerations in handling these types of OT to IT compromises:

Response will be specific to the device's purpose
If the SOC is involved, it is very important to understand the safety and health implications of the incident on the device, and work with those managing the OT to understand any life dependencies.

Limit remote response where possible
Due to the often-sensitive nature of safety and uniqueness of OT, plan to respond to incidents in physical proximity of the devices, rather than trying to respond remotely. This limits further exposure to denial of service.

Options for eradication may differ from traditional IT
Removing or rebuilding a compromised device may not be easy. Depending on the device there may or may not be a spare readily available. If the device is very expensive or limited in production, that might call for a different plan than if a replacement is readily available. And, planning for recovery as the incident is occurring is especially important with OT devices relying on availability (planning in advance of incidents is advised).

Understand the IT to OT connection is imperative
If the OT is connected to the enterprise, it is a different investigation and solution than if the OT has its own dedicated communication that was the cause of the incident.

Work with the OT device owners before acting
It is important to work with those who own the OT to understand timelines, criticality of the OT, and to understand the purpose of the OT.

Finally, consider integrating the response of OT into SOC incident response playbooks and SOPs. Some resources for assisting in how to set up an IR playbook for OT/IT integrated enterprises include these examples and resources, see [158], [159], [160], [161].

5.9 Summary – Strategy 5: Prioritize Incident Response

5.1. "Incident handling involves receiving, sorting, categorizing, and prioritizing of incoming incident reports or other requests and responding to requests and reports, and analyzing incidents and events" [5].

5.2. Effective Incident Response requires planning for what types of incidents apply to the SOC's organization, and how to respond to the events. This includes defining incident categories, response steps, and escalation paths, and codifying those into

SOPs and playbooks that are updated as necessary. Some types of incidents will be addressed frequently, such as isolated malware infections, whereas major adversarial compromises happen less frequently, and therefore should be exercised.

5.3. When triaging, investigating, and escalating incidents, team members must be given enough structure to ensure that expectations are met such as consistency, timeliness, and the removal of analytic bias, but also freedom to act on their intuition and experience.

- Investigation is usually most robust when in-depth data can be acquired from end hosts and services, however care should be taken when applying measured amount of resourcing to both forensics and attribution.
- When responding, the SOC and its constituents must act with precision and care to the threat and impacted mission; both are discouraged from falling into traps such imprecise response, responding simply because "we must do something," and actions without underlying conclusive evidence.

5.4. Choosing when and how to respond must balance several factors: knowledge of the threat, the environment, the risks and consequences of the intrusion in question, the need to limit damage vs the need to learn more about the adversary's intent, the need to find "patient zero" in more severe cases, and the finite resources that both the SOC and constituents have.

- In an ideal state, the SOC should drive clarity on the entire kill chain and MITRE ATT&CK techniques used in an incident, finding both root cause and the full extent of the incident; however finite time and imperfect data impede perfect response; experience and judgement are needed to balance these factors.

5.5. All major incidents should be followed up by robust post incident response activities. In particular, the SOC's experience must catalyze change within the SOC and across the constituency, in preparation for similar incidents in the future. A robust post incident follow-up program hastens the SOC's path to maturity, bolsters the overall cybersecurity apparatus, and is necessary to a threat-informed defensive posture.

5.6. Distributed environments require adjustments to incident response. Cloud environments involve shared responsibility for incident response and should be coordinated with cloud providers ahead of time.

5.7. Mobile devices introduce extension of organizational boundaries, and so modified tactics and techniques are used for investigations and forensics, to include examining CDR, GPS, and App data to understand incident events.

5.8. When responding to incidents involving OT, extra care should be taken in consideration of the unique equipment being used, and their connection to dependent business and mission functions.

Strategy 6: Illuminate Adversaries with Cyber Threat Intelligence

Finding malicious activity and other traces of adversaries is extremely challenging in today's complex environments, especially since it is easy for an adversary to look like a legitimate user. Cyber threat intelligence (CTI) is valuable way to augment the SOC's ability to identify adversaries and discern their movements from that of authorized users'. It moves the SOC from a per-incident approach to an adversary-focused paradigm. Analysis and tailoring of CTI and establishing context enables the SOC to prioritize the actions of detection and prevention to conserve resources, honing the effectiveness of SOC operations. For example, CTI can provide information and data to inform and assist SOCs in incident response. Incident responders focus on what is happening and how to eradicate the adversary; CTI provides the context of who are threats which enable responders to accurate and fully scope an incident's reach, leading to more effective eradication of unwanted activity.

For experienced CTI analysts who understand their constituency, context can significantly arm the SOC with information to anticipate and sometimes prevent imminent attacks. Ultimately, individual people, criminals, organizations, or governments are behind malicious activity. The distinction is important because effectively finding and anticipating threats is a dynamic, active thinking people-centric function. In cyber, how an adversary is manifested in malware or phishing changes depending on the sophistication of the person behind the keyboard.

In SOC environments, CTI can augment defenses by informing the following:

- Identifying unwanted actors in networks
- Tuning sensors and analytic systems/frameworks for better monitoring
- Prioritizing resources
- Providing context to incidents
- Anticipating adversary activities in more advanced SOCs
- Preventing or slowing down imminent attacks

In this strategy, we discuss CTI concepts and how to effectively create, analyze, and use cyber threat intelligence in the SOC. CTI is most effective when following the classic intelligence cycle, comprised of six steps: planning, collection, processing, analysis, dissemination, and evaluation [162]. Foundationally, each step is important to understand what is uniquely needed for each organization to acquire the right information and data and to filter and apply to SOC operations. This provides a starting point to navigating CTI resources and to help determine when the SOC needs CTI, and how to maximize its use.

6.1 Why CTI is Important

CTI is increasing in importance due to extended boundaries and interconnectedness of modern organizations. In addition, adversaries, from individual criminals and organized crime to activists and nation states, are expanding use of cyber as a means to meet their goals. While SOCs can react to less sophisticated and noisy attacks, responding on-the-fly to sophisticated attacks is taxing on resources, and requires fast thinking and availability of talented analysts. If the SOC wants to evolve into less reactive and more proactive, dynamic, and anticipatory defense, then producing, consuming, and fusing CTI is essential. Examining traditional SOC data such as EDR, SOAR, and network and packet capture (PCAP) traffic only allows the SOCs to see what has already occurred in their constituency. Consuming and producing CTI in context of an adversary and using tailored information about the constituency and its security-relevant telemetry, enhances anticipation ability. And, as a bonus, anticipating adversaries' movements and honing defenses to prevent successful attacks can be less expensive in the long term.

Comprehensive integration of CTI into the SOC's situational awareness and planning cycle also enhances more durable understanding of the adversary. Tools and indicators change, but generally a given adversary will not pivot every single technique across the kill chain all at once. By understanding the TTPs of a given adversary over time, the SOC can connect the dots from one campaign to the next much more reliably.

6.1.1 CTI Benefits

Most SOCs can benefit from CTI, whether large or small, and several types of decision-makers may use it indirectly. Decision-makers may use it to prioritize funding for a new security enhancement, whereas others might choose to limit access to a service. For example, a service owner may not routinely read a report about the adversary group APT3 but is very likely to be briefed by the SOC on how they need to secure their service because of APT3's recent actions. Additional examples of how constituents both inside and outside the SOC may use CTI include:

- **SOC management:** To inform monitoring and detection investments. CTI can serve as a feedback loop that sensors and those monitoring have effective signatures and anomaly detection strategies. It can also inform what else to look for on the networks. CTI provides hints to SOC staff on how to instrument networks, and for what activity to search.
- **IT executives and service owners:** CTI can inform executives outside of the SOC, as well as inside, of context of cyber incidents. For example, if similar incidents are recurring, CTI analysis can assist with why, or at least provide more information, such as what other organizations are encountering similar activity.
- **Budget decision makers:** CTI is valuable for budget decisions, especially if specific to the constituency; CTI can validate adequate money is provided for the SOC, or can be used to justify further funding, as CTI is an important input into business

and security risk decisions. CTI can shift budget toward preventative and detective capability that is more relevant to threats of greatest concern.

- **CISO or information security organization:** Timely CTI provides information about threats and how to detect and mitigate them, as well as their potential impact. CTI can assist CISO or information security organizations with:
  - **User awareness:** Improving and elevating training and informing users on threat TTPs so they can spot and report potential activity.
  - **Security engineering:** Enhancing requirements for better defensive designs and tailored defenses.
  - **Investment planning:** Informing the constituency's security investment priorities.
  - Cybersecurity defense: Enabling prioritization of network and enterprise configuration management and updates.

6.2 What Is Cyber Threat Intelligence?

Producers and consumers of cyber threat intelligence come from different fields with different goals and a single definition of CTI has not yet emerged as a standard. For the purposes of the SOC, CTI provides actionable knowledge and insight regarding adversaries and their malicious behaviors. This informs traditional and non-traditional cybersecurity defensive missions by providing information that can be correlated with SOC data and result in better visibility, reducing harm, and enabling better security decision-making through applying an iterative, repeatable process.

CTI comes from internal and external sources, and both are important. Internal sources include analysts and security researchers who curate, correlate, and analyze information about adversaries, based on sources from within the constituency; this includes incident data combined with other information and data about an adversary. External sources include the dedicated commercial threat feeds and inter-SOC or constituency threat reporting.

Combining government, industry, and academia working definitions:

CTI refers to the collection, processing, organizing,
and interpreting of data into actionable information or products that
relate to capabilities, opportunities, actions, and intent of adversaries
in the cyber domain to meet a specific requirement
determined by and informing decision-makers.[6]

---

[6] Definition adapted from combining four industry and government recognized definitions to best capture varying perspectives:
- A). CIA's seminal definition of intelligence [520]
- B). DNI's definition of CTI in the 2019 National Intelligence Strategy [180]
- C). Dragos' definition in, "Industrial Control Threat Intelligence" [521]
- D). Carnegie Mellon SEI's definition in "2019 Cyber Intelligence Tradecraft Report" [166]

In this case, the decision-makers are the SOC managers, SOC analysts, leadership, and other cybersecurity specialists. Some works relevant works to CTI include [163], [164], [165], [166], [167], and [168].

6.2.1 Evolving Capability

CTI roots stem from two fields, traditional intelligence (and counterintelligence) and cyber incident handling, with analysts varying from defensive SOC engineers to traditional intelligence analysts to those with targeting and offensive missions. With the fields merging (or colliding) in CTI, some seminal works in intelligence and structured analytic techniques inform how CTI might evolve for SOCs. These include [126], [169], [170], [171], [172], [173].

These books are worthwhile for any analyst, particularly discussions about understanding and awareness of analyst bias, and critical thinking about asking questions. Not all techniques make sense for cyber applications, and analysts must apply critical thinking to determine which make the most sense in the cyber context. In cybersecurity defense, SOC analysts typically rely on the structured analytic technique of indicator analysis. This involves analyzing and identifying those facts around a cyber compromise such as malware, connections to malicious websites, IP addresses, and other artifacts that could change with (not-so) sophisticated adversaries. This is just one technique, and it has a downside when applied to asymmetric threats, such as APTs, which is that it relies on historical (past) data which usually does not predict future activities. By definition, indicators of compromise have already occurred.

In addition to indicator analysis, other structured analytic techniques can expand what SOC analysts typically use for approaches and methods to enhance CTI. For example, key assumption checks (name and question the assumptions) and analysis of competing hypotheses (list all possible hypotheses and evaluate them) enable analysts' exploration of what adversaries may be doing, before the activities are manifested in network data and alerts, what their interests might be in the constituency, and how they might go after the constituency's important assets.

6.2.2 Is it CTI?

Because CTI is both produced and consumed across various fields, there tends to be misunderstanding about what is and is not CTI. SOC data sources are often interchanged and referred to as CTI by some; level-setting if a data source is generally considered to be CTI goes a long way towards creating a shared understanding [168].

To distinguish whether a data source is CTI, determine if the data source has adversary context built in (and if possible, details of the context). For example, depending on who you talk to, the items in both columns in Table 10 have been labeled as CTI, sometimes incorrectly. The issue with considering those in the "Not CTI" column as CTI is the context about the adversary is missing. While these data elements are still valuable to the SOC, without context, externally generated IP, malware, and network traffic are just someone's data, and when used alone they do not give the recipient knowledge about the adversary in of themselves. This does not mean they cannot be used to find malicious activity in the constituency environment. But it does means that without additional context it will be more challenging to understand the applicability to the SOC and its constituency.

Table 10. Common CTI Labeling Disconnects

CTI Examples	Not CTI*
Finished unstructured threat reporting Structured threat reporting Open-source intelligence (OSINT) Curated subscriber reports and feedback	IP Addresses Domain names Email addresses Malware samples Virus signatures PCAP captures DNS logs Intrusion detection alerts System logs Social media * These are not considered CTI unless they are associated with adversary context.

As seen in Table 10, CTI comes in many forms. Some are useful for analysts to read, and some is designed to be directly ingested by SOC systems. It is important to consider how the CTI might be used in the SOC, and multiple formats might be useful to the analysts. Example formats of CTI include:

- **Finished unstructured threat reporting:** Includes intelligence reports and analysis, sometimes lengthy, describing adversaries, observables & context analysis, e.g., multi-page CTI reports produced by many SOCs, often distributed in portable document format (PDF) and HTML format. Examples: [174], [175].
- **Structured threat reporting: Includes contextualized TTPs:** collected TTPs associated with threat actors; also includes prescribed formats including who, what, when, where, and/or why. Examples: MITRE ATT&CK®: [117], STIX-formatted CTI feeds: [43].
- **Curated subscriber reports and feedback:** Anonymized threat information from subscribers or customers. Example: Mandiant Advantage Free: [176].

CTI exemplars

Some adversary association tracking is already available in blogs, reports, feeds, and other sources. Utilizing the information identified in one organization can assist in another, as long as the context helps the analyst determine the CTI's applicability to their own constituency. There are many examples of CTI reporting SOCs can leverage. Here are some open-source exemplars, chosen for their robust format and thorough treatment of context: [177], [178], [179].[7]

---

[7] Note: These were chosen for the type of context and information and are not necessarily up to date on actual adversaries.

 Evaluating CTI Characteristics

Having a river of CTI flowing can be a great help to the SOC, but discerning which CTI to use and prioritize is important. Continual updating of instrumentation and analytics adds to the SOC analyst workload and if not prioritized, can get out of control; this leads to a lot of effort without sufficient return on the time invested. To decide what cyber threat intelligence to use, consider the following criteria for evaluation [180], [181], [182]:

- **Actionable:** Can the SOC do something constructive with the information, such as correlate with other data, create threat hunting scenarios or actions, or enact preventative protections? Is the CTI specific enough for the SOC to operationalize? Does it come in a format that is consumable and enrich a decision, while not complicating it [182]?
- **Timely:** Are events recent (in days, hours, minutes for streams, or weeks for analysis)? Are there stale data?
- **Relevant:** Does it apply to the organization and reveal unknown and possible threats? Does it come from a reputable source? Is the data volume manageable? How is the CTI ingested or analyzed? Are there application program interfaces (APIs) for feeds and platforms?
- **Accurate:** Does the content correctly describe what happened? Did the CTI include spurious or wrong data about the original attack?

All these criteria together provide means of comparison among CTI subscriptions and tools and are indicators of a CTI source that can be trusted and is likely to be of value to the organization.

6.3 Focusing Goals & Planning

One objective of CTI is to help a constituency focus on understanding their greatest threats by providing analyzed intelligence to assist cyber defenders and decision-makers in making more informed, threat-based decisions. These decision-makers are at various levels of an organization, and CTI is tailored with audiences (consumers) in mind. As with traditional intelligence and military operations, there are three levels normally applied [183], consisting of strategic, operational, and tactical levels [184].

Strategic

Strategic CTI considers business activities in context of threats to inform the decisions of executive boards and senior officers or other executives. It usually includes trends and emerging risk at the constituency, industry or sector, and business levels to provide longer horizon views. Summarized points can inform investments, and usually occur in times of large cybersecurity events directly impacting the constituency, or at a limited frequency, such as twice a year. For example, broad observations that a given group of adversaries is targeting specific companies, why, and investments to counter them.

Operational

Operational CTI encompasses the bigger technical picture and is usually tailored for middle managers, for SOC leadership, and other technology leaders. It usually includes campaign information regarding adversaries over time, such as focus areas, motivations, and including anticipating future attacks. This level of intelligence can inform defense configurations, architectural modifications, and purchase or acquiring of new cyber defense or analysis tools, or other SOC response support needs. For example: how an upward trend in a certain TTP (such as service token theft) is impacting the design, implementation, and protection of certain services.

Tactical

Tactical CTI comprises the contextual specifics about attacker methods and operations, as well as TTPs. It is the most common CTI in SOC environments and is used in writing, tuning and refining detections and analytics. Effective tactical intelligence is actionable and is used to develop defenses that are comprehensive for classes of attacks, rather than specific to indicators such as IP addresses or malware hashes. For example: information that supports creation of a new SIEM rule or Spark-based analytic operated by the SOC.

6.3.1 Integration Considerations

Being a sophisticated consumer of CTI products and services, and integrating cyber threat intelligence analysis into SOC operations, enables the SOC to be both strategic and tactical in defending the constituency. For example, CTI integration into the SOC:

- Informs shifting and prioritizing defensive actions based on imminent threats.
- Increases confidence in the efficacy and completeness of incident response actions.
- Decreases proportion of successful attacks (including APT and other sophisticated actors).
- Focuses detection to decrease time the adversary can maintain presence and avoid detection.
- Enhances SA and threat awareness through informative and thorough reporting.
- Increases context and link between incident activity and mission impact.
- Improves morale through adding value sharing threat experience with partner SOCs.
- Adds value and return on investment of SOC services through lessons learned to other areas.
- Enhances awareness of constituency's threat profile and likely targets of adversary attack.
- Provides insight into defense gaps and inspire motivation to address them.

6.3.2 Determining Approach

Effective CTI requires a sound understanding of important missions and businesses functions to assist analysts in understanding which adversaries (people, groups, or organizations) may be interested in the information the SOC is protecting and monitoring. Specifically, it is comprised of three key facets: context, analysis, and action.

The context of events provides a basic understanding of who is targeting a constituency, why, and what could happen as a result. Analysis helps to understand what has already happened and what is happening now. Action answers critical questions related to where and how something is happening and what can be done about it. Figure 13 illustrates these components, along with the outcomes associated with each and example question.

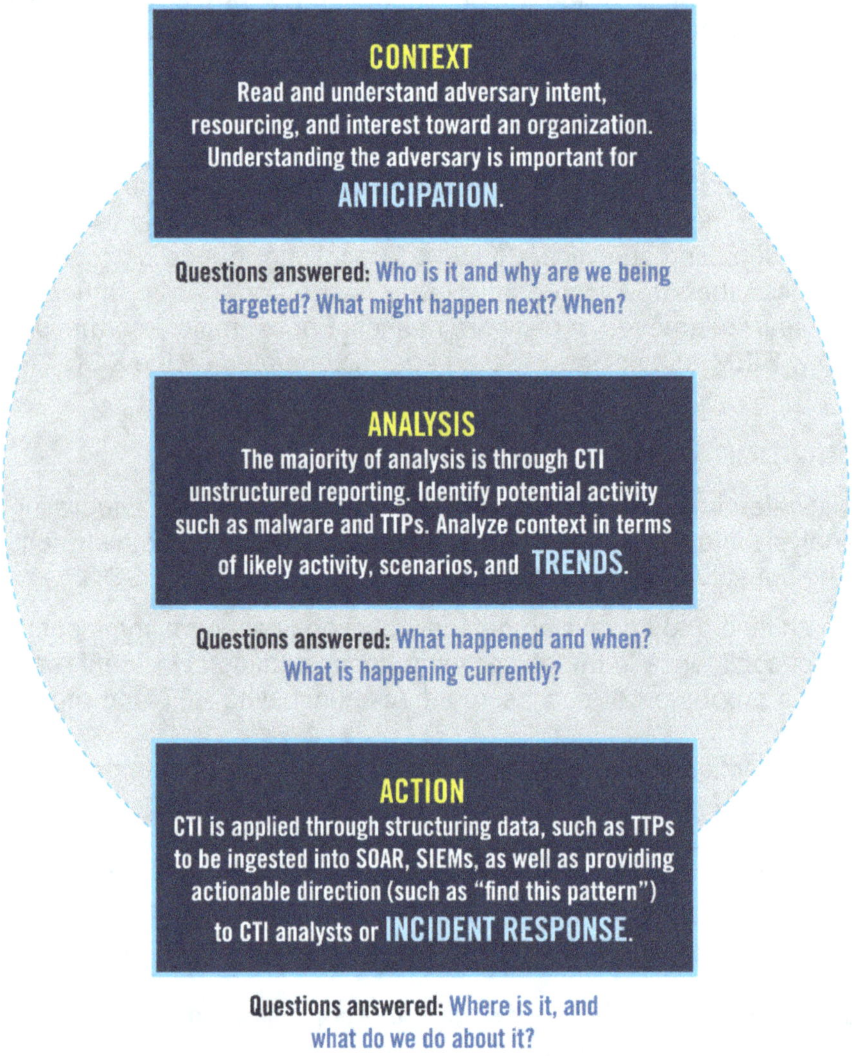

Figure 13. CTI Key Components

Using CTI effectively is an iterative process and evolves as more information is understood. CTI analysts supporting the SOC often start with the analysis key component and then move to the action key component and iterate between the two.

6.4 Getting Started in CTI

The CTI components of security operations are scalable: SOCs can start small, perhaps with just with one or two analysts working part-time or half-time with an intelligence focus. These analysts might provide new indicators, sources, scenarios, adversary searches, and reporting back to the SOC and other threat-sharing channels. This can provide immediate dividends by mitigating attacks in real time and opening paths to other threat sharing partners.

6.4.1 Choosing and Using the Right Data

Operationalizing CTI to determine what the SOC can do to deter, deny, or degrade adversary activity starts with having the right data. For a SOC to determine effective actions from CTI, three aspects of data and analysis need to be included: technical environment, adversary information, and relevancy. Technical environment encompasses data and information about the technologies in the SOC's purview, including digital assets and security relevant data, along with the data and analytic platforms (such as a SIEM or EDR). Adversary information describes everything known about adversaries and how they manifest in the technology. Relevancy is the knowledge of mission and business priorities and value of the important and relevant data to the constituency. Figure 14 illustrates the key to integrating and analyzing CTI for effective use: Actions are determined by analyzing adversary information for relevancy as applied to the constituency's technical environment.

In other words, if the SOC wants to understand how to act on CTI, it combines knowledge specific to the mission and environment to filter, prioritize, and use relevant adversary information. Depending on the size of the constituency, understanding the technical environment is possibly the most time-consuming of the three to analyze in terms of adversary information. It consists of understanding what digital assets exist and where they are, as well as understanding the state of vulnerability management and what IT and OT is connected to the constituency along with accessibility. In addition, it includes all the data used by the SOC for cybersecurity defense, which include alerts, correlated data, logs, and any other data and processing for which the SOC or IT organization have control and visibility.

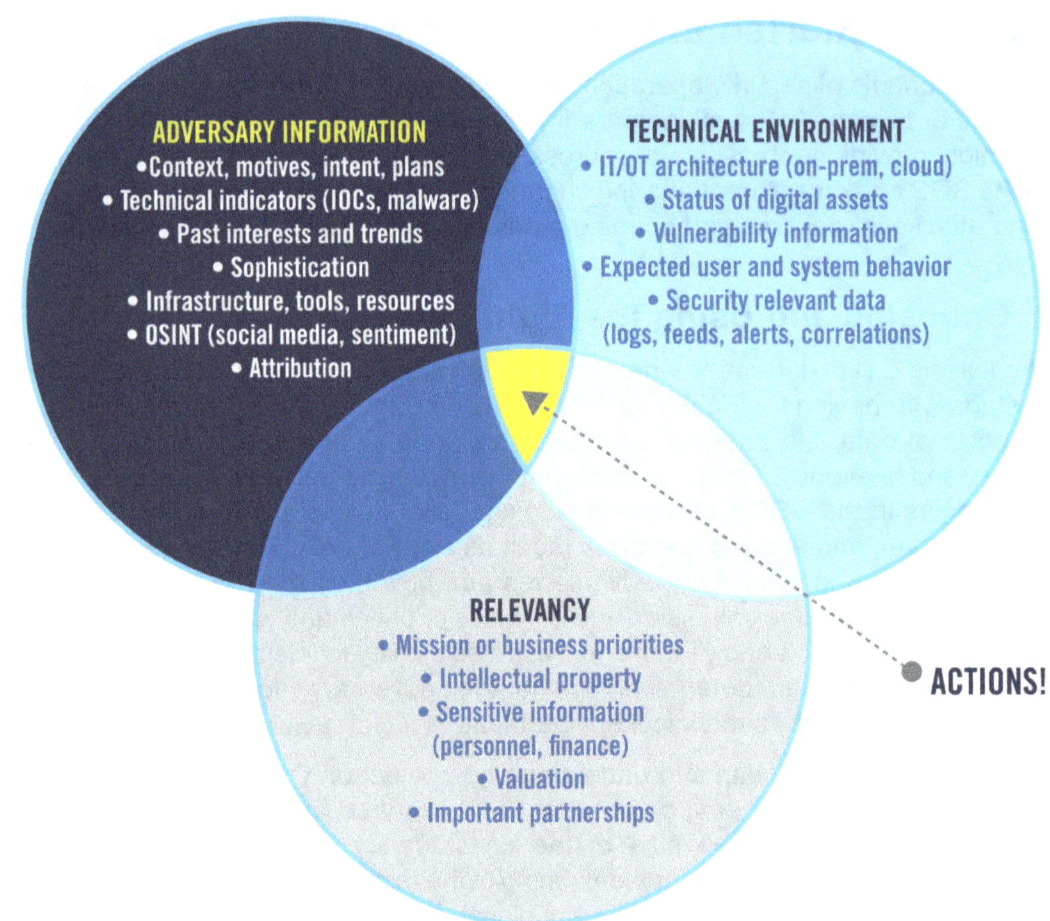

Figure 14. Developing Actions with CTI

Relevancy is the knowledge about the constituency's important data, especially information that would be devastating to the constituency if lost; also, examining relevancy from another angle, relevancy addresses if the information be interesting for an adversary. This knowledge can be challenging for the SOC, as it requires analysts to connect to other parts of the constituency to authoritatively confirm what is sensitive and important to the constituency (not to mention where it is physically housed to augment technical environment data). This can include intellectual property, privacy information, credit card processing, finance and personnel data, and mission logistics of interest. "Strategy 1: Knowing What You Are Protecting and Why," describes the details of scoping asset information which might be helpful in determining relevancy.

Adversary Information refers to the CTI technical feeds, knowledge, TTPs, and non-technology-based data about the adversary and the methods, infrastructure, plans, and motives and interests. It may include dossiers and past targets. This aspect informs:

- **Anticipation:** what an adversary might do next, and why

- **Sophistication:** how advanced is the adversary, including their ability to hide or move stealthily
- **Resources:** what tools and infrastructure are available to the adversary, as well as how many people are involved

Adversary information will depend on what the SOC is defending and can vary from information about criminals stealing credit card information to complex nation-states with complex agendas.

6.4.2 Effective CTI Correlation

Analyzing all three informative aspects of CTI in figure 14, in context of the constituency's business priorities, enables analysts to tailor CTI and choose CTI commercial and open-source services and platforms. Importantly, it provides the SOC with the actions needed to take against an adversary.

Effective CTI actions are the result of analyzing adversary information for relevancy as applied to the technical environment.

Without analyzing the three aspects together, the SOC will not be as effective. Not all CTI is useful to every environment, and SOCs cannot afford to process all omni-present data sources. For example, certain cyber criminals may be actively targeting banking and finance and credit cards. If the enterprise is not processing financial data or credit cards, perhaps the SOC will determine that cyber criminals do not have intent to target the constituency; so following CTI on cyber criminals' financial targets and public presence would not yield benefit in that case.

Analysts correlate the three aspects to populate which CTI is needed and where there may be gaps in knowledge. And, ideally, the SOC aspires to anticipate or even predict threats to the organization rather than to identify activity that has already occurred (although finding existing embedded adversaries embedded is a good starting place). To get started, some sources for enriching actions and knowledge in each of the three data aspects are described next.

Adversary information

This includes CTI as well as any other information and data about the adversary that might provide context and enable further correlation, anticipation, or other analysis. It includes IOCs and TTPs, intrusion detection alerts, interests, intent, resources, geopolitical context, past, present targets (intellectual property, organizations, etc.), and cultural norms to anticipate or predict adversary movements or targets and other information and analysis informing who and what the adversaries are targeting the constituency. This also includes impact or potential effects of the adversary on a constituency (or nation), and information on cyber actors [180].

Many sources provide both technology CTI feeds as well as written, unstructured reporting. To the analyst, the formats are used differently, which is why these are separated in the discussion of actionable CTI. Technology feeds assist in finding new instantiations of an adversary's TTPs, IOCs, etc., but do not assist in anticipation. Adversary information, with context, can assist in anticipation of adversaries, thus the distinction. Sources on adversaries include [185]:

- Organizational incident data (IOCs, TTPS, etc.) generated by the SOC and its partners
- Various Opensource Threat Feeds (31+) identified [185]
- AlienVault OTX: Adversary: Open Threat Exchange [186]
- GitHub: Awesome-threat-intelligence, A curated list of Awesome Threat Intelligence resources [187]
- Dragos: Threat Activity Groups [188]
- MISP Open-Source Threat Intelligence Platform Open Standards for Threat Information Sharing (formerly known as Malware Information Sharing Platform): [189]
- GitHub MITRE ATT&CK - Cyber Threat Intelligence Repository expressed in STIX 2.0 [190]
- Proofpoint feeds [191]
- SANS Internet Storm Center [192]
- ThreatConnect Blog: Risk-Threat-Response [193]
- X-Force Exchange [194]
- Crowdstrike Groups List Malware & Ransomware [195]
- Digital Shadows blog Threat Intelligence [196]
- FireEye APT attribution Advanced Persistent Threat Groups (APT Groups) [197]
- FireEye threat research blog [198]
- Proofpoint blog: Threat Insight Blog [199]
- Recorded Future: Adversary Infrastructure Report 2020: A Defender's View [200]

Technical environment
This is comprised of understanding what types of data are in the enterprise, what the SOC is interested in monitoring, and how it can leverage the CTI it has. For example, most constituencies keep system logs and e-mail logs, and so CTI that has relevant mission targeting that includes phishing as one of the vectors would likely be of interest. More about these types of data sources and their use and correlation can be found in "Strategy 7: Select and Collect the Right Data" and "Strategy 8: Integrate SOC Tools and Data into One Architecture." To get started, information in this area includes:

- Digital assets and connectedness
- IT/OT architecture showing design of networks, clouds, and perimeters
- User access and account information
- System logs
- Vulnerability information, including patch status
- Endpoint data
- Network sensor data and alerting such as traffic metadata
- The intel correlation capabilities of its sensors and analytic platforms

Relevancy
This includes constituency and mission critical information, specific intellectual property (IP) or other sensitive information, such as finance or personnel data that the adversary is

seeking. Essentially, build a dossier of your constituency's enterprise [201]. This information is derived from understanding the mission or business of the constituency that the SOC is supporting (see "Strategy 1: Know What You Are Protecting and Why" for more detail). Sources for determining important intellectual property (and ideally physical/virtual locations) might come from asking the following (some of this can be done in concert with CISO risk decision-making and with IT architects):

- What is the important IP or mission data?
- What does the constituency's public web site say is important? What is being sold, discussed as important, or what services are offered?
- What are the constituency's important buzz words, topics or trigger words, or organizational passions [201]?
- What is the business strategy and business objectives?
- What business or mission functions are performed?
- What is the organizational structure?
- What is the size of the various functions of the constituency?
- What matters to the constituency if deleted, stolen, or changed?
- Where is the IP or mission data maintained and is the SOC monitoring it?
- What are the major business partners?
- What does the network topology generally look like? For example, are there HR and partner network segments separate from other digital assets?

6.4.3 Analysis Through Scenarios and Sequencing

Once an analyst has collected relevant data, it is time to analyze it. Conducting analysis is a hybrid of both the SOC's standard procedures, and the individual skills, biases, experiences, and thought patterns a given analyst brings to bear. A common element to cyber analysis is that it usually consists of tracing activity from various sources and systems across various moves and sequencing events over time. Some of the activities my include:

- Discovery of new IOCs and TTPs, developed through a combination of digital artifact examination, static code analysis and reverse engineering, runtime malware execution, and simulation techniques.
- Mapping and analysis of adversary TTPs, to understand adversary activity of existing and potential threats to the constituency, as well as developing familiarity with sophisticated adversaries or criminal activity from reporting.
- Trending and reporting on activity and incidents attributed to more advanced threats to include sophisticated adversary activity or criminals.
- Tracking the evolution of adversaries and campaigns over time.
- Fusing and correlating locally derived and externally sourced cyber threat intelligence into signatures, techniques, and analytics intended to detect and track adversaries in coordination with other analysts in the SOC.
- Sharing and participating in cyber threat intelligence sharing groups, typically composed of other SOCs in a similar geographic region, similar supported organizations or industries, or both.

- Operating and populating threat knowledge management repositories, allowing SOC analysts to connect disparate but related adversary activity, incidents, indicators, and artifacts. This can be integrated as a part of incident response, or its own capability, depending on how many cyber threat analysts support the SOC.
- Matrixing into incident investigation and response activities, offering the CTI perspective during major high-severity cases and breaches

Certainly, most analysts will construct hypothetical (or real) use cases that apply to their specific environment. Once a use case is constructed, an analyst may determine how it may show up in the technology, what might be interesting from a data perspective to examine a scenario. Other options are applying some of the intelligence techniques. Building on the scenarios, analysts might consider the argument against the scenario, or otherwise lay out a devils' advocate approach to how the adversaries might attack the constituency for example. The options are limitless, and a lot like a chess game. We understand our chess board (the environment). We can study the moves of the great players (or adversaries in the SOC case), and then create scenarios and sequences based on anticipating what a player (adversary) might do. And we can apply different types of analytic techniques to determine how the adversary might manifest in the technologies. Above all, it is important to realize that context and specificity to the business or mission, the intellectual property, and the environment are key to getting good at analyzing present adversaries and moving toward anticipation.

For example, learning about an adversary's infrastructure can inform knowledge about gaps in the defenses, as well as gaps in the adversary's advantages, which can generate and link more technology-related data, such as TTPs. Open-source intelligence gained from dark or deep web resources (for more advanced analysts or purchased CTI services) can inform analysts that attackers are planning to target certain intellectual property or have been tipped off to gain access for pay, that then can prompt a constituency to place more sensors and stronger access control around the specified IP.

6.4.4 Determining Actions

When adversary information is analyzed, correlated, and corroborated thoughtfully in terms of relevancy and the technical environment, the SOC can determine a lot of different and specific actions and directions to take. And the actions can be passive or active. At times, the SOC may choose to monitor an adversary to determine the extent to which the constituency is being targeted, a passive action example. Or, based on the analysis, it may make sense to segment a network, and use active blocking for a sensitive enclave within an enterprise. It depends on what the CTI analysts in the SOC conclude in analysis.

For example, a fictitious Wiley Widget Research & Development Company (WW) values the Wiley Widget designs with secret and proprietary components, intellectual property designed over ten years, worth millions of dollars (relevancy). These designs are housed in three servers, accessible by three power user researchers (technical environment). Adversary C, at Calculating Chipmunks Laboratories, also fictitious, has been targeting widget technologies to leapfrog current state-of-the art. Adversary C uses social engineering, phishing, and specific cross-site scripting attacks and then implants command and control over a certain ephemeral port to persist and move surreptitiously around the widget industry (adversary information). So, WW's SOC would analyze the adversary information (Adversary C data) for relevancy. WW's SOC may therefore conclude WW designs would be targeted by Adversary C and are high value to WW. This would be examined as applied to the technical environment (three power users, and three servers, along with vulnerability and system logs). By combining these data points together, WW's SOC might take any or all the following actions in this example:

- Enhance phishing detection in their e-mail
- Set up more monitoring on the three servers housing IP
- Apply user entity behavior analytics on the power user profiles (and monitor lateral movement to and from the users)
- Collect and heavily monitor the system logs of the servers
- Develop pattern recognition to look for the TTPs, such as command and control over certain ports, among other actions

The options for action are endless here, but that's why CTI can assist in prioritization. And, over time, the actions will evolve depending on what is learned about the adversary and how the environment changes.

6.5 Understanding the Adversary

The SOC's defensive strategies can be greatly improved by focusing on who is targeting the organization and why. The more the SOC evolves adversary understanding, the better it is positioned to anticipate what the adversary might do. This enables the SOC to take more effective proactive actions to reduce malicious activity. In addition, the sooner the SOC is able to stop or slow an adversary, the less damage can be caused. Without focusing on and growing the adversary context, the SOC will be limited in its ability to evolve and will remain reactive, focused primarily on addressing adversaries already in the constituency.

6.5.1 Shades of Attribution – Adversary Association

The desire to attribute malicious cyber activities to their source has existed as long as SOCs existed (1988, in response to the Morris Worm) [202]. However, many practitioners opine that attribution is not an appropriate or necessary SOC function [58]. One common perception is that attribution takes resources that SOCs do not have, as it takes what few resources exist just to respond and stop current incidents that occur. Another common perception is that attribution can only be done by a law enforcement or intelligence organization that can more definitively trace back activity to its source using both technical and human methods. These perceptions may be true for many SOCs since irrefutable attribution is not necessarily a SOC goal. However, SOCs cannot predict or anticipate attacks without some understanding of the adversary [203]. Identifying characteristics of adversaries can be useful. So, there is another less rigorous approach that should be considered for use for defensive purposes within the SOC, adversary association.

Adversary association is less rigorous than full attribution yet is useful for adversary anticipation and is good enough for SOC work in most cases [203]. Responders and analysts are aware that conjectures about attribution might help to anticipate, but it is almost impossible to validate attribution from network and computer data alone. A NetFlow record may indicate that an entity from Country A is scanning the enterprise or is receiving DNS beaconing from a compromised host. Is it really someone in that country or is that just the next hop out in the network connection? Or an audit log is stamped with user Pamela may not really be Pamela sitting at the keyboard. It might be Alexis who compromised Pamela's account, or perhaps it was automated activity using Pamela's identity. Most times, an incident responder can only propose theories and suggest a degree of confidence about who is behind a given set of malicious or anomalous activities. Unless there is direct proof of who is sitting at the keyboard, user attribution is theory and not fact, and thus why adversary association is more appropriate for SOC defensive activities.

Adversary association aides the SOC in providing context to data. Adversary association includes linking adversaries to likely attributes, such as IOCs, behaviors, and TTPs across the kill chain, which can then provide clues about what else the adversary has done and start to identify why. Patterns of behavior such as frequency, tools, targets, locations, and sectors can be identified to assist in anticipating adversaries' movements [204]. This linkage can be constructed by analyzing bodies of knowledge about adversaries and previous events and drawing conclusions about what else the attacker may do or how else they may be detected, without getting caught up in the details of exactly who is behind the keyboard. It is useful for adversary anticipation and other cyber defense functions. With the vast amount of data available on the Internet, and through sharing with others, some semblance of adversary association, such as an adversary's activities, including IOCs and TTPs, is viable even in small SOCs with few resources.

There are varying aspects of adversary association analysis that can be useful to the SOC; for example, sometimes understanding the high-level motivation of an adversary can help a SOC anticipate what an adversary might do next [205]. Information such as the industry that is being targeted, relationships to other threat groups, or correlation with events, can also assist analysts [205].

Generally understanding who attacked a constituency can benefit most SOCs; for example, the association can lead to motives, and a better understanding of the why, which then can provide insight into what an attacker might attempt next [205]. This can also lead to more potential TTPs that might be used on others and in the past [205]. Anticipating enables responders to develop and adjust defenses and proactively defend, rather than respond and clean up after an incident occurs.

6.5.2 Adversary Frameworks

Today there are multiple frameworks to help the SOC, and CTI analysts, express what is known about an adversary and how they act. This is important as it allows analysts to discuss and share CTI in a repeatable and consistent way. Some of the frameworks include:

- Lockheed Martin, the Cyber Kill Chain® model [206]
- MITRE ATT&CK [117]
- U.S. Department of Defense, Defense Science Board Cyber Threat Taxonomy tiered model [207]
- The Diamond Model of Intrusion Analysis [74]

Each framework represents adversary actions, and other related information, in slightly different way. Selecting the model or models to use will depend on the intended outcomes and needs of the SOC, and they may sometimes be combined. For example, MITRE ATT&CK provides a more detailed look at adversary behavior across the cyber-attack lifecycle. The curated knowledge base is based on real-world adversary activity. It is a continually evolving globally accessible knowledge base of adversary tactics and techniques based on real-world observations [117]. ATT&CK provides a common taxonomy for both offense and defense and has become a useful conceptual tool across many cybersecurity disciplines to convey threat intelligence, perform testing through red teaming or adversary emulation, and improve network and system defenses against intrusions. It includes matrices for enterprise usage (Windows, macOS, Linux, and various cloud infrastructures) as well as Mobile and ICS.

At a high-level, ATT&CK is a behavioral model that consists of the following core components:

- Tactics, denoting short-term, tactical adversary goals during an attack
- Techniques, describing the means by which adversaries achieve tactical goals
- Sub-techniques, describing more specific means by which adversaries achieve tactical goals at a lower level than techniques
- Procedures, describing the specific implementation adversaries use for techniques or sub-techniques, as observed
- Documented adversary usage of techniques, their procedures, and other metadata.
- Matrices for enterprise usage (Windows, macOS, Linux, and various cloud infrastructures) as well as Mobile and ICS.

Table 11 lists the ATT&CK v10 Enterprise Tactics.

Tactic Name	Tactic Description
Reconnaissance	The adversary is trying to gather information they can use to plan future operations.
Resource Development	The adversary is trying to establish resources they can use to support operations.
Initial Access	The adversary is trying to get into your network.
Execution	The adversary is trying to run malicious code.
Persistence	The adversary is trying to maintain their foothold.
Privilege Escalation	The adversary is trying to gain higher-level permissions.
Defense Evasion	The adversary is trying to avoid being detected.
Credential Access	The adversary is trying to steal account names and passwords.
Discovery	The adversary is trying to figure out your environment.
Lateral Movement	The adversary is trying to move through your environment.
Collection	The adversary is trying to gather data of interest to their goal.
Command and Control	The adversary is trying to communicate with compromised systems to control them.
Exfiltration	The adversary is trying to steal data.
Impact	The adversary is trying to manipulate, interrupt, or destrooy your systems and data.

For more information on ATT&CK, the following resources guide on getting started and its use [208], [209], [210], [211].

6.6 CTI Tools

For CTI analysts, choices on which tools and techniques to use are usually fusion of personal preferences and team consensus. Chosen tools depend on several factors, including skills of the analyst such as if the analyst likes to write scripts or program, preferences for open-source analysis, and types and sources of data preferred. No matter what tools or other unifying technology is used, getting the most out of CTI tools takes care and feeding. Some best practices in use of CTI include the following:

- **Have a good feedback loop:** Is the CTI actionable such that it can be integrated into the various SOC tools? Is it yielding useful results to the SOC team?
- **Quality control:** If the data is inaccurate, feedback should be used to adjust the SOC reliance.
- **Track the pedigree:** Is the CTI derived or evolving over time? Where did it originate? What type of data was used to derive it? Is it a reliable source? Also, ensure any caveats are tracked (such as proprietary, cannot be forwarded, etc.).
- **Define and maintain standards for attribution/association:** Attribution is difficult, adversary association is more attainable: know ahead of time what is "good enough" for the SOC and drive consistency in language and thresholds used to articulate analytic certainty.

- **Define and maintain standard for annotation and action:** Understand ahead of time when to watch based on activity, and when to act based on hits.

6.6.1 Leveraging a Cyber Threat Intelligence Platform

SOCs that evolve a deep and sustained investment over time in CTI often start with an analyst or two and manage incident data and indicators and intelligence in an incident response management system, file shares, spreadsheets, log management, and other tools such as node analysis and correlation software. CTI analysts keep track of history of campaigns, adversary activities movements and trends, IOCs, and other methods typically through tools they are comfortable using. In fact, it is important to emphasize that experienced CTI analysts are the key to effective use of CTI, and technology can assist, but is not a replacement.

As SOCs mature in CTI and/or gain more resources CTI analysts and money, they typically identify a need for a purpose-built platform supporting correlation, linkage, tracking, and knowledge of CTI. These platforms, also called threat intelligence platforms (TIPs), ingest, organize, connect, correlate, and use high volumes of adversary-related data, including IOCs, to enable the SOC to perform more effective threat knowledge management beyond incident-focused tracking. CTI platforms are generally most powerful when they are integrated with the SOC's other high-scale data processing environments such as a SIEM or big data platform, thereby supporting correlation, data enrichment, and workflow. A CTI platform can enable the SOC to better answer questions including:

- Has this adversary been seen before, and when?
- What adversary activities were exploited in the past in the enterprise?
- Who is reporting on this TTP?
- What reports might be related to the current activity?

There are many platforms available. Many companies offer products that function as CTI platforms. Evaluating and choosing the right one depends on several factors and specific requirements [212], [213].

6.6.2 Evaluating a CTI Platform

In considering these platforms, the SOC should evaluate the following criteria and tips to select and leverage the right platform for the enterprise:

Workflow and organization
The CTI platform should support collaboration and repeatability around CTI artifact handling, correlation, and organization, particularly around adversary campaign tracking and indicator linkages. This means that the tool supports functionality similar to a case tracking tool (case open/close, analyst notes, and other structured knowledge capture). In fact, case management systems can also be used, and configured around actor and campaigns as an option. One of the benefits of choosing a CTI platform is to be able to organize and link by adversary activity and by campaign (for example, different adversaries may use the same activity, and this is interesting to track across campaigns). The most common mechanisms

to do this is allowing extensible category tagging and tracking that supports association and attribution between different knowledge elements.

Data integration
The CTI platform should be able to ingest, persist, correlate, and interface with many other CTI and other relevant tools, including open-source CTI feeds, commercial CTI feeds, and the SOC's analytic architecture(s) (SIEM, SOAR, big data, etc.). This means the CTI tool should support both open CTI standards (STIX/TAXII) as well as the APIs of the tools the SOC favors, such as their SIEM/SOAR. This also means that the threat intel management tool supports both batched and NRT data automation in and out of the tool.

Off-the-shelf feeds
A good CTI platform will offer a "menu" of integration and ingestion from paid and open-source CTI feeds. At the same time, it should also enable and automate deduplication across the feeds ingested.

Feedback and confidence scoring
Analysts should be able to tag, vote, or otherwise score various CTI according to their assessment of the quality and usefulness of the CTI in question. This will support not only an internal feedback loop of the SOC's own products, but more importantly, scoring against the feeds it receives from others. This should also enable the binning and filtering of IOCs and other intel so that the SIEM/SOAR can be set to only alert on the top-quality CTI sources, according to SOC's assessment of CTI pedigree.

Access management and identity integration
The CTI platform should support identity integration for the SOC, such that the CTI platform is insulated from risk by general constituency compromise and yet enables analysts to correlate user identity mapping to other workflow activities (email, chat, ticketing, and others).

Scalability
It is very likely for the SOC to encounter IOC feeds that measure in the millions of distinct IOCs. Accordingly, the CTI platform should be able to persist and query this data at scale, usually on timescales measured in seconds or minutes.

Confidentiality and source tagging
SOCs that ingest CTI data from many different sources often have data sharing agreements and data handling caveats to preserve and maintain. A good CTI platform should support various tagging and metadata for handled artifacts (IOCs and otherwise) that enable the SOC to be clear on the pedigree and handling caveats of source data, and label data sharing for intel product. This should include sharing protocols, such as Traffic Light Protocol [214].

User pivoting
The CTI tool should enable analysts to follow or trace through adversary activities and create threads of associations, make correlations, and pivot to new threads of associations and links easily, while maintaining a history.

Auditing and change management

The system should support tracking of who did what, and ideally, the ability to revert changes. This is particularly important when one analyst will add attribution or association in a given IOC, campaign or adversary record that could in turn trigger downstream correlations and alerts to be fired. This also helps avoid losing days of analysis or work.

6.7 Analysis Reports & Products

Cyber threat analysts produce a set of deliverables and artifacts on a routine basis. Some of these deliverables are easily recognized briefings or papers, whereas others take the form of short reports, or as input to an online knowledge base or updates to tools or technologies used by the SOC.

Table 12 lists some of the written artifacts a cyber threat analyst is likely to produce. Notably, the types of reports will overlap quite a bit with incident reports. Cyber threat reports put context around the incident artifacts and data. Any of these work products can be for internal or external consumption. They consist of formal or informal information and reports from SOCs, commercial vendors, independent security researchers, and independent security research groups that discuss information about attempted or successful intrusion activity, threats, vulnerabilities, or adversary TTPs, often including specific attack indicators and vectors.

Table 12. Cyber Threat Analyst Artifacts

What	Discussion	Typical Frequency
Cyber threat intelligence reports	Cyber threat intelligence reports can range from monthly to annual and often summarize cyber threat activities for the constituency, specific enterprise, or for a business sector	Monthly-Annually
Case management notes and reporting	Incidents that are targeted in nature or are related to a known APT may be referred to cyber threat analysts for in-depth analysis. The working notes, activities, recommended follow-up, and other analyst-to-analyst communications are recorded in the SOC's incident case management capability and/or CTI platform.	Daily-Weekly
Formal incident write-ups	Particularly notable incidents may deserve formal documentation or presentation outside the scope of what is captured in the case management tool. This may take the form of presentations, written reports, or sometimes both, authored by cyber threat analysis or co-authored with SOC incident responders.	Monthly-Quarterly
Cyber threat tipper	Short, timely information "tipped" to a cyber threat sharing group within minutes or hours of identifying activity as likely relating to a targeted intrusion attempt. The information may be as simple as the sending IOCs with a short bit of informal context derived from malware analysis, log analysis or other digital artifacts.	Daily-Weekly

What	Discussion	Typical Frequency
Short-form malware report	Depending on the SOC resources, the report may be a brief multi-page report that provides some indicators and information regarding an observed piece of malware. Usually stems from malware that took one or two days of static or dynamic analysis to understand and co-authored with SOC malware analysts.	Weekly-Monthly
Long-form malware report	Three-plus-page report that provides detailed indicators and reporting on an observed piece of malware co-authored with SOC malware analysts for major attacks. Generally, stems from a deep-dive reverse engineering effort that took several days or weeks to accomplish. Typically includes a full description of the malware sample's functionality, any encryption used, and its network protocols used for command and control. It may include additional tools and techniques developed alongside the analysis, such as malware network protocol decoders, and ways to unpack and extract encryption keys and other indicators from malware samples within the same family. Commentary about attribution or adversary association and comments on the malware use across the kill chain, such as actions on objectives may also be included.	Monthly-Quarterly
Adversary and campaign reports and presentations	Briefs that discuss the TTPs, intent, activity seen, incidents, etc., stemming from a named adversary or adversary campaign. Usually combines activity seen from multiple incidents and/or several months of reporting. Oftentimes these will be framed in a larger context of geopolitics, economics, or other motives of the parent nation state or non-state actor.	Quarterly- Annual

For advanced or large well-resourced SOCs, cyber threat analysts are likely to apply substantial efforts toward non-traditional tangible work products, as detailed in Table 13.

Table 13. Other Cyber Threat Analysis Work Products

What	Discussion	Update Frequency
Trends of adversaries, campaigns, or other incidents	Cyber threat analysts are tracking the bigger picture of adversaries and incidents across an enterprise, and therefore are in the position to conduct trend analysis, pattern recognition, and make associations across activities that responders may miss. This may be cumulated in comprehensive annual (or quarterly) reports.	Quarterly -Annual
Indicator lists and TTPs with context	Part of cyber threat analysts' job is to aggregate, correlate, and associate various indicators of compromise (suspicious IP addresses, domains, email addresses, etc.) from external cyber threat intel reporting and its own malware reverse engineering. These indicator lists are primarily used to generate signatures and other detection content in the SOC's tool set (network-based intrusion detection systems [NIDS], SIEM, EDR, etc.). They may also be housed inside—and generated from—the CTI platform.	Daily-Weekly

What	Discussion	Update Frequency
Sensor and analytics enhancements	CTI analysts will frequently write or enhance SOC detections and analytics themselves or pass off technical details to a team member to create them. In either event, the SOC's CTI apparatus should routinely impact detection and analytic work product.	Weekly-Quarterly
Custom tools or scripts	Cyber threat analysts may uncover activity that highlight gaps in capabilities that cannot be satisfied through Free and Open-Source Software (FOSS) or COTS solutions. Quarantining and observing the adversary, simulating command and control traffic, or ingesting foreign sources of data into a tool are three examples where custom code may be needed. Custom tools are spun off on an irregular basis, usually developed very quickly, and do not always reach full maturity before they are no longer needed.	Irregularly; Monthly-Quarterly

6.8   Organizational Relationships

Cyber threat analysts' relationships with personnel outside the SOC will vary depending on the nature of the relationship. For instance, most users and the IT help desk need to know that potential cybersecurity incidents should be referred to the SOC; they see the SOC as one unit and have no visibility into specific functions like CTI. Other parties, however, may recognize and interface with cyber threat analysts directly, due to their special role in operations. Intel analysts are the SOCs early warning system, when they get reports of incoming attacks from threat sharing partners, they relay that information in real time to the IR analysts to take defensive actions. These relationships are depicted in Figure 15.

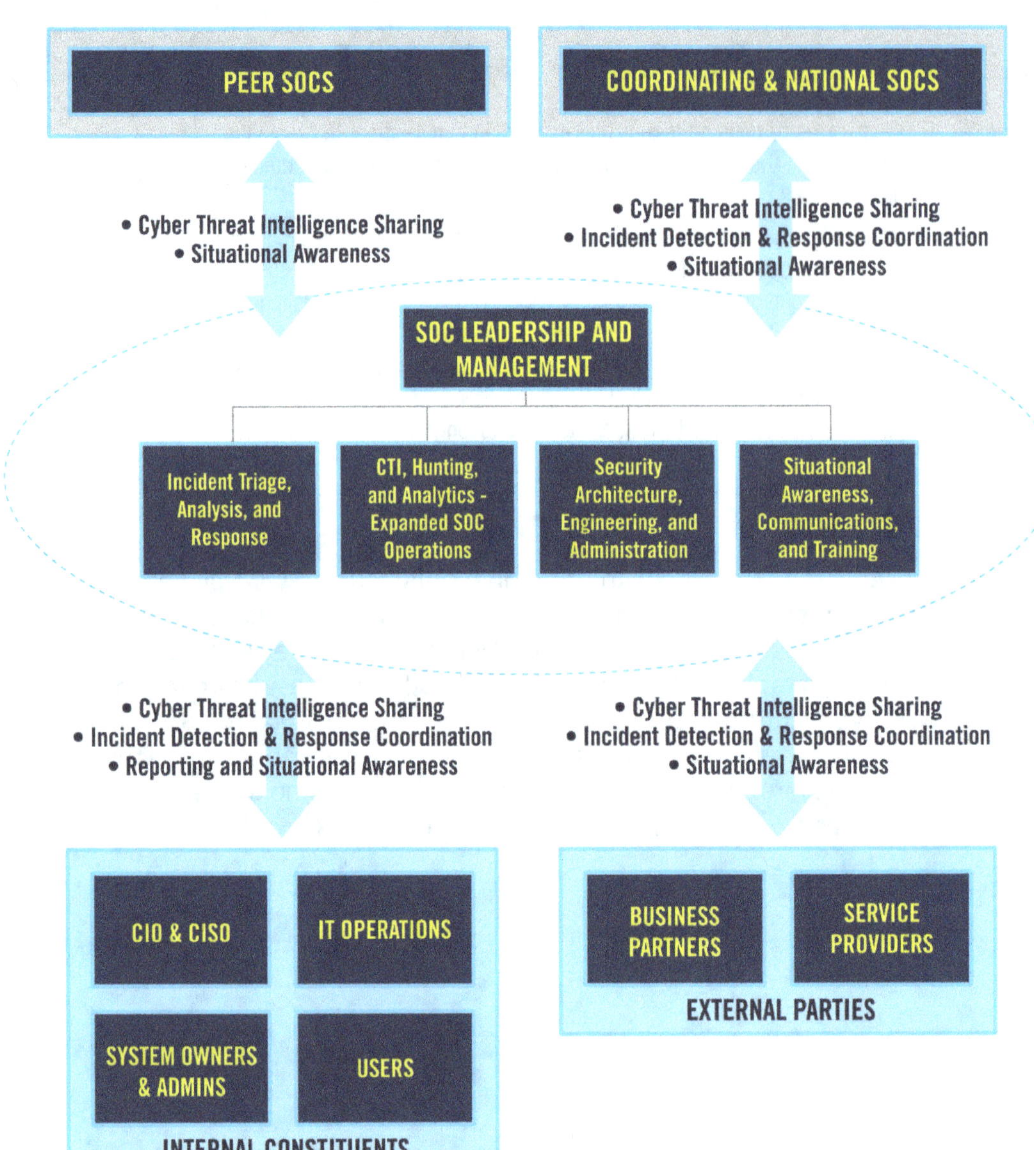

Figure 15. Cyber Threat Analysts and SOC Relationships to Other Organizations

Cyber threat analysts' knowledge of the adversary will likely be of specific interest to IT and security executives, such as the parent organization's CIO, CISO, and CSO. The SOC can expect that cyber threat analysts will provide monthly or quarterly threat briefings to interested executives. Providing these briefs is important, even if they are just informational: these briefings build trust and familiarity with the SOC and help justify its budget. If the SOC's parent organization has any parties that must maintain strong awareness of cyber threats, such as industrial counterespionage or insider threat groups, the cyber threat analysts should consider collaborating with those groups as well. The analysts will also require direct liaison with certain IT and security personnel.

When it comes to sharing cyber threat intelligence with other SOCs outside the organization, cyber threat analysts take the lead. In some cases, this can be pairwise sharing with one other SOC. However, nearly all cyber threat analysts participate in cyber threat intel sharing groups external to the constituency. These groups usually consist of a handful to several dozen other SOCs with some common attribute—usually geographic region, nationality, or business function such as government, industry sector, or education. Most typically, these relationships are reputation-based, brokered at the analyst-to-analyst level. There must be a mutual sense that each participating SOC has something to add and that indicators and TTPs will be protected; hiring dedicated and experienced cyber threat analysts is the best way to gain substantive entry to such sharing groups.

During focused adversary engagements and deception operations, cyber threat analysts have a goal to gain intimate knowledge of the adversary in the context of the impacted systems, mission, data, and users. This usually requires the SOC to make ad hoc instrumentation of enclaves and hosts at the edge of the network, and potentially redirection of adversary activity. Ideally, the CTI analyst will work with other SOC analysts, who will have a close working relationship with system owners and IT operations for the duration of the engagement. CTI analysts will also frequently interface with business and mission specialists, particularly as it related to mission impairment, IP theft, potential fraud, and potential abuse activity.

Similarly, cyber threat analysts of larger or well-resourced SOCs might also cultivate relationships or be acquainted with relevant law enforcement organizations empowered to investigate cybercrime. The SOC and its champions might consider enabling cyber threat analysts to have direct liaison authority with these outside parties, with broad clearly defined authorities and guardrails regarding collaboration and disclosure.

6.9 Common CTI Challenges

Using contextualized information about adversaries can greatly enhance SOC ability to anticipate, detect, and respond to incidents and unwanted adversary attention. It should be added once basic SOC functions are up and running. The point of CTI is to look beyond exact signatures, hashes, IP addresses, or other static and easily changed characteristics of an adversary's actions. Adversaries avoiding detection will move around and attempt to evade sensors, which means changing static characteristics in technology data. Some common pitfalls and misconceptions to consider and avoid when integrating CTI functions and services into the SOC include:

CTI is not all created equal
Some CTI sources repackage open-source feeds and apply proprietary filters or apply other value-add secret sauce. Ask questions about the provenance and data sources. Also, CTI is not defined standardly, so there is a wide range of CTI; some provide lists of indicators of compromise IOCs, malware, or other data as CTI, while others provide finished, long reports that must be analyzed and applied to the SOC enterprise.

Too much CTI can result in information overload
If too much CTI is streaming into the SOC, there may not be enough analyst time to review and process the information, and it may introduce more "noise" than assistance to SOC analysts. CTI should provide clarity, not just more traffic.

Not every CTI source is useful for every SOC
If the constituency processes payments for rental units, criminals that steal credit card information and targeting the housing sector are of interest, including TTPs, geo-political data on groups targeting, etc. But perhaps adversaries targeting the electric grid are less relevant. The SOC should shape its CTI ingest and focus around those relevant to its constituency.

Cost does not necessarily translate to great CTI
Sometimes SOCs can purchase tools and set them up to run automatically. CTI is different. If a team does not use or integrate purchased CTI, money spent is not worthwhile. Large, advanced teams with many CTI analysts might prefer freely available open-source CTI feeds and reporting from partner SOCs. In contrast, an advanced but small team (e.g., two analysts) with resourcing, might opt to purchase commercial CTI services.

Malware and CTI are not equivalent
Malware can be one focus of CTI; however, the SOC needs to ensure broad coverage of adversary TTPs that eschew extensive use of malware. For example, if a SOC is just looking for malware hits and malware pedigree, they miss other pieces of the puzzle, such as adversaries "living off the land" (residing in an enterprise, pivoting through trusted resources, and using resources). Malware is a tool which adversaries use to further their cause, and it is important to understand the bigger picture of adversary targeting and TTPs, and where available, motives/intent to anticipate how to design the defenses.

IOCs and CTI are not equivalent
In traditional intelligence analysis, indicators are facts that inform the analysis, but are not the intelligence analysis. Ensure CTI answers decision-makers questions and contains context[8] [215].

6.10 Final Thoughts

Because CTI is an important aspect of today's cyber defenses, we can expect the field to crystalize and form around some standards in definitions. In addition, attribution analysis and certainly the more attainable adversary association techniques, and analysis from intelligence will increase in SOC use to become important in addressing cyber adversaries for efficiency and effectiveness. Finally, investing in CTI analysts can save a constituency and its SOC money over time. Effective CTI analysts can assist the SOC in understanding which cyber defense products are worthwhile, and which are not worth the money. They understand how adversaries behave and manifest in the technology, and therefore, which security products are effective. With all the security products available on the market, SOCs can spend endless

---

[8] Homeland Security Digital Library (hsdl.org): A Compendium of Analytic Tradecraft Notes provides definitions of common elements of intelligence, including what indicators are and how to use them.

---

amounts of capital trying to be secure. So, it is worth the investment to hire great cyber intel specialists to assist in prioritization of resources.

6.11 Summary – Strategy 6: Illuminate Adversaries with Cyber Threat Intelligence

6.1. Benefactors of the CTI are not only all aspects of the SOC itself, but also stakeholders elsewhere in the constituency ranging from IT executives to cybersecurity partners.

6.2. In this book, CTI is defined as the collecting, processing, organizing, and interpreting of data into actionable information that relates to capabilities, opportunities, actions, and intent of adversaries in the cyber domain to meet a specific requirement determined by and informing decision-makers.

- Data is considered CTI only when it is clearly connected to contextual information about the adversary behavior it describes. For example, an IP address by itself is not CTI.
- Good CTI is: Actionable, meaning the SOC can how to apply the information and it is accurate; Timely, meaning events are recent (and thus still applicable); Relevant, meaning it is applicable to the constituency; and it is Accurate, meaning the content accurately and correctly describe what happened.

6.3. CTI can be applied at the strategic, operational, and tactical levels of cybersecurity and cybersecurity operations. CTI is comprised of three key facets: context, analysis, and action.

6.4. CTI becomes truly actionable at the intersection of: Adversary Information, Technical Environment, and Relevancy:

- Adversary Information: adversary information outside of the constituency.
- Technical Environment: data from inside the constituency, and the constituency's capacity to do something with that CTI (such as IOC matching in a SIEM or a custom detection in an EDR).
- Relevancy: information about the constituency regarding as what is important to protect, including intellectual property, privacy information, credit card processing, or mission logistics of interest.

6.5. SOCs cannot predict attacks without adversary association, or some semblance of attribution and are limited in anticipation. There are different shades of attribution, and the SOC may not need absolute certainty in attribution to form associations or act on their conclusions. There are several frameworks useful for understanding the adversary, including MITRE ATT&CK, Lockheed Martin's Cyber Kill Chain, and The Diamond Model.

6.6. SOCs that have a sophisticated CTI capability, such one where they are synthesizing several sources of CTI or producing their own CTI may invest in a Cyber Threat Intelligence Platform, sometimes knows as a TIP. A CTI platform can quickly become central to the workflow of many analysts, and thus should be acquired and supported with care.

6.7. The SOC is likely to produce myriad products that fuse CTI, ranging from daily updates in a CTI platform to annual threat intelligence summaries.

6.8. CTI analysts are better informed through sharing and are often the best to take the lead in SOCs for outreach and collecting and sharing CTI externally and internally to the constituency.

6.9. The SOC should exercise care and discrimination toward what CTI it ingests and how it processes that data against factors discussed earlier in this chapter.

Strategy 7: Select and Collect the Right Data

Gaining visibility into cyber activities requires thinking strategically about the data and feeds that best render a complete picture of those activities. There are many types of data the SOC will find valuable. For example, "Strategy 1: Know What You Are Protecting and Why" includes information on gathering and maintaining asset information, and "Strategy 6: Illuminate Adversaries with Cyber Threat Intelligence" delves into the cyber threat intelligence data available. In comparison, this strategy focuses on the sensor and log data collected by network and host systems, cloud resources, applications, and sensors, wherever they may reside. This data makes up the largest percentage of data the SOC collects and persists.

7.1 Planning for Data Collection

With so much data to choose from, it is necessary to identify the most useful data for the job up front to avoid missing important events. As shown in Figure 16, this can occur either from collecting too little data and not having the relevant information available to detect intrusions or from collecting so much that tools and analysts become overwhelmed. Additionally, designing, acquiring, deploying, and maintaining sensor and log collection capabilities can take tremendous resources depending on the scale and complexity of deployment. Tradeoffs must be made regarding ingestion rate and storage versus cost.

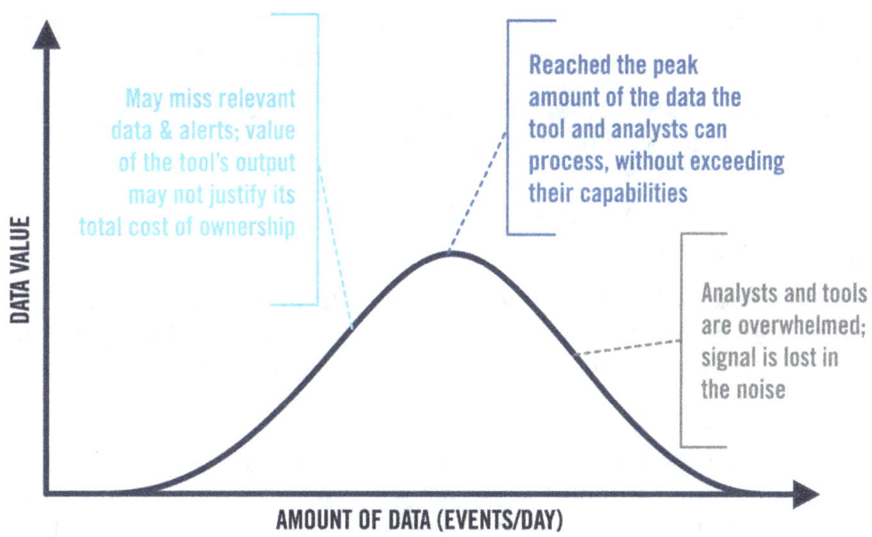

Figure 16. Balancing Data Volume with Value

This strategy begins by broadly introducing considerations for data collection including planning for, tuning, maintaining, and retaining those collections. It then dives deep into host and network collection and monitoring as these are at the core of the SOC's data collection strategy. It finishes up by discussing considerations for cloud and OT environments, along with summarizing considerations for a data collection instrumentation approach. The goal of this strategy is to gather the right data in the right amounts from the right places in the enterprise, with an economy of effort and expense.

One of the most frequent questions posed by SOCs is, "What log and sensor data should we gather?" Every organization is different, so the answer will vary, however there are some common approaches all organizations can take to identify the right data for their constituency. Most importantly, the SOC needs to be deliberate in their planning, not just taking in any data they can, but selectively targeting the data that is most relevant and ensuring appropriate polices are in place to allow them to collect that data. They also need to ensure they have the appropriate policies and budget approach in place to support collection.

7.1.1 Mission Drivers and Use Cases

To get started on selecting the right data to collect, the SOC should pick a handful of business use cases which are important to the constituency and identify the top threats the SOC wishes to defend against. This will then drive which feeds are necessary to build out those use cases. There are several benefits to this approach:

- The SOC has a clear set of requirements and goals to meet.
- There is a concrete set of deliverables and outcomes.
- The set of data being collected is seen as finite and therefore manageable.

When considering business use cases, there are several drivers for collection of security-relevant telemetry, many of which overlap between the SOC, other elements of the cybersecurity apparatus, and traditional IT operations. They include:

- Defending networks, systems, cloud resources, and other digital assets
- Insider threat monitoring and audit collection
- Performance monitoring
- Maintenance troubleshooting and root-cause analysis
- Configuration management

When considering threat-based use cases, it is helpful to combine threat intelligence with the use of a threat framework, to guide collection choices. More on this topic can be found in "Strategy 6: Illuminate Adversaries with Cyber Threat Intelligence."

Use case: Log audit review
One non-defensive use case that may be of interest to many SOC is the use of the SOC's log collection architecture to support log audit review, especially as mandated by law or regulation. There are many advantages to fusing SOC data collection and mandated audit efforts. The SOC will have access to a large set of audit data as the assigned collector, and

the logs can be brought into a single place while serving several use cases. To minimize burden of serving multiple purposes, here are some tips to consider:

- The SOC's mandated audit log collection systems may be subject to multiple legal collection and retention requirements and the SOC must be resourced appropriately to meet these needs.
- The SOC and its partner security and IT organizations should set clear expectations and division of labor around full-scope audit review, especially when compliance, regulation, or laws are in play. Often, security officers (in government, ISSOs) and sysadmins are responsible for comprehensive, widely scoped reviews of the enterprise's audit logs, whereas the SOC's role may be limited or operate at a higher level of cross-service or cross-enterprise visibility. These expectations must be clarified and agreed to by all key stakeholders.
- Those granted access to audit logs (e.g., cybersecurity specialists outside the SOC) should only be given access to the portion of logs reports necessary to fulfill their job. Widespread access to all logs by people outside the SOC may lead to conflicts in incident identification and escalation and risks compromise of sensitive insider threat cases.
- When appropriate, use existing tools to segregate non-SOC audit data. For instance, SIEM or log aggregation agents/collectors can extract audit data once and transmit it to separate log management warehouses in parallel.

7.1.2 Data Sources

Once the SOC understands the situations for which it wants to collect data, it next needs to consider the types of data available. Each SOC will choose data feeds to best illuminate the enterprise for preventing and detecting intrusions and other monitoring activities. Figure 17 shows one way the SOC could leverage host, cloud, and networking sensor data and logs to support detection and investigation activities.

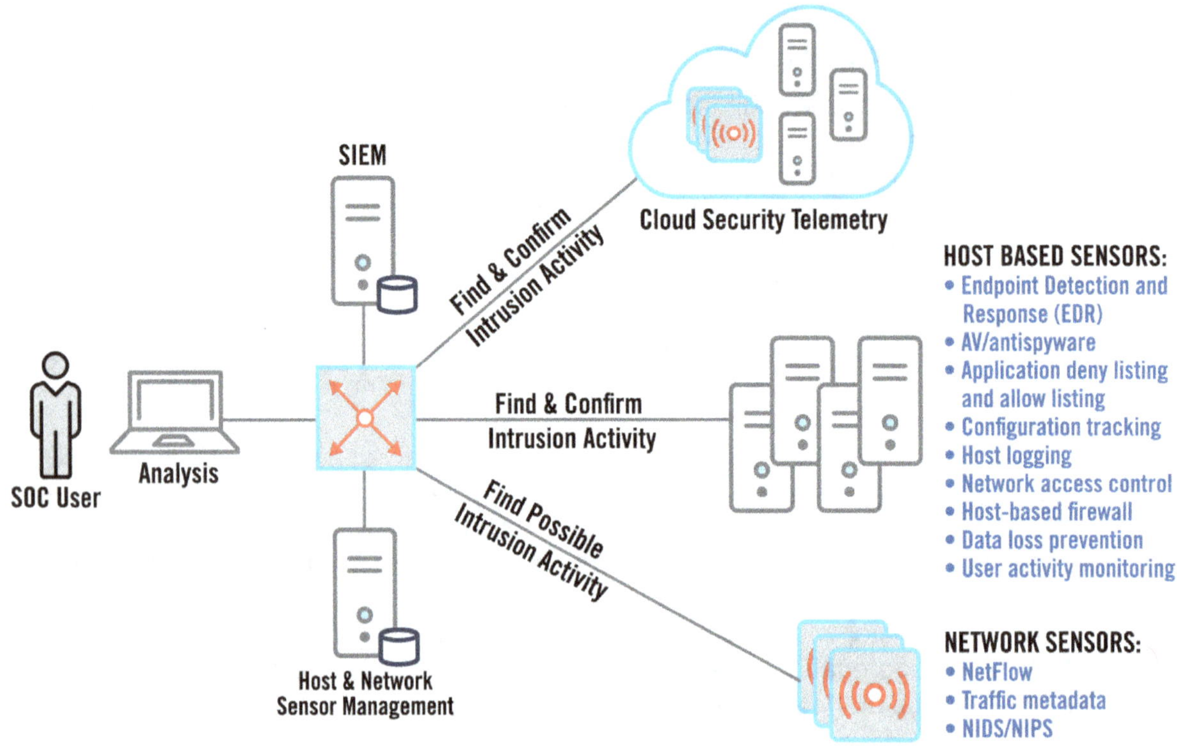

Figure 17. Notional Sensor and Log Data Sources in Intrusion Context

The data the SOC needs will come from a variety of sensors and logs; understanding these potential sources is a key step in determining what data can support what use case and what data to prioritize collecting. Appendix E provides a more detailed look at common data sources collected by SOCs. The appendix table includes information about (1) summary of what the data source reveals to analysts, (2) rough order of magnitude event volume, (3) volume dependencies, (4) general value of data, and (5) common fields of interest.

In addition to the information contained in Appendix E, the SOC will want to consider the following when deciding on data to collect:

- How does the quality, visibility, and attack life cycle coverage are offered by purpose-built host or network sensors on or near the hosts in question complement the log feeds themselves?
- Quality and clarity of logs produced:
  - Human and machine readability; does the log use cryptic messages and obscure encoding, or can a human easily read it?
  - Does one log entry correspond to one logical event, or is human or machine correlation needed to piece together disparate audit records?
  - Are the clocks synchronized on all systems considered for monitoring?
  - Are system owners willing to adjust their logging policies for SOC needs?

- Availability of logs:
  ◦ Are logs written in a vendor-neutral, open format (e.g., key/value pair, Open-Source Security Events Metadata (OSSEM), or Common Event Format [CEF]), and/or can they be interpreted by audit collection tools or SIEM currently in use?
  ◦ If no existing parsers exist, what resources are needed to integrate them in a data aggregation or analytic framework?
  ◦ Are system owners willing to provide direct or mediated access to log data in their original format and in real time, instead of feeding all of them to the SOC?
  ◦ Is there a natural aggregation point (such as a message bus, syslog collector, or data warehouse) that can be capitalized upon?
  ◦ What is the overall volume of collected logs? Will the networks and systems that connect the source system with the SIEM or log aggregation node support the requisite bandwidth and disk space?
- Coverage of logs:
  ◦ Will a given log feed cover a wide portion of the enterprise, such as Windows domain controller or Web proxy logs?
  ◦ Or, will the logs enhance coverage for a specific high-value application (e.g., a financial management system) that deserves deep visibility and detailed monitoring use cases?

When considering what data to collect, it is also important to recognize the technical impact audit data collection places on constituent systems. There are several tips for minimizing the impact of collecting security telemetry:

- Minimize the number of agents deployed, especially on end systems.
- Carefully tune performance-related parameters of the agent, such as the polling frequency for events, the number of alerts retrieved in each poll, and CPU/disk/network/memory footprint limitations.
- Leverage existing collection points (such as syslog aggregation points and management servers) where they exist, provided they meet the following criteria:
  ◦ Data is delivered to that collection point without substantial loss in original fidelity and detail.
  ◦ The SIEM or platform system has an agent for the collection point.
  ◦ Data is delivered quickly enough that they may be correlated with other related events.
- Leverage assured delivery where this option exists:
  ◦ Wherever possible, use TCP-based protocols instead of UDP to avoid event loss.
  ◦ Placing the agent close (logically or physically) to the source systems can minimize the "distance" events travel before encryption and delivery guarantees are put in place.
  ◦ Many message bus technologies such as Apache Kafka provide positive acknowledgement and "at least once" or "exactly once" event delivery along with batching, encryption, and compression; these techniques should be leveraged, when possible.

- Consider using a SIEM or log management system that can transmit events from one agent to multiple destinations, thereby supporting redundancy and COOP, if needed.
- Consider the stability and resilience of telemetry collection software or service; does it suffer from frequent crashes? How resilient is it to being circumvented or shut down by an attacker? The SOC may need to make careful choices in finding resilient and reliable data collection technologies, especially if they need to reach out to an external party in the event of crashes.

7.1.3 Data Collection and Monitoring Policies

In conjunction with identifying available data sources, the SOC needs to ensure it has the proper authorities and agreements in place for the data it is going to collect.

While quick and informal agreements with system owners and sysadmins can get results quickly, they may not be durable, due to personnel turnover. In medium to large constituencies, an MOA or MOU may help when setting up major set of data feeds or targeted monitoring engagement. Constituents often perceive better service when their specific monitoring needs are formally documented and the SOC can establish service expectations, such as regular cyber threat or awareness reporting.

A data collection memorandum usually includes the following:

- Technical POCs and Management POCs
- What data is being gathered
- What it is being used for, such as targeted use cases, general monitoring, insider threat, etc.
- Who is responsible for a regulatory or compliance-driven audit review and long-term log retention, if applicable?
- Whom to contact if the data feed goes down or changes in any major way
- How data will be secured, including steps taken to protect users' privacy or data confidentiality
- Reference to important authorities such as a SOC charter or CONOPS
- The digital assets being used to collect the data
- Additional expectations of the SOC and system owners, if needed

In addition to MOUs with specific stakeholders, the SOC should leverage exiting process and policies. For example, if the constituency has an engineering or CM process that supports timely delivery of services, the SOC should work in concert with that process. Additionally, the SOC can look to inject collection requirements be articulated in an IT policy for engineering of new systems and services. The SOC, however, must be sensitive to how these policies impact organizational resourcing for new and ongoing projects. When authoring policy mandating monitoring coverage for the enterprise, it is best for the SOC to work with other stakeholders not only to set governance for instrumentation and data collection, but to have standard service offerings to satisfy that governance. This most likely takes the form of pre-packaged host and sensor suites, whose deployment, costing, and servicing are standardized.

7.1.4 Budget

As the SOC is developing its data collection plan, it needs to consider how it will pay for the transport and storage of the data. However, the SOC needs to maintain awareness of how the collection volume may change over time. For example, as enterprise applications move to the cloud, the volume of data flowing past perimeter taps and through firewalls increases, which also increased the amount of NetFlow or firewall events captured. When these types of changes are being considered, the SOC needs to be a part of early planning.

In general, it is simplest for SOCs to maintain their own budget supporting comprehensive monitoring coverage and staffing. Moving to a fee-for-service model or "tax" can become challenging: many programs and projects will not budget for new capabilities, and the SOC's year-to-year budget planning will become overly complex or subject to third parties that the SOC will have a hard time influencing. Avoiding a "tax" on lines of business helps maintain the perception of the SOC as a source of value rather than a sunk cost. An alternative is for the SOCs to look to large projects to help fund monitoring tools at initial deployment time and then build in recap and staffing costs for out years into the long term SOC budget.

7.1.5 Data Volume

Figure 18 illustrates some of the data sources available to an IT enterprise; these potential data sources vary in value and volume for prevention, detection, or analytics/forensics. For example, some data, such as PCAP, is extremely resource intensive, whereas traffic metadata collection analysis, given its comparatively lower volume, provides improved bang for the buck in both detection and analysis. When utilizing this diagram, a couple caveats should be considered:

- Data volume estimates assume wide coverage and saturation of each technology.
- This is not a strictly stack-ranked list of sources; it is possible to make a single email gateway produce far more data than a given network firewall, for example.
- This is not a complete list of all security-relevant data sources; for a more comprehensive list, see Appendix E.
- Even though a given data source may not be shown to cover a given function (such as analysis or forensics), it does not mean it will *never* support that goal; for example: Wireless Intrusion Prevention System (WIPS) and AV telemetry certainly have been used for forensic incident timeline reconstruction by some SOCs.

In addition, some data is better for determining what happened after an event, and therefore may better inform forensics which in turn can help prevent future attacks. This strategy captures common considerations and presents a pragmatic, operations-driven approach to prioritizing what data to gather.

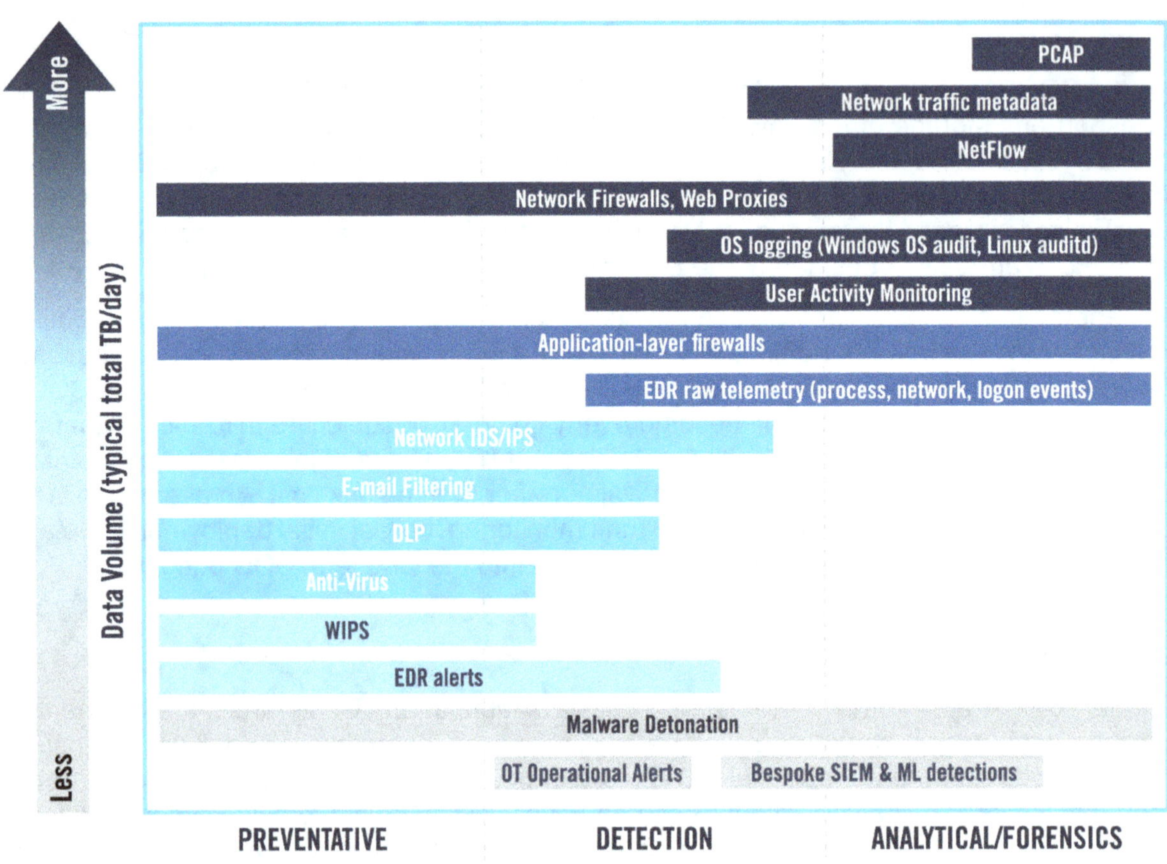

Figure 18. Data Source in Context of SOC Usage and Volume

7.1.6 Tuning

Once the SOC has selected data feeds for collection, how should it pare down its volume? If the SOC over-tunes its data, it will miss important information that could help it identify malicious activity. If it under-tunes its data, it can result in inefficient processes or missing the important activity due to all the noise (and very high IT/cloud bills).

Tuning approaches
There are two classic approaches that SOCs may take in selecting and tuning data sources: tune up from zero or tune down from everything. This section also includes a third, somewhat orthogonal approach: leverage data in place. Table 14 identifies the pros and cons of each approach.

Table 14. Approaches to Tuning Data Sources

Approach	Pros	Cons
Start with the entirety of a given data feed and tune down to a manageable data volume that meets common needs.	• Requires little foreknowledge of the data being gathered. • Easiest to implement. • Enables SIEM tools to leverage full scope of data features and event types offered.	• May overwhelm tools and analysts if data feed is too voluminous. • If methodology is used for many data feeds, poses exponential risk of "data overload." • "Default open" filtering policy toward data collection may pos long-term risk to data aggregation systems as feeds change over time.
Start with a candidate data feed, and tune up from zero, focusing only on what is deemed useful or important.	• Keeps data volume low. • Focuses systems and analysts only on what is deemed to be of interest. • Less problematic for SOCs with limited budgets.	• Carried to its extreme, limits value given time/effort granting SOC access to given data feed. • Analysts blind to features of data feeds not explicitly set for input into data collection systems. • Approach may require more labor to implement.
Leverage data closer to its source, such as in an intermediate log management or big data platform, rather than ingesting it directly into the SOC's data pipeline.	• Keeps latency, performance low. • Context of local data is retained. • Usually lower costs due to fewer copies of the data and less impact on network utilization.	• Can be complicated to configure and maintain; as traffic changes, local processing needs to be updated. Recommended for large enterprises and datasets. • Analysts still need to pivot into the remote data store, which can vary from easy to impossible, depending on the specific scenario. • Analysts need to keep up to date access to remote data; in the presence of dozens or 100 disparate stores, this can become error-prone and time consuming. • This data usually only supports forensics and not detections, as ability to process that data for detections is usually predicated on centralized collection and processing like in a SIEM. • Ensuring chain of custody and anti-tampering of the data may be a concern, depending on where it has kept, who is control, and surrounding security controls, such as data left on end hosts.

Reducing volume for lower value data

No matter which of the tuning approaches is taken, one of the first things to do is to tune out any data that is known to have a lower value. For example, in network sensing, filtering out encrypted packets or known-good high-bandwidth sources such as security cameras can reduce data that are unlikely to be used.

Local versus centralized processing

There are many options for determining which data is collected and processed locally, compared to bringing data to a central SIEM for correlation. In general, when tuning data sources, larger, geographically distributed constituencies design collection with a combination of local and central techniques for processing and collection.

Processing locally can greatly assist in limiting network traffic and bogging down centralized systems; on the other hand, it can also be implemented in a way such that the SOC does not benefit from the data. Using local collection and retention is most frequently used in large enterprises with multiple regions and diverse data lakes with many stakeholders. Local retention does not necessarily mean leaving it on the source host or cloud service, but rather pulling it to a log store local to the region, application, or service in question. This is particularly the case as a) SOCs leverage more sources of data that were not originally meant for security purposes and b) more services, applications, and cloud resources have a local logging store. By gaining remote access to other services' data warehouses and leveraging that data in place, the SOC stands to gain some of the same benefits as collecting all that data but at a fraction of the technical expense and political disruption. It is best if the technologies chosen support querying multiple disparate datasets simultaneously through techniques such as federated search [218], [219] and geographic sharding, "geosharding" [220].

Data processing tools versus data collection volume
The tools the SOC brings to data analysis changes as the volume, variety and velocity of that data grows. As shown in Figure 19, the value of the data divided by the relative cost of implementing a given analytic framework is shown on the y axis, as a function of the volume of the data and the number of different feeds on the x axis. Whereas manual tools might be appropriate and effective for 10,000 lines of logs, it is not for 10 billion. These are discussed further in "Strategy 8: Leverage Tools to Support Analyst Workflow."

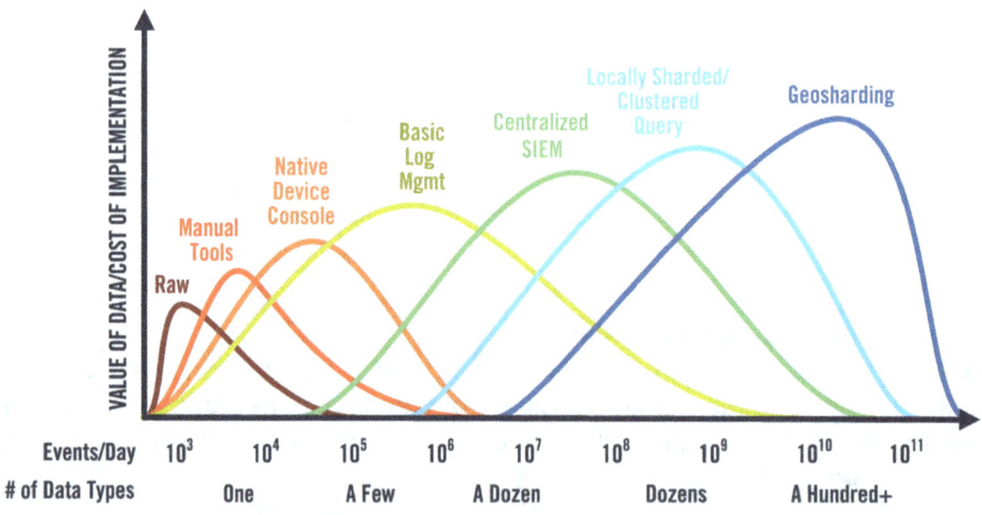

Figure 19. Leveraging Data at Many Scales

Tuning failure and success auditing

It's also important to avoid a common pitfall when defining audit policies: generating messages only on a "fail" but not on a "success." Failure events include users typing in the wrong password or being blocked from visiting a website. Failures mean a security control did its job: it stopped someone or something from doing what it should not do, which is *usually* a good thing. Successes, such as file modification granted, file transfer completed, and database table insert, are often where the SOC is most interested when performing investigation and analysis. This leads to an important point:

> *Do not log just the "denies"; the "allows" are often more important.*

This is because in most situations, a "deny" is an attempt by definition; it did not get through, at least on this attempt. An "allow" is either a legitimate transaction, or it is an attacker or unwanted activity that got past some access controls. Consider situations in which "allows" are often more important than "denies" such as malware beaconing, RATs, data exfiltration, and insider threat. With only failure attempts logged, the SOC will not understand what happened. Failure, block, and deny events are frequently an analytic dead end. Successes events are necessary for both investigation and correlation.

Summary

Most SOCs end up leveraging all three of these approaches. No SOC can consume every single piece of data it will ever need. Rather, what is more realistic is making judgements about how data to backhaul vs how to leverage in place. Similarly, few SOCs have the time and discipline to tune every data feed up from zero. Every SOC is encouraged to be both thoughtful, disciplined, and pragmatic about how to proceed with choices around tuning, given finite time, compute, storage, networking, and people resources.

7.1.7 Long-Term Maintenance

Sensors and log feeds require long-term care. For SOCs with large constituencies and a variety of data feeds, this is a daily activity. Constituency systems and services are constantly being installed, upgraded, migrated, rebooted, reconfigured, and decommissioned; with cloud computing, this rate of change tends to be even more pronounced. These changes frequently present blind spots in monitoring coverage. One of the most important aspects of monitoring the enterprise is:

> *Sensors and log feeds require their own routine monitoring*
> *to ensure they are performing as expected.*

Here are a few tips to keep sensor maintenance manageable:

- Enforce robust but not overbearing configuration management to track changes to monitoring data feeds. Some SOCs maintain a list of systems and technical POCs from which they receive logs.
- Maintain regular contact with sysadmins and engineering process change boards to track changes to systems. Regular SOC representation may help to maintain awareness of new projects.
- Check data feed status daily or every shift. Just because an agent is green does not mean the data feed is online. It may just mean the agent software has not crashed.
  - Consider performing regular checks against feeds from high-value targets to ensure no interruptions.
  - Perform regular checks against SIEM content dependent upon key data feeds; is a dashboard blank because there are no attacks today, or because the feed is down? Either is a possibility.
  - Include data feed health and coverage statistics as part of routine SOC metrics program and with key stakeholders as necessary.
- Use built-in sensor monitoring capabilities to alert on changes. For example, some capabilities measure average data feed rates, calculating a moving weight average, and alert when feed rate skews far from that weighted averaged.

One virtue of maintaining vigilant watch over data feeds is feedback to end system owners the SOC is active. Not only does the SOC minimize downtime in event feeds, but if outages are caught in real time, system owners can be contacted and asked, "Hey, what did you just do?" This minimizes the time needed to track down the changes that caused the outage.

7.1.8 Data Retention

The length of time the SOC needs to retain data is driven by a combination of legal and regulatory requirements, the risk profile of the constituency, and financial constraints. Table 15 suggests guidelines for minimum online log retention within the SOC, recognizing the distinct needs of SOC triage analysts, SOC forensics/investigations analysts, and external audit and investigation support. These can be used as a starting point for the SOC to evaluate how long it believes it needs to keep data readily available for query in the context of its own operations and mission. These time frames are primarily based on retention within the constituency's environment. Note that bulk long-term retention of PCAP data is no longer regarded as widely necessary, given the rising importance and comparative value of traffic metadata and host telemetry.

Table 15. Suggested Minimum Data Retention Time Frames

What	SOC triage	SOC forensics & investigations	External Support
EDR, network sensor alerts, and SIEM-correlated alerts	2 weeks	6 months	2+ years
NetFlow & traffic metadata logs	1 month	6 months	2+ years
Full-session PCAP	as needed*	as needed*	as needed*
System, network & application audit logs	2 weeks	6 months	2+ years
Emails	2 weeks	2 years	As needed

The most common standard policies set audit data retention at 12 months or more; some venues such as government agencies may mandate 60 months' retention. The most onerous requirement usually stems from supporting external investigations and regulations, such as law enforcement asking for logs on a given subject as far back as available, possibly several years. This can be a challenge. Also consider this: if the SOC stores a given feed for several years, how long will it take for the SOC to extract data on a given person or host from that full span of time? This calculus should also be factored into long-term retention plans and architecture; all too often, a SOC may store some data for years, but it has no reasonable means to "carve" out the old data needed for rare but critical legal and administrative proceedings.

7.2 Intrusion Detection Overview

Many SOC sensing technologies follow a similar basic pattern at their core:

- Knowledge of the environment and the threat is used to formulate detection policies and mathematical models that define known good, known bad, normal, or abnormal behavior.
- A detection engine which consumes a set of cyber observables (e.g., events or stateful properties that can be seen in the cyber operational domain) and compares them against a detection policy. Activity that matches known bad behavior or skews from typical "normal" behavior causes an event or a series of events containing details of the activity to be fired.
- Events are sent to an analyst in near-real-time, stored for later analysis, or both.
- Feedback from the events generated will inform further tweaks to the detection policy, known as tuning.
- Events can be filtered in several places in a large monitoring architecture, most notably: before the events are displayed to the analyst or before they are stored.

There are two classical approaches to intrusion detection [219, pp. 87-88]:

- **Misuse or signature-based detection:** Where the system has a priori knowledge of how to characterize and therefore detect malicious behavior, such as with IOC matching

- **Anomaly detection:** Where the system characterizes what normal or benign behavior looks like and alerts whenever it observes something that falls outside the scope of that behavior

Each of these approaches have pros and cons. In practice, finding tools that rely exclusively on one approach or the other is rare; most modern monitoring and detection products integrate both techniques. The most important thing to recognize is that signature-based detection requires the defender to know about adversary techniques in advance, which means the tool cannot alert on what has never been seen. On the other hand, an anomaly detection tool suite is most effective when it can cleanly differentiate "normal," from "abnormal" which can be challenging to attain, especially in highly heterogenous environments.

Although EDR, network sensors, anti-virus, and SIEM operate at different layers of abstraction, they all generally fit this model. The network sensor observables are network traffic; host observables feed EDR. Network sensor alerts and logs feed SIEM and SOAR, which treats these events as cyber observables, as the sensor did. One can even use the same mental model to understand ML and User Entity Behavior Analytics (UEBA) at a high level. Figure 20 show SOC detection data and tools in the abstract.

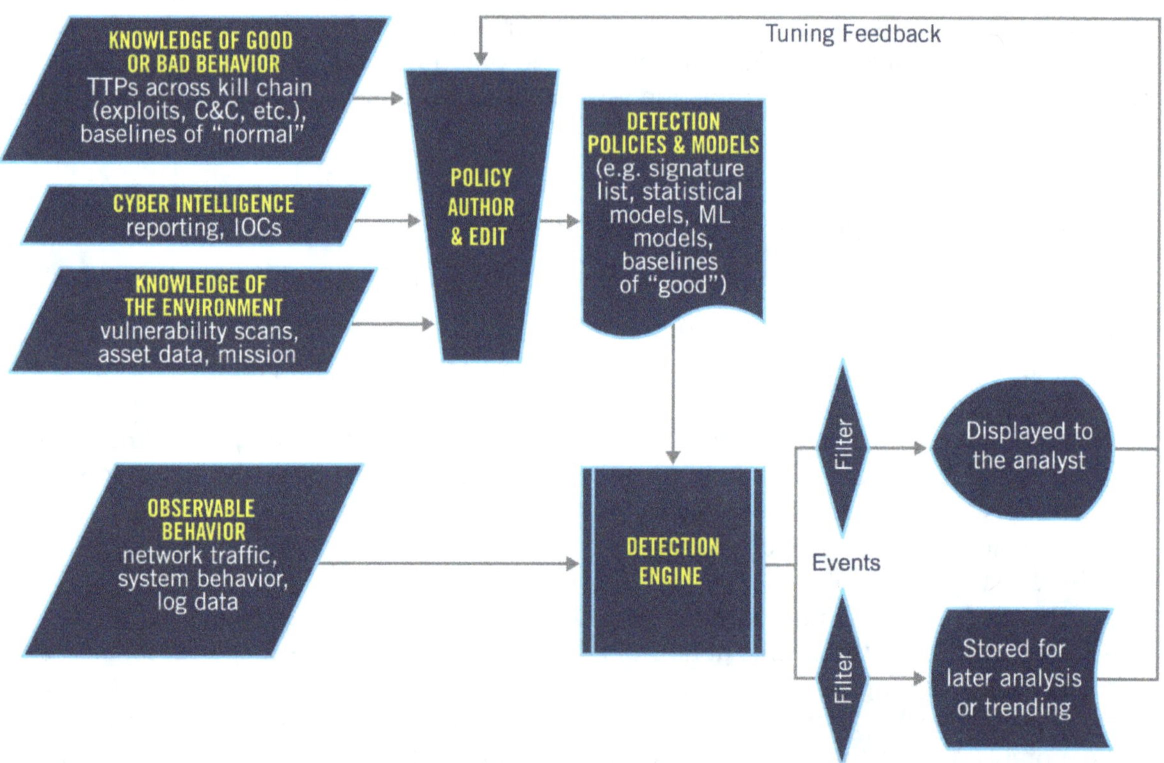

Figure 20. SOC Data and Detection in the Abstract

Each of the different types of sensors has its strengths and weaknesses. These major points of comparison are summarized in Table 16.

Table 16. Advantages and Disadvantages of Intrusion Detection Elements

Characteristic	Type	Advantage	Disadvantage
Detection method	Behavior-based, including anomaly detection and machine learning	• Behavior-based detection can detect previously unknown attacks and misuse within a session, prior to a specific attack being publicly known (e.g., with "zero days").	• They are complex and prone to false positives. • May require longer ramp-up times to learn baseline system behavior. • Networks or systems with frequent changes and activity surges may be difficult to profile. • Their behavior may be harder to understand, particularly by analysts without background in machine learning.
	Knowledge-based: Known as misuse or signature-based detection. Includes IOC matching	• Signature-based detection is fast and *sometimes* has a lower false-positive rate than behavior-based detection. • Signature-based detection can find known attacks immediately.	• Signature-based detection can only alert on known attacks. • If signatures are not updated, new types of attacks will most likely be missed, putting attackers and defenders in a game of "cat and mouse." They are prone to false positives. • They may be especially prone to circumvention by content obfuscation or protocol encryption.
Source[9]	Network: Detect activity from network traffic at perimeter or core monitoring points Traditionally: NIDS/NIPS; often part of "next gen" firewalls	• A network sensor can monitor a large range of systems for each deployed sensor. • NIDS should be invisible to users.	• A network sensor can miss traffic and is prone to being attacked or bypassed. • A network sensor often cannot determine the success or failure of an attack. • In absence of SSL/TLS "break and inspect" decryption, network sensors cannot examine encrypted traffic. • Some malware may encrypt its traffic.

---

[9] These same concepts extend into cloud monitoring to a degree, especially in the IaaS context. For a treatment of cloud monitoring as a scenario distinct from host and network, see Section 7.5

Characteristic	Type	Advantage	Disadvantage
	Host: Monitor OS and interactive user activities. Sensors are often software agents deployed onto production systems. Part of EDR	• A host sensor will not miss attack traffic directed at a system due to packet loss or encrypted obfuscated traffic, assuming the detection is based on locally observed host behavior. • A host sensor can help determine the success or failure of an attack. • A host sensor can help identify misuse by a legitimate user. A host sensing product often bundles myriad capabilities such as host integrity/ assurance monitoring.	• Host monitoring software could be disabled or circumvented by a skilled attacker, such as at the Basic Input/ Output System (BIOS)/Extensible Firmware Interface (EFI) level or with certain rootkits. • A host sensor often requires privileged access to the system to prevent or block misuse. • Incorrectly configured host monitoring capabilities can easily interfere with correct host operation.
Response mode	Active: Active IDS, called IPS, react by terminating services or blocking detected hos- tile activities.	• IPS are well matched with signature-based IDS because of the need for well-known attack definitions. • IPS can prevent or reduce damage by a quick response to a threat or attack. • No immediate operator intervention is required.	• IPS require some control of services being protected. • IPS require careful tuning in order not to block or slow legitimate traffic or host activity.
	Passive: Passive IDS react by sending alerts or alarms. These do not per- form corrective actions.	• Comparatively less risk to deploy and use as false positives do not negatively impact enterprise.	• Requires operator intervention for all alerts. This adds time to interpret, determine corrective action, and respond, which could allow more damage to occur.

As shown in Figure 21, a sensor is most valuable between the time an exploit is put in use by the adversary and when the exploit is patched against. This figure also shows the gap inherent in signature-based detections where signature implementation lags behind when an exploit is in use. Because a signature-based detection is usually more precise in spotting an attack, once the corresponding signature is deployed, a signature-based detection may be regarded as more valuable than a heuristics-based detection with respect to the exploit in question. As use of the exploit wanes, the value of that detection with respect to the particular vulnerability also diminishes.

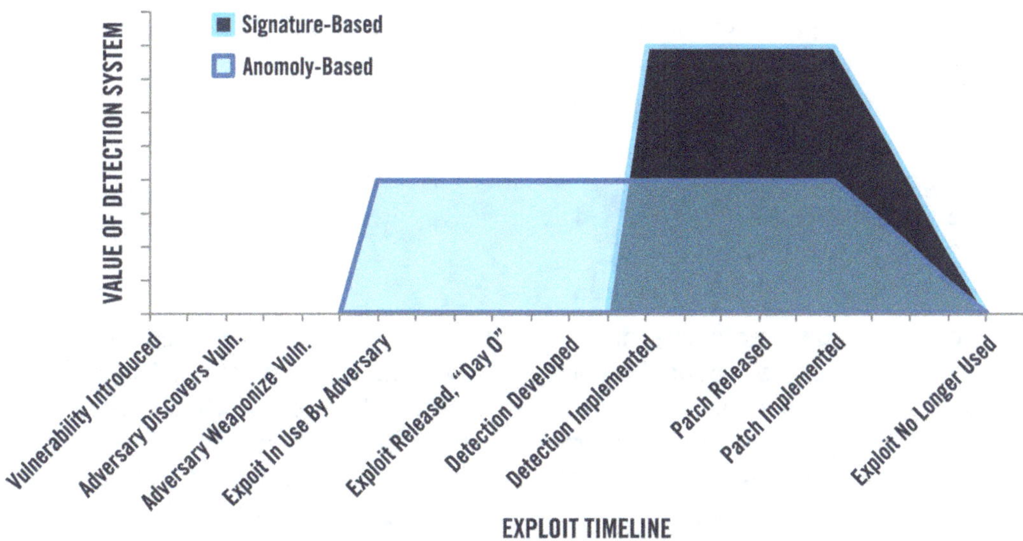

Figure 21. IDS Signature Age Versus Usefulness in Detection

7.3 Host Monitoring and Defense

This section encompasses the scope of host sensor instrumentation used by the SOC to detect, analyze, understand, monitor, prevent, and respond to security incidents. These tools generally take the form of a software agent installed on the host connected to a central management system.

In the early days of intrusion detection and incident response, there tended to be a huge emphasis on network-based sensing. Network sensors have many virtues; one sensor provides situational awareness and tip-offs for potential incidents across thousands of systems. But insight is only as deep as what can be seen in network traffic. With the expansion and maturation of host-based monitoring, along with the proliferation of network traffic encryption, emphasis has shifted to host-based instrumentation and prevention. In general, if you are trying to positively confirm an attacker was successful in hacking an account, generally data retrieved from end point sources, such as EDR, will be more effective than network traffic sources such as NetFlow. In contrast, network traffic can be better than EDR at providing hints such as "where else do I have a problem" and "which end nodes to I look at first" in a situation where EDR coverage is incomplete.

Data from an endpoint is generally more informative than network traffic data for confirmation of intrusion.

A few different host level capabilities can provide data to the SOC:

- EDR
- Application allow listing and deny listing
- Executable integrity checking
- Host-based firewall
- AV/antispyware
- Data loss prevention
- User activity monitoring

Different host products emphasize different elements of the above capabilities, and it is increasingly rare to find one product that only performs one single function. The SOC will want to carefully review the functions of any host-based capability it deploys and look to reduce the number of agents deployed on end systems. When possible, it is ideal to deploy exactly one security agent that performs all local monitoring and detection functions. Not only does this reduce the amount of computing resources the agent uses, but tools from different vendors have also been known to cause conflicts, making coexistence of specialized tools a challenge.

Regardless of the tool selected, almost all of them can leverage a set of observables contained in system memory, CPU, disk, peripheral contents, or a combination of these. The differences among tools are in their targeted features and the techniques that leverage various observables. Therefore, this section begins by discussing these observables in some detail. These observables are data that can be collected by the SOC.

7.3.1 Host Observables and Perspective

Observables can be gathered either on an ongoing basis or on demand and synthesized in many ways. To start, the building blocks include:

- **From mounted file systems and any other storage:**
    - OS version, installed service pack(s), and patch level
    - Installed applications
    - Resident files, modification times, ownership, security permissions, contents, and summary data (size and cryptographic hash value) [220]
    - Author, date, header, hash, and other qualities of executables and libraries such as Portable Executable (PE) files, Dynamic Link Libraries (DLLs), ELF binaries, etc.
    - File system "slack space" containing deleted files and recycle bin/trash contents
    - Contents of the entire physical disk such as a bit-by-bit image
    - OS and application logs
    - OS and application configuration data (e.g., Windows registry hive contents)
    - Browser history, cache, cookies, and settings
- **From system memory and processor(s):**
    - Application process identification number (PID), creation time, executable path, execution syntax with arguments, name, user whose privileges it is running under, parent (spawning) process identified, and cryptographic hash (user context), CPU time used, and priority

- ◦ Actions and behavior taken by running processes and threads, such as execution behavior and system calls
  - ◦ RAM contents and memory map
  - ◦ Clipboard contents
  - ◦ Contents and disposition of CPU registers and cache
  - ◦ Logged-in users or applications acting with privileges of a remote user such as with a database or custom application.
- **From attached devices and system input/output (I/O):**
  - ◦ Network flow (sometimes known as "host flow") data, possibly including enrichments that tie process name to the ports and connections it has open
  - ◦ Content of network traffic
  - ◦ User keystrokes
  - ◦ Actions from other input devices such as mice, touch pads, or touch screens
  - ◦ Screenshots
  - ◦ Connected devices, potentially including details such as device type, driver info, serial number/ID, system resources, addressable storage or memory (if applicable), and insertion/remove events

To paint a complete picture of what is happening on the host, SOC analysts frequently examine all three described elements (on disk, in memory, and attached device I/O). For instance, focusing on just the local file system will blind an analyst to malware operating exclusively in memory. The host monitoring package must also implement its data collection in a manner that does not obligate a large portion of system resources, especially CPU and random-access memory (RAM).

Host-based monitoring defense perspective:
Depending on where in the system the host monitoring resides, the data available to collect varies widely. For example, most host monitoring tools reside on disk, and they run in memory like any other program. In this configuration, the tool must verify its code has not been compromised when it starts. By leaving a permanent presence on disk, it automatically runs each time on startup, but it is rendered easier for malware to recognize its presence and circumvent detection. Furthermore, malware that resides at the firmware, BIOS, or hardware has the advantage at defeating disk resident monitoring packages, at least in part, as is the case with "bootkits" [221].

Some host monitoring approaches leverage permanent storage or a root of trust at the hardware level. Rootkit detection, for instance, could be driven by specialized monitoring tools implemented as a peripheral component interconnect card in a system but comes at prohibitive cost [222]. Techniques leveraging a hardware-based root of trust, such as with trusted platform module (TPM) [223] and trusted boot [224] help ensure both the OS and other components match expected code.

When considering any sort of host-based monitoring package, the SOC engineer should choose tools that can defend themselves against the direct attack or circumventions. For instance, a tool that detects rootkits is useless if it runs with user-level privileges and therefore is easily subverted by rootkits running with system-level privileges [225]. In this case, the monitoring package would likely be fooled into seeing the system's content manipulated by what the rootkit wants it to see.

7.3.2 Endpoint Detection and Response

Purpose-built, commercial host monitoring systems have been around for roughly 20 years. Around 2013, a new market segment was coined, "Endpoint Detection Threat Response" [226], and later shortened to simply "Endpoint Detection and Response." While using a mix of signature- and host-based techniques to detect and block attacks had been around for a while, new themes emerged or were emphasized:

- Leveraging more perspectives in the operating system to detect presence of the adversary, particularly adversaries who leave few traces on persistent storage
- Allowing the user to interactively collect host state and other details on demand, and to interact with that rich telemetry in a manner that goes beyond alert triage
- Stronger coverage across the cyber-attack life cycle, combined with an increased integration and focus on high-fidelity cyber threat intelligence

Since then, techniques consistent with EDR capabilities have become an indispensable tool for the SOC. EDR capabilities can be achieved by buying a single commercial product, by building a solution from disparate tools, or a combination thereof. Essential elements of EDR include [227], [228]:

- **Strong detections:**
  - Uses large library of detections that spread across the cyber kill chain and ATT&CK framework [229].
  - Leverages a combination of detection techniques behind the scenes, including machine learning.
  - Looks at evidence of adversary activity both on disk and in memory.
  - Provides deep linking to cyber threat intelligence.
  - Enacts automated response capabilities on the box to quarantine or evict the adversary.
  - Uses observables seen from one system to inform detection and hunt capabilities across other nodes, particularly with cloud-based systems.
- **Rich telemetry:**
  - The system collects streaming telemetry generally considered essential to incident investigation and hunting: file touches & creations, DNS queries and responses, network connections tied to host process, logon/logoff, and process creation & termination, etc.
  - The system collects OS configuration and state, like configuration changes (such as registry on Windows; /etc. /boot on Linux), DNS cache, network adapter settings, logged on users, current running processes, items typically associated with adversary persistence mechanisms, installed programs and patches, known system vulnerabilities, and so forth.

- ○ The system will vary the information it streams to backend systems to manage network bandwidth.
  - ○ A user interface (UI) is optimized for interacting with this data at scale.
  - ○ Users can pivot between entities associated with alerts, including hosts, users, processes, files, and named adversaries through a rich UI.
- **Arbitrary/interactive command and response:**
  - ○ The system will gather additional information from the end host under different conditions: a fired detection will trigger the EDR to pull down addition running process or file details.
  - ○ The user can, on demand, initiate information retrieval or configuration sweeps, such as from files, processes, network connections, and memory.
  - ○ The user can open an interactive shell or command session on remote hosts, allowing them to perform actions like deleting a file, terminating a process, terminating a network connection, or isolating a host on the network.
- **Integration with managed services:**
  - ○ If contracted and configured, the user can push a button, getting almost instantaneous help from the EDR vendor; using screen and data sharing, the vendor can provide immediate expertise with incident analysis or response.
  - ○ If contracted and configured, the vendor will be hired on a routine and ongoing basis to perform threat hunting directly on the customer's data.

There are several competitors in the EDR marketplace. Before purchasing an EDR product, here are some areas of differentiated value between different product offerings:

- Features discussed above; in particular: kill chain and ATT&CK framework coverage
- True positive, false positive, and false negative rate for detections, and the lag between adversary activity and when detections fire
- Richness and pedigree of threat intelligence supporting detections, and integration with the product
- Telemetry quality and completeness, including how telemetry is throttled and shaped before download
- Coverage for targeted systems, including not only Windows desktop/laptop, but Windows Server, Linux, macOS, iOS, and Android
- System utilization footprint, and the optimization needed to get the agent working on production workloads (if any)
- If this is being used as a standalone capability (no SIEM or SOAR), whether built-in alert and case management is "good enough"
- Controls put in place on the agent to ensure it is not circumvented by the adversary

Alternatives to an EDR solution

It is possible to compose many of the same capabilities as an "all in one" EDR from disparate host telemetry scanning and capabilities. Sysmon, OS Query, and GRR provide hugely rich host SA. Even ordinary Windows Event Collection and Forwarding (WEC/WEF) and Linux auditd provide tremendous insight into host activity. The advantages to building a custom capability is typically increased flexibility and lower initial acquisition cost. However, they also require an increased time investment to develop and deploy, and usually do not have the benefit of technical support. SOCs looking to take this approach should also observe the following considerations:

- A best-of-breed EDR will ship with a library of thousands of curated detections; it is virtually impossible for a single SOC to achieve this same level of detection coverage and sophistication from scratch.
- The EDR graphical user interface (GUI) is optimized for working with EDR data; building this using other available tools may not yield the same user experience.

Ensuring EDR success: An analyst view

A good EDR will enable the analyst with all kinds of robust functionality. As with any other response capability "on the box," some key safety tips should be observed:

- Have a standard triage process using the EDR for common threats like phishing and non-targeted malware hits.
- Enact and refine a process for proper usage of automated response capabilities:
  - Leverage built-in role-based access control (RBAC) as appropriate.
  - Establish criteria and testing process for use of "heavy" response actions like running expensive Yet Another Recursive/Ridiculous Acronym (YARA) signatures [230] or pulling artifacts from many hosts, also taking care that there may not be time to thoroughly test a YARA signature before it is deployed in response to an emergent threat.
  - Draft communication templates in advance, for informing system owners when certain high-risk response actions are being taken, taking care to mitigate the risk of the adversary having access to the same communication path (e.g., email).
- As the SOC is able, consider measuring key metrics Mean Time to Detect (MTTD) and Mean Time to Recovery (MTTR) before and after EDR is deployed widely.
- Consider using the EDR as an essential element in purple teaming; also consider measuring its effectiveness with precision after a red team activity.
- For advanced users, be clear about how the EDR is selecting and throttling telemetry like process creation and file touch events.
  - Specifically, an EDR may download file "touch" and process creation events but that may be less than 1% of all process creations or file modifications that occurred on the host. These events are far too voluminous in many cases to download every instance of a given event.
  - The SOC should be highly cognizant of how these mechanisms work, and how they will influence detections, analytic results, and downstream data dependencies.

Ensuring EDR success: An engineer's view

Widespread deployment, operations, and maintenance of a best-of-breed EDR solution can be a time-consuming endeavor. Moreover, integrating with server workloads, desktop/laptops and mobile systems can be tricky. The SOC engineering team should consider the following tips when planning and executing EDR deployment and operations:

- **Be clear about both initial and out year cost model:** consider implications of cost model particularly around per seat and how painful vendor lock-in may be.

- **Consider the vendor's coverage and roadmap:** especially for non-Windows systems.
- **Weigh the advantages and disadvantages of on-prem versus cloud-based EDR:**
  - An EDR with a cloud-based backend may be much faster to deploy and leverage, but the SOC should evaluate this in context with its other tools, risk model, and security model.
  - A cloud-based EDR solution may be especially helpful for SOCs and constituencies with a remote worker population and/or a large existing cloud investment.
- **A large portion of deployment will be semi-technical; consider allocating non-engineering resources to support:**
  - Scoping and staying clear of systems may be too old, too specialized, or too small to tolerate a commodity EDR product (Internet of things, Industrial Control Systems (ICS)/SCADA, embedded systems, legacy mainframes, etc.)
  - Deployment process phasing, rollout, and updates
  - Service and mission owner communications
  - Measuring coverage
- **Test the EDR agent on pre-production environments, and in particular, critical workloads like:**
  - Point of sale terminals, embedded systems, human machine interfaces (HMIs) for ICS/SCADA systems
  - Domain controllers, SharePoint, and Exchange servers
  - Databases and high-transaction systems
  - Anything used for development, build and test automation
  - Endpoints used by executives and their assistants
- **Build and implement mechanisms for agent deployment that make it "zero touch":**
  - Consider building automation that can be linked at the highest node in the management hierarchy, such as the root in an AD GPO tree.
  - Ensure the vendor has automation that copes with different OS versions and environments and works correctly if old versions of the agent are already present.
  - Be clear about the EDR's code update model, it may pull updates directly from the cloud; this may be a concern with some high-security enclaves; Implement an offline agent update model as security and services demand it.
  - If other SOCs are present in the enterprise, build in logic for agent provisioning that automatically deconflicts deployments.
- **Have agreement with system owners in advance:**
  - Demonstrate the processes built around how analysts will exercise automated response and data collection.
  - Establish clear expectations and facts around issues of control and visibility that may scare or threaten system owners.
  - Show past "wins" with the product that show mission and business owners the value of the EDR.
  - If the EDR supports it and it makes sense for the SOC to do so, consider giving system owners a view into their "slice" of the data coming from their own systems.

- **Establish a strong understanding of EDR performance management and resource utilization:**
  - Ensure the vendor implements controls around utilization and provides that in writing.
  - Be careful around systems with extremely high disk I/O like databases, or endpoints with very low throughput to the backend EDR management systems or cloud, like mobile, remote workers, and 4G/5G- or satellite communications (SATCOM)-connected systems.
  - Test the EDR product with any other monitoring agent technology present, in particular: vulnerability scanning, anti-virus, system performance monitoring, deception, and automated security test validation systems; try to avoid agent sprawl.
- **Implement a health management process that balances completeness with resourcing constraints:**
  - Monitoring EDR agent health on a per-host basis by the SOC is challenging. Deployment and updating mechanisms sitting outside the agent framework should detect and correct most agent issues, such as due to broken upgrades, re-imaged systems, or corrupted files. EDR health and coverage/saturation metrics should be pushed down to service and system owners, like any other hygiene metric.
  - See "Strategy 10: Measure Performance to Improve Performance" for more on measuring coverage.
- **Be clear about different ways data can be moved out of the EDR:** Any good EDR will support near-real-time alert export, but it is best to gain clarity on how to move bulk telemetry out of the EDR, if desired.
- **Harmonize deployment and instrumentation strategies with any other log collection and SIEM:**
  - Some EDRs may be "good enough" to replace OS-layer syslog, auditd, WEC/WEF [231], etc. collection, but care should be paid to the depth and breadth of coverage provided, such as equivalency to Windows event IDs (4688, 4624s, etc.).
  - If bringing EDR data into a SIEM, evaluate whether the EDR bulk telemetry will "light up" other host monitoring use cases, analytics, detections, and content in the SIEM due to peculiarities and differences in schema.
- **Consider synchronization opportunities:** If the EDR provides a case management capability, it may be appropriate to synchronize alert triage between the EDR console and an external case management capability, such as in the SIEM, SOAR, or ticketing systems.
- **Recognize that the EDR tool itself can be a channel for adversary attack:**
  - Evaluate the agent against externally facing services or APIs, and for advanced SOCs, the resilience to direct attack or circumvention by rootkit or other RAT.
  - Protect the console and backend infrastructure; consider red teaming the tool.

7.3.3 Application Allowlist, Denylist, and Integrity Checking

Often built into operating systems, application deny listing is a technique whereby an OS module or protection agent blocks unwanted processes running on the end host [179]. Similarly, application allow listing policy uses a default deny approach. Sysadmins must define which programs and software publishers are authorized for execution; all others are blocked from running either by the OS or by the allow listing/deny listing client. For example, Gatekeeper in macOS and AppLocker in Windows [232], can be used to limit which users use what programs, and with which permissions. For Windows, a more extreme security control of application allow listing component is CodeIntegrity. On later versions of Windows (version 10 at the time of this writing), this component can enforce "S" mode, allowing only those programs from the Windows Store to run [233]. Similarly, Apple has released Gatekeeper, an allow listing component to keep out unnotarized (vetted) applications and malware [234].

SOCs wishing to pursue application allow listing or deny-listing technologies should consider the additional management overhead involved in tracking allowed or denied applications on the enterprise baseline. To implement allow listing, all monitored hosts should adhere to a known OS and application baseline, and the SOC must continually maintain consistency with that baseline (to ensure legitimate applications or services can run). This can be problematic with a complex enterprise baseline or decentralized IT administration. As a result, many constituencies struggle to achieve full saturation of application allow listing or deny listing. This is particularly true for workgroups that perform development or are regarded as free form "labs." *Generally*, application allow listing and deny listing are most successful for high-risk users that have a finite software baseline and/or stick to software from a known set of publishers or app stores without much divergence.

More traditionally, Tripwire is used on end hosts to detect changes to key configuration files and can alert on changes that may be an indicator of malware or a malicious user [235]. Changes that are detected in monitored files and settings can be detected and reversed by the administrator.

7.3.4 Host-Based Firewalls & Antivirus and Antispyware

Traditional security devices still play a role in overall security and SOC success. Although firewalls are most widely recognized as appliances that filter traffic crossing between two or more network boundaries, host-based network traffic filtering capabilities can be found in virtually all popular varieties of UNIX, Linux [236], and Windows, and are integrated in some EDR products [237], [238].

Antivirus was one of the earliest host-based defensive capabilities. It is a program that inspects file system and memory contents, leveraging a large signature pool and heuristics to find known malware or malware techniques. Antispyware capabilities are often included in most AV suites. They add to malware detection capabilities by examining Web browser specifics such as stored cookies, content, extensions, and stored cache.

A common criticism of AV tools is that their system resource utilization, RAM footprint, and regular disk scans outweigh their diminishing benefits. Historically, various sources have quoted percentages of anywhere from 15 to 40 percent [239], [240]. AV on non-Windows platforms such as Apple, Linux, and UNIX are regarded as unnecessary by some defenders, whereas on Windows, AV still provides some value, especially for malware detection resulting from Web surfing. However, AV indicated "cleaned" infections can be deceiving, sometimes leaving adversary tools and persistence on the system. Today, it is most common for a SOC to leverage traditional AV as part of a larger EDR suite.

7.3.5 Data Loss Prevention

For many constituencies, there is significant concern about the exfiltration of sensitive data from the enterprise. This can include anything from sending sensitive documents over personal email to downloading HR data to a thumb drive. One feature set of certain endpoint products, including purpose-built data loss prevention (DLP) solutions is to monitor, detect, or prevent loss of confidential or sensitive data [241]. This can range from healthcare records to financial data, to PII.

No matter how implemented, the host is often the only place where the SOC can expect to clearly see this activity (e.g., through network traffic, clipboard, file copy, print activity and system call observables). As a result, many of the enterprise host monitoring suites listed in this chapter include functionality that will scan and report on data transferred to local removable media as well as website and email postings (known as intellectual property or DLP). Some DLP packages can also be used to block or limit user access to removable media, enhancing functionality already present in Windows domain GPOs.

Alternatively, some adversary engagement and deception products [242] can leverage techniques like honeytokens, or bogus records, datasets, or other data of no value, are often set to entice intruders. When altered or exfiltrated, alerts are sent to the SOC.

7.3.6 User Activity Monitoring

In some enterprises, there is a significant concern over the actions of portions of the user populace. These constituencies must follow a policy of "trust but verify," whereby users are given latitude to perform their job functions, but their actions are heavily monitored. These may include any constituency that handles large amounts of sensitive or high-value data, such as defense, intelligence, or finance.

In such cases, security, counterintelligence, or intellectual property loss prevention may require full scope user activity monitoring, primarily from monitoring on the host. Typically, these capabilities involve comprehensive capture of user activity on desktops, where users' actions can be monitored in real time or replayed with screenshots and keystrokes. The efficacy, ethics, and legal issues surrounding use of such software are outside the scope of this book.

7.3.7 Host Sensor Placement

Although host-based sensoring is both scalable and frequently used, not all SOCs are resourced to put an agent on every constituency host. Table 17 provides some considerations and examples for where to prioritize host monitoring deployment.

Table 17. Host Monitoring Placement Considerations

Prioritized Placement	Example(s)
Host, service, application, or workload mission criticality	Key enterprise database servers, financial systems, manufacturing automation control, systems containing PII, systems under regulatory/compliance controls
Number and strength of trust relationships between that system and other hosts, especially hosts residing in other enclaves	• Web servers directly exposed to Internet • Web services systems forming a business-to-business (B2B) relationship with a partner company • Remote access VPN or webmail servers
Number of, and privileges wielded by, users on that system, especially users residing in other enclaves	Web-enabled financial application server; call center ticketing system
Vulnerability and attack surface exposed by system(s) of concern	Any server that cannot be regularly patched for whatever reason (legacy, operational demands, fragility, etc.)
Stability, maturity, applicability of protection mechanism(s) to that platform	Commodity, non-embedded IT such as Windows, Linux, and macOS systems

Not all monitoring tools are applicable to all hosts. In some cases, the most important systems in the enterprise may not be well suited for a typical host sensor suite, such as legacy mainframe systems and embedded OT. The SOC may depend on other tools like configuration checkers, robust logging, and native OS host firewalling.

7.4 Network Monitoring

Although there is a strong move toward host-based monitoring, network-based monitoring is still used by many SOCs. Network-based monitoring technologies can sometimes be the most cost-efficient and simplest means by which SOCs can gain visibility and attack detection coverage for a given enclave or network, especially in cases where they have no other visibility.

NIDS and NIPS history

For SOCs first stood up in the late 1990s, network intrusion detection systems (NIDS) often dominated any discussion of monitoring and detection. In 2003, Gartner, Inc., declared NIDS dead [243]. They made the point that NIDS' prodigious quantity of false positives renders them not worth the trouble. Although NIDS did not completely disappear, NIPS gradually came to replace them. NIPS are built on the same concepts as NIDS although NIPS also have the capacity to actively respond in real time.

Unfortunately, many SOCs never get their NIPS into an in-line, blocking mode. There are several potential causes for this: (1) the SOC does not have enough organizational authority and operational agility, (2) the SOC is not confident enough in its signature tuning, or (3) the SOC simply decides that in-line blocking mode is not worth the perceived risk of a self-inflicted DoS.

Attacks detected by NIDS, NIPS, such as exploits executed across the network (most notably remotely exploitable buffer overflows), no longer comprise the majority of initial attack vectors. Client-side attacks, such as phishing, have long become far more prevalent, giving way to the content detonation and analysis devices. As a result, signature-based methods by themselves (e.g., AV and traditional NIDS) are no longer sufficient for finding attacks and defending a network. Further compounding this, many cloud-based services consumed by many enterprises do not support the deployment of traditional NIDS/NIPS due to their network topology.

Additionally, the existence of single capability NIDS/NIPS capability is diminishing. Vendors such as Cisco and Palo Alto merged their firewall and NIDS/NIPS capabilities into single product suites years ago. It is increasingly difficult to find a firewall without an IDS/IPS feature set, and vice versa. The term "Network Detection and Response" or NDR is often used to refer to products with NIDS/NIPS functionality.

Today, the focus for network sensing is often around a) merged function network security devices, sometimes referred to as "next gen firewalls," b) NetFlow and traffic metadata collection, and c) malware detonation, though d) some dedicated network sensing/analytic platforms are also available [244], [245], [246]. All these techniques are discussed in this section.

7.4.1 Network Intrusion Prevention Systems

As mentioned above, NIPS can both detect malicious activity and respond to the activity in real time. This means there are a number of deployment considerations and options.

False positives
There is a high false-positive rate associated with IDS technology, NIPS administrators are justifiably cautious. Consider that each false alarm results in blocked traffic. If not careful, the NIPS administrator can inflict a very serious DoS. As a result, many SOCs will be careful about which NIPS signatures they turn to block, doing so only after several days or weeks of use in alert-only mode. This places a great deal of emphasis on choosing a NIPS with a robust protocol analysis and signature detection engine.

Signature details and context
Like any other detection technology, a good NIDS/NIPS will provide rich contextual data to the analyst or operator. Having access to raw signatures is highly desirable, as this, married with full session PCAP, gives the user clarity on why an alert fired, and how prone that detection is to false positives. Seasoned analysts tend to favor sensor platforms that offer open, published

signatures. By contrast, historically, many systems with closed "black box" signatures tend to be disregarded by some analysts as unactionable.

Response choices

Different network sensing technologies offer different means to respond to malicious traffic. One common method is to send a TCP reset to both hosts involved in a network connection, which, unfortunately, is not a good idea. Attackers, being malicious, probably expect this behavior and will take advantage of it in two ways:

- They can simply ignore the reset and continue the conversation.
- They now know there is probably an IPS in front of their victim.

Some platforms will implement blocks by automatically updating a firewall or router's access control list (ACL), blocking the network communication in question. This is problematic because they can make a mess of router and firewall configurations. The best protection capabilities support blocking by dropping the offending packet and every subsequent packet between the attacker and victim host(s).

Response actions

When an IPS blocks packets in an offending network connection and takes no other actions such as TCP resets, questions include how long should the attacker be kept from communicating or with whom? and just the victim or anyone on the network? Architectural aspects of the network, such as NATing, can make this more difficult. Imagine an IPS that sees attacks as originating from a Web proxy or NATing firewall. The attacker may appear to the IPS as the firewall, whereas, in reality, it is a host somewhere on the other side. But, because the IPS is blocking traffic, it is now dropping packets to a remote firewall, thereby inflicting a DoS against a much larger swath of traffic. Careful placement of the IPS to affect correct response action is critical.

Presence

A NIPS should not advertise presence to systems on either side of the network connection. This means not sending out traffic to the attacker and target, and not even having a MAC address. In other words, the device should simply appear as a "bump on the wire." That said, even the best IPS may disclose its presence simply by doing its job. Skilled attackers can detect an in-line sensor by using very old attack methods against a target network. Ordinarily, these attacks will fail because the targets are well patched. However, the sensors will do its job by blocking the attacker from any further communication to the targets. As a result, the attacker now knows a sensor is present and can change attack techniques. Therefore, the SOC may choose response actions that only block specific attacks rather than banning attacker IPs completely.

Latency and bandwidth

Being in the middle of network traffic, a NIPS may have an undesirable impact on network performance. An incorrectly implemented or undersized NIPS can introduce latency into network communications or inadvertently throttle bandwidth or traffic that passes by. The SOC must be careful which sensor products it chooses for a given network connection to avoid this problem, especially for high-bandwidth links.

Cost of decoding

A NIPS may operate at its advertised speeds only with a synthetic set of protocols or with certain decoder modules turned off. For instance, a NIPS advertised as 40 gigabit (Gb)-capable may operate at only 10 Gb with full application traffic decoding and inspection turned on because that is more computationally expensive [247]. Or a NIPS may advertise compatibility with an open signature format (such as Snort) but use of this capability will slow the NIPS's decoding engine. Moreover, a lot of sensors offer "packet shaping," "traffic shaping," or firewall-like capabilities. Use of these features may introduce latency into network traffic. The SOC may wish to assess whether this latency has a meaningful impact on network services.

Single point of failure

If the NIPS is in-line, what happens if it breaks or loses power? Good NIPS will have features (or accessories) that allow them to fail open (i.e., even if it malfunctions or loses power, traffic will continue to flow). Use of these features is highly dependent upon the use case and presence of other blocking capabilities, such as a separate firewall.

Involvement in network operations

When the SOC deploys any sort of in-line capability anywhere in the enterprise (NIPS, EDR, or otherwise), it becomes a primary player in IT operations. It is common for the SOC's equipment to be blamed for any problems (e.g., network outages or slow performance), even if there is no relationship between the SOC equipment and the actual problem. A SOC must be vigilant in watching its device status and data feeds to catch issues, and it must constantly work with IT operations to ensure its equipment is well behaved.

Price

In-line NIPS performance or availability problems are more impactful than those of a passive IDS sensor. NIPS vendors are, therefore, compelled to build robust, sometimes custom hardware platforms into their products. High-speed NIPS/next gen firewall devices are typically priced in the $10s of thousands range and above. Blanketing an enterprise with these devices can be expensive. Is a dedicated NIPS sensor necessary or will a single combined firewall/threat prevention appliance suffice?

Despite these challenges, some SOCs find NIPS to be useful tools. The gap between exploit release and patch deployment presents a period of serious risk to the enterprise; sometimes this is measured in hours, other times it is measured in weeks or months. A few well-placed sensors may provide protection during periods when systems would otherwise be vulnerable.

7.4.2 NetFlow

Whereas some sensors examine entire contents of network traffic, it can also be useful for the SOC to have a capability that succinctly summarizes all network traffic. One data source complementary to sensor alerts are NetFlow records (often referred to as flow records or flows). Rather than recording or analyzing the entire contents of a conversation, each flow record contains a high-level summary of each network connection. While the development of the NetFlow standard can be attributed to Cisco [248], it is now used in various networking hardware and software products and is an Internet Engineering Task Force standard [249].

A NetFlow generation device is like a telephone pen register. A pen register is a device that produces a listing of all phone calls that go in and out of a specific phone line, including the time, duration, caller's phone number, and recipient's phone number [250]. However, a pen register is just a summary of what calls were made and does not include the contents of the call, or what the people discussed. NetFlow is to PCAP collection as a pen register is to a telephone transcript. NetFlow does not include what was said, it simply indicates a conversation took place.

NetFlow Characteristics

While different NetFlow generation and manipulation tools are available, each flow record generally provides the following information:

- Start time and end time (or duration since start time)
- Source and destination IP
- Source and destination port
- Layer 4 protocol—TCP, UDP, or Internet Control Message Protocol (ICMP)
- Bytes sent and bytes received
- TCP flags set (if it is a TCP stream)

Whereas the contents of a network connection could be gigabits in size, a single flow record is less than a few kilobytes. The power of NetFlow, therefore, is found in its simplicity. NetFlow record collection and analysis is regarded as an efficient way to understand what is going on across networks of all shapes and sizes. It is critical to understand that NetFlow records do not generally contain the content of network traffic above OSI layer 4. This is a blessing, because (1) little processing power is necessary to generate them, (2) records occupy little space when stored or transmitted, (3) they are agnostic to most forms of encryption such as SSL/TLS, and (4) just a few flow records can summarize terabits worth of network traffic. On the other hand, this is a curse, because the flows capture nothing about the payload of that traffic. Whereas an IDS consumes significant processing power to alert on only suspect traffic, NetFlow generation tools consume little processing power to summarize all traffic.

Interesting clues can be generated from NetFlow alone. For example, they might indicate answers to, "Why do I see email traffic coming from a Web server?" or "One workstation was seen transferring vastly more data out of the enterprise than any other workstation." By combining flow records, knowledge about the constituency, and NetFlow analysis tools, an experienced SOC analyst can find a variety of potential intrusions without any other source of data.

NetFlow Devices

Flow records can be generated by different devices, including:

- Routers and switches
- Some sensor products, in addition to normal stream of intrusion detection alerts
- Some EDR tools record not only flows seen by the local host, but also tie the flow to the OS process transmitting or receiving on the port in question (sometimes known as hostflow), enriching the contextual quality of the data at the potential expense of extremely high volume if widely deployed

- Software packages purpose-built for flow generation, collection, and analysis, such as SiLK [251] and Argus. As discussed in the next section, Zeek can also generate functionally similar telemetry (and much more)

SOCs that leverage purpose-built tools for their flow generation and collection needs do so because they can operate and control them directly, vice routers or switches. More important, however, the SOC can place flow collection devices where it needs them. This gives the SOC an advantage when analyzing how an advanced adversary moves laterally inside the network, something a border device would not see.

Like other sensor architectures, NetFlow tools are split into two parts:

- One or more flow generation devices that monitor network traffic and produce NetFlow records in real time
- A central component that gathers and stores flow records and provides analyst tools via either command line or a Web interface

Some NetFlow tool suites accept flow records generated by third-party systems such as routers or switches just like a native producer of flow records. Argus, for instance, collects flows in Cisco NetFlow standards.

Many SIEM tools and log management systems are more than adequate at consuming and querying flow data. SIEM tools are useful in analysis because they can process and alert on NetFlow records in real time, whereas, in traditional flow analysis, tools like Argus and SiLK *by themselves* are usually not meant for persisting and querying billions of Flow records, nor NRT streaming analytics.

NetFlow Limitations

NetFlow is not without its limitations. Among these are:

- A NetFlow sensor cannot generate records on traffic it does not see. Therefore, as with other network-based monitoring, careful placement of NetFlow sensors is important.
- Classic NetFlow records do not record content of network traffic above layer 4 of the TCP/IP OSI stack. A flow over port 80, does not necessarily mean its contents are legitimate Web traffic.
- Under a heavy load, a NetFlow sensor is likely to resort to sampling a portion of the network traffic that passes. If this occurs, records generated will contain incorrect packet and byte counts, thereby skewing derived statistics and potentially fouling downstream use cases.
- Like other network-monitoring capabilities, NetFlow analysis can be partially blinded by frequent use of NAT and proxy technologies throughout the network. For this reason, it may be prudent to collect flows from both sides of a proxy and/or the proxy logs themselves.
- NetFlow is sometimes used to perform analysis on encrypted connections because going deeper into the network stack is not useful. That said, the nature of the protocol must be carefully analyzed, because some combinations of tunneling and encryption can render NetFlow analysis marginally useful (e.g., the case with some uses of Virtual Routing and Forwarding/Generic Routing Encapsulation [VRF/GRE] and VPNs).

- Because they generate a record for each network traffic session, NetFlow records can dwarf most other data feeds collected by a SOC, especially if flows are collected from the end host.

When SOC analysts look at a sensor alert, they see only something potentially bad about one packet in one network session. The flow records for the source and destination hosts involved in that alert bring context to the analysts. What other hosts did the attacker interact with? Once the alert fired, did the victim start making similar connections with yet other hosts, indicating a spreading infection? These questions can be answered with flow records and the tools necessary to query them. As will be discussed later, full-session capture can also support these use cases, but the advantage of NetFlow is the amount of data needed to draw these conclusions is often vastly less, affording the analysts greater economy of data and speed.

7.4.3 Traffic Metadata

With all the data sources, and particularly combining host and network data, parsing can be unreliable and cumbersome when trying to manage full content analysis. Fortunately, network traffic metadata tools take NetFlow one step deeper into the TCP/IP stack, providing analysts with rich network-based situational awareness. Metadata is roughly as voluminous as NetFlow, in terms of the number of records generated on a busy network link, but it can serve as an enhanced source of potential intrusion tip-offs.

For example, given a known set of websites that are hosting malware, a bad-URL list can be matched against incoming traffic metadata to look for potential infections. Sensors can collect metadata on DNS requests and replies, which can be used to look for malware beaconing and covert command and control of persistent malware. As another example, these sensors can look for known bad SSL/TLS cipher suites being used on the network, such as those associated with known malware.

Collecting metadata in the right places on the network allows the SOC to be more selective in what is collected. Thereby, it presents less of a performance burden on both the SOC's collection systems and network services such as DNS servers or mail gateways. More bluntly, many DNS servers will crash with full DNS logging turned on, whereas a traffic metadata sensor is designed precisely to record every DNS request and reply seen on the wire at very high speeds.

Some tools such as Yet Another Flowmeter (YAF) and SiLK [251] and Zeek (formerly "Bro" IDS) [252] provide robust metadata generation and analysis capabilities. There are some commercial vendors, such as Corelight [246], who sell a pre-packaged Zeek capability, freeing the SOC of the laborious task of build high-scale network monitoring sensors. Regardless of implementation details, Zeek has a vibrant community and ecosystem of plugins and analytics; more can be found at [253].

Traffic metadata can be even more voluminous than NetFlow—making it one of the top producers of security data for the SOC. That said, because network traffic metadata contains the most critical data the SOC would otherwise get from full session traffic capture, and in a structured format, it can be a comparative bargain compared to widespread, routine "full PCAP" collection discussed next.

7.4.4 Full Session Capture PCAP and Analysis

When analyzing a serious incident such as one that requires active response or legal action, the SOC requires concrete proof of what happened. This confirmation comes from host data. Having a complete record of network traffic can also be helpful, especially when host telemetry is not available or untrustworthy. What is the content of this suspicious beaconing traffic? What was that user printing out at 2 a.m.? What was the full payload of this exploit, and what did this infected host download after it was infected? Full session capture can answer these questions, but the SOC must be careful to scale traffic collection and analysis platforms effectively and efficiently.

Traffic capture is typically done on major perimeter connections, and in an ad hoc manner near systems that are suspected of compromise, such as with adversarial engagements and other incidents. While the SOC can filter out traffic that has no value of being recorded past the header (such as SSL/TLS sessions), or in extremely long-running flows (e.g., "elephant flows"), the SOC can still face scalability challenges in all but the smallest deployments.

The majority of full-session capture and analysis support input and output in binary libpcap format, ranging from open-source tcpdump [254] command line utilities to some commercial-grade appliances. There are many excellent sources [255], for information on tcpdump and associated protocol analysis tools such as WireShark [256], which provide details. At a high level, WireShark is a GUI-based utility that can record and display PCAP data in graphic format to the analyst. This allows the user to view how each OSI protocol layer is broken down for each packet. There is also a text-based version, tshark, that records and displays PCAP data, but without the fancy interface. Most SOCs leverage WireShark as a core-analyst tool because it is easy to use, decodes PCAP, sniffs low-bandwidth connections well, and is free.

The biggest challenge with full-session capture is volume. Consider an office building that connects to the Internet through an ordinary 10gigabit/s Ethernet connection on its way to a VPN and ISP. At an average of 50 percent utilization, the SOC would collect this volume in a 24-hour period:

10 Gbit/s * 60 sec * 60 min * 24 hours * .5 utilization/8 bits per byte = 54 TB

A single NAS can easily store this, and 10-Gb network adapters are easily acquired. Depending on other data sources, such as EDR, some analysts may need at least 30 days of online PCAP for analysis and response purposes. In this example, that is 1.62PB (30 * 54 TB). This is significant for a moderately sized connection and modest retention.

Specific traffic collection and analysis tools include RSA NetWitness Network [257] and Arkim (formerly Moloch) [258]. Commercial tools are almost always PCAP–interoperable but, in some cases, will record data in their own proprietary format. Whereas open-source tools are free, some traffic capture vendors license their products on a per-TB basis, making retention of large amounts of captured data expensive. The biggest advantage of these types of tools is that they both record data from high-bandwidth connections and present metadata about captured traffic to the analysts, allowing them to pivot and drill down very quickly from many days or weeks of traffic to what they need. One way to save money on the entire solution is to leverage FOSS tools to collect large amounts of PCAP cheaply and then run them through a commercial tool on an as-needed basis. This provides much the same usability but without the high price of a COTS tool.

7.4.5 Malware Detonation and Analysis

The rise of phishing and waterhole attacks in the early 2000s gave way to a new set of network protection technologies dubbed "detonation chambers." Targeted primarily and email attachments and visited websites, the purpose of these devices is to open suspect files in an artificial virtualized environment, enabling a) confirmation of whether the file or website was in fact malicious based on resulting behavior and b) provide detailed analysis on the nature of the malware, if it was in fact malicious. Popular example vendors include FireEye [259], Broadcom [260], and Palo Alto [261], and sometimes may be integrated as a feature of next gen/all-in-one firewalls.

Malicious files will usually behave in suspicious ways like making privileged system calls, beaconing out for command and control, or downloading additional malicious packages. Malware detonation systems look for this sort of activity, but generally without sole reliance on signatures that define a specific attack or vulnerability. As a result, they are better tuned to ongoing detection of zero-days and specially crafted malware. While the attack vectors may change, the outcome does not, and that is the focus of detonation. Even if malware is obfuscated or packed inside a binary file, the detonation chamber can typically still detect that it is acting in an anomalous or malicious way—something traditional IDS is not usually capable of.

Generally, on-prem content detonation systems come in two varieties:

- **Offline:** The first type accepts file uploads by users in an "offline" manner. This is particularly useful for incident responders and malware analysts who need an on-demand capability that can provide quick details on whether a file is likely to be malicious or not. This capability can automate several hours of manually intensive malware reverse engineering. Best-of-breed products will provide specifics on system calls, network connections made, and files dropped. Historically these were only on-prem capabilities but of course can be bought and consumed as cloud-based services. Some malware will try and detect whether it is running inside a virtual analysis environment and then change its behavior—a good content detonation system should also detect and mitigate this, at least to some extent.
- **Real time:** The second type is a device that can scan network traffic (usually Web or email) in real time, pull out malicious files, and detonate them fast enough to block the malicious content from reaching the victim user or system. This kind of device is targeted toward use cases where a SOC wants to detect or block client-side attacks

in real time, thereby catching a lot of the more easily detectable malware traversing the enterprise perimeter. Some systems can also take the "bad" files they have seen and automatically produce IDS signatures that can be leveraged to see whether the same file appeared elsewhere on the network. Most best-of-breed, cloud-based email services have this kind of capability as part of the product offering.

Whether it is on-prem, in the cloud, a dedicated detonation appliance, or an all-in-one network security appliance, automated malware analysis today can be thought of as a spectrum. On the one end of the spectrum is ordinary signature matching as with traditional IDS. From there, things get more sophisticated: "carving" files out of traffic, matching them against hashes, scanning the files themselves, and running ordinary anti-virus on them. Moving on, the sophistication of content detonation grows, starting with very simple lightweight environments that will only catch some malware, and terminating in sophisticated multi-OS "rigs." Finally, illustrative purposes, by-hand static code analysis and reverse engineering is considered of course the most labor-intensive means of dissecting malware originally caught on the wire. See Figure 22.

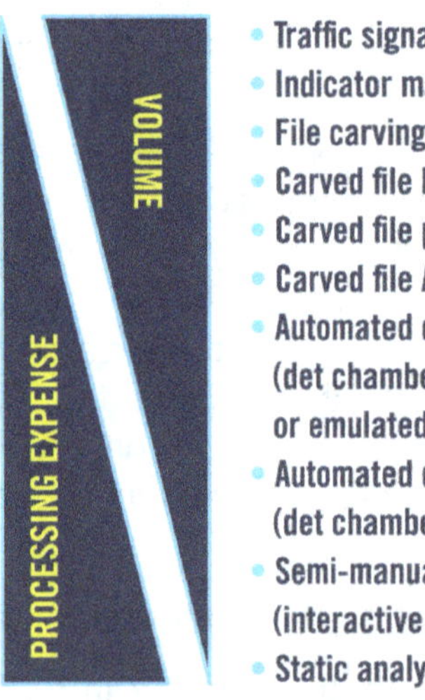

Figure 22. Malware Detonation Cost Versus Volume

Each of these techniques have their place, given the volume and complexity tradeoff. First, when considering use and tuning of any malware detonation capability, SOC operators should consider optimizing use of these different techniques, given finite compute, storage, and network capacity. When considering commercial products, the SOC is well-advised to gain clarity from prospective vendors about which of these techniques they are employing. Second, many modern network protection appliances will perform some of this in the cloud; whereas in the past malware would be detonated in the on-prem appliance, now some on-prem appliances will simply "ship" new and novel files to the cloud for analysis. The SOC's own situation and risk appetite may inform whether this is desirable. Finally, as with the convergence of firewall and NIPS appliances, some will feature malware detonation features. Again, in these cases, it is worth gaining clarity from the vendor about how deep the "detonation" features go, and where that detonation occurs-on device or in the cloud.

In addition to the above, when working with malware detonation capabilities, either on-prem or in the cloud, the following advice should be heeded:

- **Some malware requires user interaction before it will execute:** The detonation chamber's ability to interact with malware is often limited, thereby resulting in a false negative. The SOC should give care to whether the det chamber is "high" or "low" interaction and what automation it provides in this department.
- **Some exploits such as heap sprays require very specific arrangements of an OS's components in memory:** Because on-prem malware detonation systems try to fit as many VMs as possible in a modest platform, there may not be sufficient memory to fully simulate what would run on a real end host.
- **A lot of malware is less than perfect and may not successfully exploit a host every time it is executed:** A content detonation device most likely only has few opportunities to detonate a file that is potentially malicious. As a result, the malware may not be caught in the single time it is run, resulting in a false negative.
- **The content detonation device will time out after a certain number of CPU cycles or seconds:** Some malware authors will specifically write their malware to delay execution to avoid detection by AV engines or content detonation devices.
- **The content detonation device will open files within some sort of VM or sandbox that is meant to match common corporate desktop configurations:** However, these configurations may diverge significantly from what is being used in the enterprise. The operators of the content detonation device should make sure their VMs/sandboxes match the OS, browser, browser plug-in, and application revision, service pack, and patch level as used on their corporate desktop baseline. Failure to do so may result in either false positives or false negatives.
- **If used in in-line blocking mode, the content detonation device or cloud service may serve as an additional point of failure in email or Web content delivery:** If redundant mail or Web gateway devices exist, the SOC may consider also dedicating a content detonation device to each gateway device, thereby preserving the same level of redundancy but making deployment more expensive. For most SOCs, this means that content detonation device should fail "open," allowing traffic to pass if it fails.
- **SA provided by a content detonation device will make it very clear how much malware hits an enterprise takes in a given day or week:** Without additional context of whether these hits were mitigated or blocked, great alarm could result, even though concern is not warranted. The SOC should be certain to provide assistance in interpreting these results.

7.4.6 Wireless Intrusion Prevention

The use of wireless networks, and therefore the use of laptops, smart phones, and other portable devices, as well as non-traditional IOT devices in many different physical locations is increasingly the norm in many constituencies. With this convenience, wireless network access has become important to many constituencies, and so has monitoring points of access for unauthorized access, radio frequency (RF) attacks, and suspicious network activities.

Because wireless local area networks (WLANs) are an RF-based signal medium, they can unintentionally expand beyond the physical boundaries of a building. This often-unknown network presence outside the physical perimeter can attract drive-by attacks and eavesdropping, therefore considerations for the SOC. WIPS can assist with distinguishing between neighboring WLANs and unauthorized WLANs as well.

WIPS assists the SOC in focusing on detection of the following:

- Rogue access point detection
- WLAN attacks, including Man-in-the-middle attacks
- Rogue/unauthorized endpoint access and hotspots
- Unauthorized network connections
- DoS detection/prevention

In addition to these, some WIPS, such as Cisco Adaptive can also alert on authentication or encryption hacking attempts [262], [263].

7.4.7 Open source Versus Commercial Network Sensors and Sensor Virtualization

Enterprise-grade purpose-built network sensors and all-in-one firewall/network appliance products can retail anywhere from $2,000 for a modest hardware appliance to over $100,000 for an appliance capable of multiple 100 GigE connections. However, they offer stability, scalability, enhanced user interfaces, and other benefits that come from a managed device. In contrast, free and open-source software for network detection and prevention can give the SOC more flexibility and insight into exactly how their device is performing, but they require a much higher level of technical knowledge and additional management time. Commodity network protection appliances tend to bundle many technologies in one package. This level of full integration for FOSS (sensing, detection, firewall, file carving, file detonation) can be done, but tends to require significant up-front engineering cost.

Table 18 summarizes some typical differences between best-of-breed FOSS and COTS network sensing and prevention platforms.

Table 18. COTS Versus FOSS for Network Monitoring

Characteristic	COTS	FOSS
Code base	Closed	Open
Signatures	Closed. This can be very frustrating when an analyst wants to understand why an event fired; the number and robustness of custom signatures are sometimes very limited.	Open. Analysts understand exactly what is being detected and why alerts fire. Detection authors can harness the full capabilities of the system; there is no inaccessible "secret sauce." The system can run thousands of custom signatures as necessary.

Characteristic	COTS	FOSS
Protocol detection engine robustness	Usually very good. A major value-added as it is a vendor's "secret sauce."	Also, can be very good, but is up to the community to keep pace with evolving threats and protocols.
Predisposition for false positives	Varies widely depending on the individual signature and detection engine.	Varies widely depending on the individual signature and detection engine.
Availability of signatures for critical new vulnerabilities	Usually within hours	Usually within hours
Bandwidth	Capable of handling most any network connection, including those 40gigE and beyond, for a price.	Most solutions can handle 10gigE monitoring with little issue on modest hardware; scaling to 40 GigE and beyond typically requires attention to hardware and software optimization [264].
Management complexity	Almost always point-and-click, which makes training new staff straightforward, but some systems can be deceivingly hard to manage due to several layers of complexity; some COTS solutions become difficult to manage with fleets of hundreds of sensors.	For the novice, this can be a daunting task; experienced Linux/UNIX sysadmins can usually automate management of large fleets of sensors with minimal sustained labor by leveraging tools native to the Linux/UNIX environment.
Overall suitability for in-line prevention use	Depends on the product implementation; usually very good to excellent.	Manual configuration of commodity hardware and OSs can make this problematic if the system is not built from the ground up to be fault tolerant.
Combined cost of software and hard-ware to implement full solution	Moderate to very expensive, depending on bandwidth requirements; expect yearly maintenance costs to be around 20–25 percent of initial acquisition.	Assuming the availability of a few talented Linux/UNIX administrators, this can be comparatively inexpensive, especially with deployments of 50+ sensors; the only outyear cost is hardware maintenance and staffing.
Integration with other capabilities: firewall, proxy, file carving, file scanning, malware detonation	Most best-of-breed products offer all-in-one products that bundle many capabilities together, greatly enhancing value and ease of deployment.	Building and integrating these capabilities with FOSS can be done but tends to become ornate and requires highly skilled Linux experts when done at high throughput and high scale.

Both FOSS and COTS sensing platforms offer compelling features and drawbacks. Mature SOCs should consider their resourcing and tradeoffs when weight COTS versus FOSS for network sensing.

7.4.8 Building an Open-Source Network Monitoring Capability

The SOC has different options for which platform may serve as a basis for network monitoring capabilities. Each option offers different performance, scalability, and economic advantages. Here is what you can get for $20,000 and one unit of rack space:

- Two CPUs with many hyperthreaded cores
- Several hundred GB of RAM
- Ten high-capacity spinning hard drives or solid state drives (SSDs) that can be managed either through a hardware redundant array of independent disks (RAID) controller or a RAID-aware file system such as Z File System (ZFS) [265] or software RAID such as Linux mdadm [266]
- Many embedded Ethernet ports supporting 10 GigE speeds and beyond, and PCIe expansion for more ports and faster speeds (40gig, etc.)

That is a large amount of computing power in a small amount of space. Blade systems offer even higher density computing; however, storage and networking become more complicated, especially when considering networking monitoring applications. While the above specs obviously will become less impressive in the years following this book's writing, the point is that dedicating one server to one application is a thing of the past.

If the SOC needs collocated NIDS monitoring, NetFlow, or (and sometimes full PCAP) collection, these tasks may be accomplished at scale with FOSS tools such like Suricata [267], Zeek [253], and tcpdump [254], with scripting to glue them together. Some SOCs have bolted on additional functionality like file carving and file YARA scanning as well. However, as mentioned above, this can become complex.

From a hardware perspective, integrating all this equipment into one rack space is relatively straightforward. From a software perspective, integration is a bit more challenging. There are two potential approaches.

One approach is to run all monitoring software on top of the same OS, directly installed on the system as in "bare metal." In the case of multiple monitoring taps, each detection package will usually be running its own instance with a unique identifier, separate set of configuration files, and data output destination—be it physical disk, RAM disk, or syslog.

Modern virtualization affords the SOC the ability to take another approach—each sensor becomes virtualized. If the SOC wishes to use multiple incompatible monitoring programs (e.g., both COTS and FOSS), they can be separated into different VMs. These can run on top of typical virtualization technology such as VMware ESXi Server [268] or Xen [269]. While adding some complexity, this approach adds a great deal of flexibility not found in a bare-metal OS install.

Figure 23 illustrates a high-level architecture of such a virtualized arrangement. This example folds six disparate sensors onto one hardware platform—even greater consolidation is possible in practice. It leverages commodity hardware, free or cheap virtualization technologies, and FOSS tools to collapse the monitoring architecture and maximize the hardware resourced at the SOC's disposal.

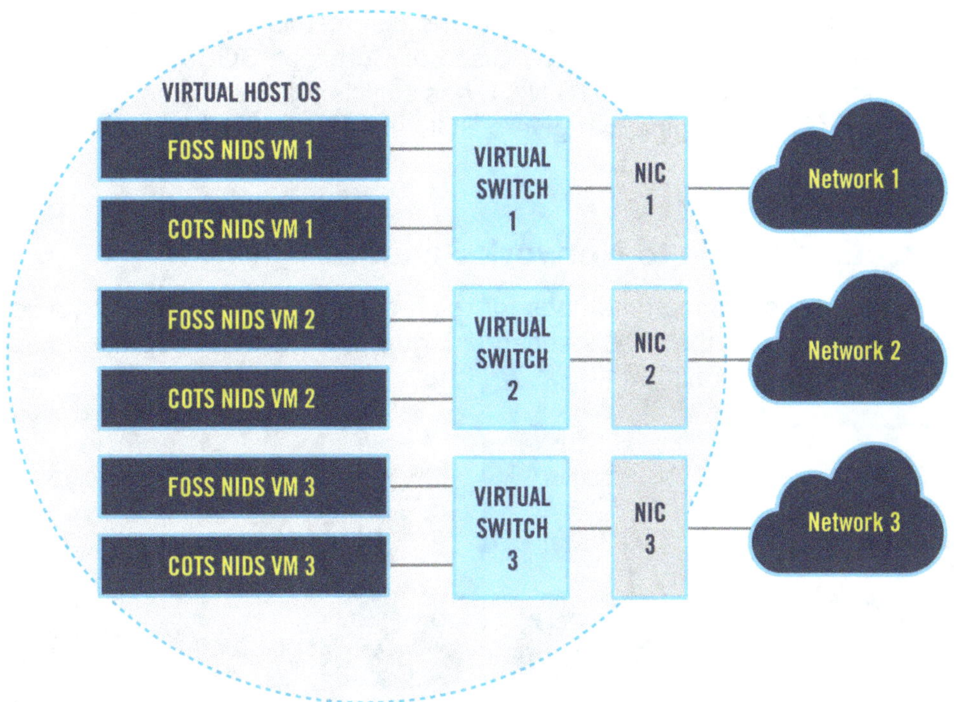

Figure 23. Virtualizing IDS

If the SOCs wants to pursue this approach, they should pay careful attention to optimizing their virtualization platform to their monitoring needs, being keenly attentive to any issues that may lead to packet loss. For example, using host virtualization of any sort may not work in very high bandwidth scenarios, due to its potential performance overhead.

Zeek, Snort, and Suricata are regarded as de facto standards when it comes to performing custom network sensing. Some commercial vendors have integrated these technologies directly into their platforms and/or provided interoperable signature and scripting formats. When considering use of Snort-, Zeek-, or Suricata-based custom signatures, the SOC should consider:

- What proportion of the sensor fleet will need to fly these custom signatures?
- How many custom signatures are needed—a few dozen or several hundred? Some sensors pay a higher performance penalty for custom signatures than native ones.
- How does the COTS network sensor support custom signature implementation? Is it actually running Suricata, Snort or Zeek, and, if so, how old is it? If it is emulated, how good is the emulation and are all functions supported?
- If the SOC already has a COTS network sensor with custom signature support, does that custom syntax support its needs? Is it willing to spend time translating signatures provided by other SOCs into this custom signature syntax?

Some SOCs, which favor widespread use of bleeding edge signatures, may choose to use native versions of the sensing software in question. Other SOCs, which do not have strong

UNIX/Linux expertise, may choose to go down the commercial path—either with the COTS version of the same capability such as by Cisco or Corelight. SOCs can avoid running an extra sensor fleet if they feel their COTS NIDS has sufficient signature support. Having more than one network sensing engine can give a SOC extra options when facing an elusive or targeted threat.

7.4.9 Directing Traffic to Network Sensors

This section addresses common methods for redirecting copies of Ethernet traffic from the constituency's networks to the SOC's sensors. Figure 24 illustrates popular approaches to making copies of network traffic.

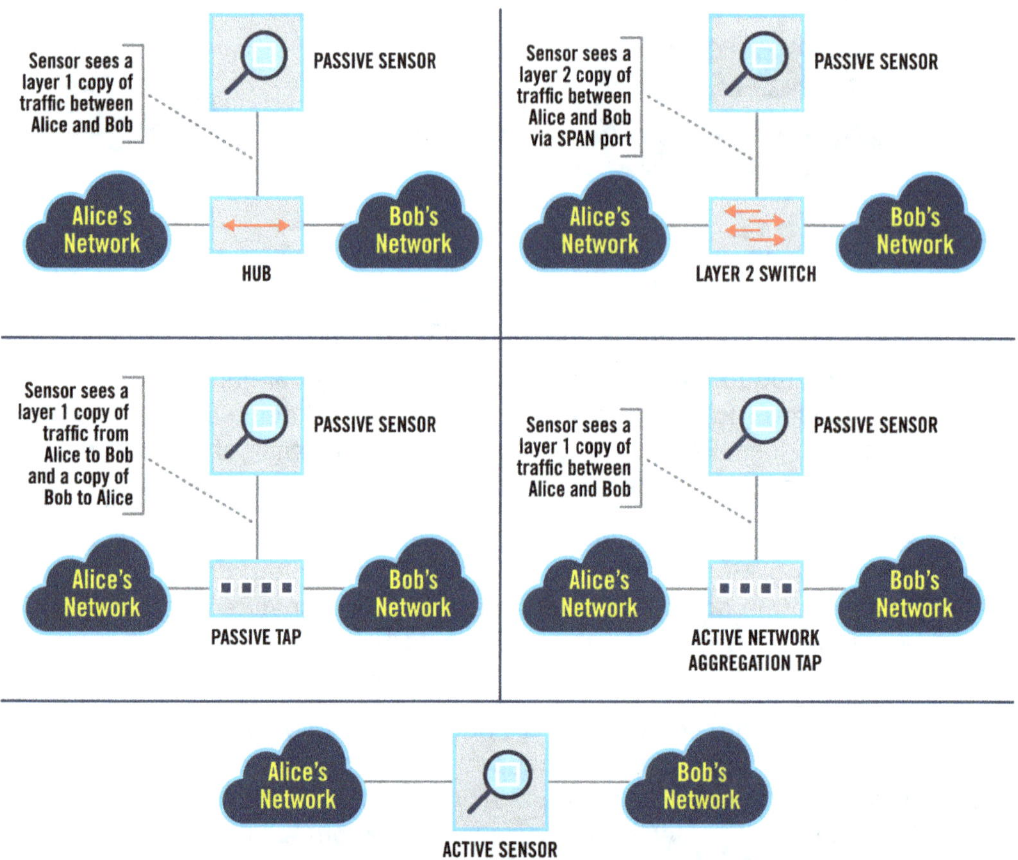

Figure 24. Copying Network Traffic

At the top left of the image a network hub is being used to copy network traffic to a passive network sensor. This is the most straightforward approach. By inserting a layer 1 Ethernet device between Alice's and Bob's networks, the sensor will see a copy of the traffic passed. However, in an age of widespread 10gig+ networking, this is almost never used. Packet collisions and packet latency can become a serious problem. By using a hub, packets will be dropped, and the sensor will miss traffic. In addition, most modern edge networks operate at

1 Gb/s or 10 Gb/s speed; hubs essentially do not exist in speeds faster than 100 Megabits/s. Finally, hubs are generally not very fault tolerant. Thus, network owners are unlikely to approve the placement of a flimsy device between two networks.

At the top right, the hub is replaced with a layer 2 or layer 3 network switch. This switch is configured with a switched port analyzer (SPAN) to copy or "span" traffic from one or more source ports or virtual LANs (VLANs) to the port hosting a network sensor. This approach offers most of the benefits of the hub approach, by using enterprise-grade network equipment that probably is already in operation across the enterprise. The major caveat to this approach is that the SPAN must be set properly, and that its configuration must continue to match the intended source ports and VLANs down the road. Also, utilizing span ports can impact switch performance. As network topology changes, respective SPAN configurations must be altered to match.

In the middle of the diagram, a network tap is used. A network tap is essentially a device inserted between two network nodes that makes a copy of all network traffic flowing between them. On the left, there is a passive network tap. In the case of copper, it simply makes an electrical copy of the traffic flowing. Or, in the case of optical taps, it uses a mirror or prism to split the transmit (TX) and receive (RX) light beams. With "dumb" or passive network taps, the SOC must use a sensor that has built-in logic that recombines the RX and TX lines into one logical stream of traffic suitable for decoding. On the right, an active network aggregation tap does this work but, at the same time, has the disadvantage of saturating its monitor port if both Alice's and Bob's aggregate bandwidth exceeds that of the monitor port. Notably, some active network taps that pull data from multiple locations (e.g., aggregation taps) have traffic deduplication features that eliminate redundant copies of the same traffic. This feature can be a big cost saver if used in the right places. Popular manufacturers of network taps include Network Critical [270], Netscout [271], and Gigamon [272]. Network taps are not generally subject to the same range of misconfigurations that switch SPANs are. However, even network sensors certainly get disconnected from time to time.

Finally, a sensor is placed directly in-line between Alice's and Bob's networks (e.g., a NIPS). Most modern network security appliances offer a monitoring mode where the sensor can sit in-line and passively listen to traffic without any intentional interference. Less robust sensors may "fail closed" (e.g., if an error occurs or the device loses power) and disconnect Alice and Bob. However, most best-of-breed products take great care to avoid this problem.

In every case discussed, the network sensor must be physically near the network devices it monitors. This usually means that the network sensor is in the same rack, server room, or building as the monitored network segment. While it is certainly possible to send a copy of network traffic to a distant sensor using a remote SPAN, long-range optical connection, or even a WAN link, doing so can become prohibitively expensive. Generally, the longer the connection, the less sense this makes. Physically placing sensors within close proximity to their monitored network segment is almost always the cheapest option; as a result, effective remote management is essential.

All these traffic redirection options have implications for how to prevent the network sensor from compromise or discovery. First, the monitoring port or ports should not have an IP address assigned to them. This will minimize the likelihood that it will talk back out on the network or bind services to the port.

The pros and cons of approaches to traffic redirection are summarized in Table 19.

Table 19. Traffic Redirection Options

What	Pros	Cons
Hub	• Inexpensive • Easy to install • Can attach as many monitoring devices as there are free ports	• Sensors will miss packets due to collisions. • Almost never an option: modern networking is usually 1gigE and up, whereas hubs only work on 100mbit and below.
SPAN	• Free to use if monitoring points already have managed switches in place, which is very likely. • RX and TX are combined; one network cable off a SPAN port can plug right into a sensor. • Straightforward for monitoring traffic from any device hanging off a switch (such as a firewall, WAN link, or cluster of servers). • Can attach as many monitoring devices as there are free ports (and switch SPAN capacity). • Can be used to monitor network core, such as spanning multiple ports off a core switch or router.	• An adversary with access to the enterprise network management platform can disable monitoring feeds to the sensor. • Some older or cheaper switches support only one or two SPAN ports per switch, limiting options. • When spanning traffic from multiple source ports, the destination SPAN port may become oversaturated if the source ports' traffic aggregate bandwidth exceeds the SPAN port's speed. • Changes to VLAN or port configurations after initial SPAN configuration can partially or completely blind the network sensor without the SOC necessarily realizing it.
Tap (include both passive and active)	• Invisible from a logical perspective. Only operates at the physical layer, meaning the adversary does not have an obvious target to exploit or circumvent. • Should not alter packets in any way. • Active network regen taps support multiple monitoring devices. • Active aggregation taps with packet deduplication can reduce the total network sensor count and total sensor capacity requirement.	• An additional device (albeit usually well-built and simple) that can fail is introduced into critical network links. • Only appropriate when observing conversation between two networked devices (as opposed to many with a network switch SPAN), as is often the case in perimeter network monitoring. • Every monitoring point requires the purchase of a tap device. • With a passive tap, RX and TX lines need to be recombined; some sensor technologies do have the internal logic to do so. • Passive network taps only support one monitoring device.

What	Pros	Cons
In-line	• Sensor can actively block traffic, depending on rule set.	• If sensor goes down, it may cut off communication unless resiliency features are built in (e.g., "fail open"). • Some sensor technologies introduce packet latency or packet reordering, which in turn can sometimes degrade network quality of service or make the sensor detectable. • More than one monitoring device means serial attachment of devices in-line, each being a separate point of failure.

7.4.10 Network Sensor Placement

There are variety of network monitoring technologies that can be placed throughout the environment, including:

- Dedicated network sensors
- Combined security devices that feature IDS/IPS features such as next gen firewalls and all-in-one security perimeter protection
- NetFlow and/or traffic metadata record generation
- Sustained and ad hoc full PCAP collection

Even if a SOC completely eschews classic signature-based NIDS, it should consider other strategies like network traffic metadata collection. Regardless of the technology chosen, Table 20 features some tips for sensor placement.

Table 20. Network Sensor Placement Considerations

Goals	Placement Example(s)
Gain visibility into systems important to constituency mission.	• Servers hosting custom mission applications and sensitive data placed behind sensor
Provide coverage for systems that are of especially high value to adversaries.	• Systems behind sensor contain trade secrets, source code, or confidential records • An Internet-facing email gateway serving a large user population
Achieve greatest "bang for the buck" by picking locations that host a large number of network connections (e.g., network "choke points").	• All network traffic between two major corporate regions transit sensor, covering 10,000 systems
Protect systems that sit on the trusted side of a controlled interface (e.g., a firewall).	• Sensor is between university dorm networks and the university's registrar's office. • Company A's servers communicate with Company B's servers across a private link

Goals	Placement Example(s)
Have complete insight into the traffic being observed (e.g., it is not encrypted and uses protocols the sensor understands).	• On the unencrypted side of a VPN termination point or SSL accelerator • On both sides of a NATing firewall or Web proxy
Leverage passive monitoring as a compensating control for systems that lack critical security features or have serious unmitigated vulnerabilities.	• Legacy or proprietary ICS/SCADA or mainframe enclaves or network segments

An enterprise's Internet-facing gateways are typically a first choice for sensor placement. This sensor placement at these locations meets most of the goals discussed: (1) mission-critical systems usually connect through it, (2) a large proportion of the entire enterprise's traffic passes through, (3) systems on the other side of gateways are untrusted, and (4) it is expected that many enterprise systems will expose various vulnerable services through it. Many organizations that do not even have a SOC will often choose to place an all-in-one network protection appliance at their Internet gateway, with monitoring duties falling upon their general IT department or NOC.

Beyond major enterprise perimeter points, the choice of network sensor deployment becomes more complicated. The SOC must analyze where they will get the most value from a sensor deployment. One way to do this is to work with network owners to acquire throughput statistics for proposed sensor point of presence prior to hardware acquisition and deployment. This allows SOC engineers to plan for hardware resources needed, accounting for tuning that will filtered out some protocols from the monitored stream. Consider the following criteria for sensor deployment:

- **Maintainability and access:**
  - Is the monitoring location physically located where sensor(s) have good connectivity and bandwidth back to core SOC systems?
  - Can SOC sysadmins physically access the equipment, or can a representative at the site perform touch maintenance when needed?
- **Existence of Other monitoring capabilities:**
  - If the SOC cannot put a sensor at a given location (for whatever reason), what alternatives such as robust log feeds can offset this blind spot?
  - What value will the proposed sensor provide above and beyond other log feeds?
- **Ownership of assets:**
  - The SOC may have restricted ability to place monitoring capabilities on a network, based on system ownership issues.
  - These issues may bring up such situations as (1) comingled assets belonging to different owners, (2) outsourcing, (3) B2B connections, or (4) coexistence of multiple SOCs within a larger organization.
  - The SOC may consider partnering with other network operations teams and SOCs to ensure networks of mixed or ambiguous ownership are monitored by someone.

- **Active network sensors, firewalls, and content detonation:**
  - When considering active network sensors, inclusive of all-in-one firewall devices, the SOC will leverage the same criteria and considerations for passive sensors but with more selectivity, considering the higher costs associated with acquiring them.
  - The most obvious place where the SOC would want to put a NIPS is where they wish to block direct network attacks, perform bandwidth throttling or "packet shaping," or filter out specific content such as games in social networking sites. This would usually be accomplished at an Internet gateway, but major transit or interconnection points within the network or WAN also may be considered. Blocking an adversary as they move laterally is one example of when this makes sense.
  - The choice of where to put content detonation devices is perhaps the most straightforward—at any gateways where the constituency exchanges email or Web traffic with the Internet or between large networks of differing trust levels. Ideally the enterprise is designed to limit these, and, therefore, few devices are needed. The SOC may also choose to host its own out-of-band content detonation device in its enclave for the purpose of ad hoc malware analysis.
- **Application-specific monitoring devices:**
  - Rounding out the conversation of network-based sensors, are application-specific monitoring and prevention devices such as EXtensible Markup Language (XML), database, or Web Application Firewalls (WAF). Use of these devices is very straightforward. Basically, any system that serves semi-trusted or untrusted users with a corresponding protection technology is a candidate for such a device. For instance, a Web services interface between a government agency and many private corporations' business-to-government (B2G) connections might be a good place for an XML firewall. An externally facing Web server that allows members of the public to access health or financial records might be well served by a database firewall. And WAFs can provide application-level visibility in their logs for systems they are in front of. If allowed by the constituency they can also break SSL and feed packets to a network sensors. For web servers, the visibility provided by the logs is almost as good as PCAP and they remain a lower cost option when compared to PCAP systems.
  - While these are certainly intrusion monitoring and defense capabilities, they are also tightly embedded in key applications. Deployment and tuning of these systems should be done in coordination with respective sysadmins.

7.5   Instrumenting Constituent Systems in the Cloud

The use of cloud computing has become common for constituencies both large and small. But just because an asset is in the cloud, does not mean the cloud provider has fully secured or monitored that asset. The SOC mission carries on for cloud, as it does for on-prem systems.

There are however some new complexities for how to properly instrument cloud systems for detection and response, which are explored in this section.

7.5.1   What is the Same as On-prem?

Many (if not most) concepts from on-prem monitoring, detection, and incident response carries forward to the cloud. For those familiar with on-prem, the following aspects should be familiar:

- SOC analysts still need to know what it is that they are defending.
- Analysts still need both good detections and good supporting data; without both they will be lost.
- The SOC's existing approaches for on-prem monitoring and defense will likely work for IaaS.
- Cloud resources developed and released by familiar vendors (for example firewalls) are likely to provide the same or similar telemetry, detections, and visibility as their on-prem counterparts.
- Constituents still need good guidance that makes expectations clear and easy to follow, with as much automated onboarding to security monitoring and scanning as possible.

7.5.2   What is Different in the Cloud

For IT assets in the cloud, the SOC has some new and changing realities to consider:

- Utilizing the cloud gives the SOC a strong advantage in understanding the constituency. Within cloud environments that are sanctioned by the constituency, all assets and resources should be easily discoverable or known. In ensuring those assets are monitored, the SOC will have strong starting point for monitoring.
- The number and type of resources will increase faster than the SOC can keep up, and usually faster in comparison to on-prem, both in terms of instrumentation and writing good detections. The SOC (and larger security org) have three choices:
  - Limit constituents to utilizing cloud assets, resources, and providers that the SOC has established security controls such as monitoring and log collection.
  - The SOC will be in constant catch up mode, coping with constituents buying cloud resources that are beyond the SOC's reach, or for which no monitoring/defense capability exists.
  - Strike a balance between the two.

- Cloud asset topology is generally more distributed, meaning a further shift away from network monitoring and toward monitoring endpoints and services.
- It can be easier to make fatal mistakes in cloud, particularly with Internet-facing resources; getting compromised can be that much faster and easier than on-prem.
- Measuring coverage and applying controls will need to be more tightly integrated with other forms of security compliance and automation, because they are frequently intertwined.
- The SOC will need to become conversant in the telemetry sources and security mechanisms indigenous to the cloud provider(s) being used, in particular for non-IaaS resources.
- Forensic data recovery and evidence preservation are likely to be cloud vendor-specific and require advanced planning to be effective.
- Because security monitoring and detections can be enabled on a per-deployment and resource level, those costs may be more frequently borne by the constituent service or business unit, than in on-prem situations where the SOC is responsible for buying free-standing equipment.
- Perhaps most importantly, monitoring non-IaaS assets in the cloud will look substantially different than on-prem, challenging SOCs that take an attitude of "slap a sensor on it and walk away."

7.5.3 Areas of Cloud Visibility and Coverage

One big challenge for a SOC supporting a constituency moving to the cloud is harnessing telemetry types and detection capabilities that are new, different, or arguably more important than on-prem [273], [274], [275]. In on-prem scenarios, monitoring coverage frequently boiled down to network and OS-level monitoring, and perhaps applications and IoT devices. With cloud, the variety of asset types, and thus telemetry types, has exploded. Many of these are not unique to cloud, but are even important or prevalent for the cloud environment than they were on-prem.

- Monitoring, detection, and telemetry scenarios generally unique to the cloud:
    - Cloud provider control plane, such as asset deletion, asset deactivation, access patterns, use of high-privileged roles like Azure subscription ownership
    - Cloud identity providers and cloud to on-prem identity federation
    - "Serverless" code execution/Function as a Service (FaaS) (e.g., Azure functions and AWS Lambda)
    - Cloud Email and productivity such as G Suite and Office 365
- In addition, there are several prominent monitoring, detection, and telemetry scenarios that in cloud take on a different shape, use different techniques/technologies, or are simply more important than they were on-prem. These are usually PaaS and SaaS services that have both cloud and on-prem incarnations:
    - Identity providers and identity integration, such as AD integration
    - Key/certificate storage and secrets management

- Storage, particularly internet-facing, including cache devices, service bus, and message buses like Kafka
- Structured data storage, including Relational Database Management System (RDBMS), big data platforms, ML frameworks, Hadoop stacks, etc.
- Containers and containers as a service, such as Kubernetes and Docker
- Web infrastructure as a service: web server, middleware, proxy servers, and load balancers
- Continuous integration/continuous development (CI/CD) platforms that form a trust relationship and command and control (C2) channel for downstream services, particularly cloud services

Finally, there are services (usually Software as a Service (SaaS)) that the constituency may consume from the cloud that may or may not be particularly critical, serve as a point of entry to the enterprise, or otherwise present risk. In said cases, the SOC should engage in a conversation whether monitoring is required:

- HR and annual ratings systems
- Training as a service
- Knowledge management and knowledge stores
- Social media platforms
- Business-specific applications

> *There are hundreds of cloud resource types; having a security "story" for all of them can feel hopeless at times; it is best for the SOC to carefully prioritize its efforts.*

One of the best ways for the SOC to cope with the ever-expanding set of cloud resources is to prioritize them based on just a couple of factors: risk and criticality level, and the popularity of that resource type amongst the constituency overall. Meaning, for example, if a constituency has 10,000 distinct cloud resources, the resource type that has only 2 instances may get pushed down the priority list, all other things (such as risk and criticality) being equal.

If the SOC can keep a stack ranked or prioritized list of major cloud investment areas, that will help it from feeling overwhelmed. Accordingly, combining these initiatives with overall constituency cloud migration and security cloud support should help.

7.5.4 Cloud Success Considerations

- **Be clear about what the cloud provider is and is not doing** [276]:
  - The SOC and cloud provider roles should follow typical Infrastructure as a Service (IaaS), PaaS, and SaaS alignment.
  - The SOC will need to consult the cloud provider's specific technology stack for how to monitor non-IaaS assets.

- A full-stack cloud provider should have a security help desk, and built-in monitoring tools and layer of both automation and security data consolidation; the SOC should leverage these capabilities where appropriate [277], [278], [279], [280].
- **Keep the list of sanctioned cloud providers to a manageable size:**
  - Each different provider will likely mean a different monitoring, defense, and response stack.
  - It is helpful if the constituency can limit to one or two major full stack cloud providers, plus a handful of other specialized SaaS solutions and point products.
- **Expect that constituents will stand up services faster than the SOC can track through manual or human-to-human configuration management practices:**
  - Have a process for measuring and enforcing security compliance that is highly automated.
  - Use cloud governance and policy framework to measure and enforce monitoring and scanning coverage, other typical controls.
- **The SOC's monitoring approach should be integrated with a larger investment in automation around measuring security hygiene and compliance:**
  - Provide templates and cloud resource automation to allow system owners to make implementation of security controls easy, if not automatic, right from the start.
  - Be thoughtful about the access rights needed when integrating data sets and ensure least privilege is used.
  - To help measure coverage and automation success, marry the logs for the automation against inventory and activity logs for customer cloud assets.
- **It is extremely unlikely the SOC will be able to write detections and alerting on all these cloud resource types by itself:**
  - Consider picking a cloud provider and solution providers that offers easy controls for instrumenting and detections on their non-IaaS offerings.
  - For advanced SOCs, consider red teaming popular resource types and detection services to evaluate their effectiveness.
- **Integrate cloud-source monitoring telemetry into an enterprise-wide situational-awareness capability:**
  - Be aware of how cloud resources generate and store security telemetry; by default, each enclave and subscription may have its own default security telemetry storage.
  - Pursue cloud automation strategies to consolidate security data across disparate customer cloud resources into one system (perhaps the SIEM), as is the case for on-prem assets.
  - Be aware of the storage, compute, and networking costs of backhauling security telemetry to the SOC's incumbent monitoring architecture; it may be necessary for the SOC to leave some security telemetry in place.
  - Due to the emphasis of application- and service-level telemetry, expect that logging (and associated logging costs) for cloud-based services may increase steeply; it helps to have SOC representatives "plugged in" to cloud investment planning.

- **Be careful with expectations regarding network sensing:**
  - Because cloud resources are more disparate and do not usually flow through a small number of perimeter choke points like on-prem, network sensing will be even more disperse.
  - Not all cloud providers have virtual tap capabilities, meaning integrating sensors into virtual network fabrics may be more challenging.
  - Virtual network sensors and tap capabilities *may* be more limited in performance and throughput than their on-prem counterparts.
  - Putting network sensors in customer environments can complicate costing for those sensors; it may be difficult to break out the network sensor cost in a way that the SOC can easily pay for it.
  - Cloud Access Security Broker (CASB) is an approach to overcoming some of these challenges; consider a combination of CASB, security application logs such as from a WAF or firewall, and application logs to achieve some of the same outcomes as a traditional on-prem network sensor.
- **Constituents are more likely to bear the cost of some security protections and instrumentation:**
  - The cost model for implementing monitoring will likely be metered the same way that cloud resource itself is metered, plus the cost to store and/or collect the telemetry generated.
  - The SOC should pay careful attention to how security monitoring and protection controls cost as a ratio to the total cost of the cloud service in question; if that ratio is especially high, perhaps 20% or 30% of the total cloud resource cost, constituents may balk or deactivate those controls.

7.6 Monitoring in Zero Trust Environments

Traditional perimeters provide little protection to many of today's breaches by themselves; moreover, movement to cloud-based services has continued the dissolution of the enterprise perimeter. Zero trust, or the practice of not trusting users or endpoints, regardless of whether they are inside or outside the enterprise (gateway perimeter), consists of reauthenticating at various points [281]. There are different ways of configuring zero trust, and typically, micro segmentation is used. Micro segmentation is breaking up networks into smaller zones to control more granularly what is secured in applications and user access. No matter how they are rolled out, designers must think about what to reauthenticate, and when or performance of the application, access, and system can take a performance hit, if not carefully configured. Because there are not standard practices at the time of this writing, approaches to implementing zero trust vary.

Just like setting up zero trust, monitoring in these environments can be challenging. To move toward zero-trust, the SOC is integral. Several security products can be combined for monitoring. For end points, EDR provides monitoring of devices. For Data assurances, DLP monitoring can be used. For application portions of zero-trust monitoring, a combination of monitoring user access, such as Single Sign On (SSO), plus monitoring isolated application resources can suffice. In zero trust environments, robust logging and detections at the user to service and service user impersonation become that much more critical, in contrast to other trust and authentication models. The goal of monitoring at the application level is to set boundaries around the applications and data in question, and then monitor access through that boundary. Finally, for moving information across (transport and session) monitoring, virtualizing traditional non-secure applications can assist (and monitoring the isolation). For more on zero trust, see NIST 800-207 [282] and [283].

7.7 Monitoring Operational Technology

OT systems, or the hardware/software for controlling and monitoring physical systems, is increasingly a security monitoring consideration for the SOC. Before dismissing OT because the constituency is not running a nuclear power substation or public transportation, consider that OT also includes WiFi-enabled remote cameras, physical security monitoring cameras, building management and automation systems, lighting controls for internal and external facilities, energy monitoring for building environments, security and safety systems ... and the OT list continues to grow.

Operational technology, such as building controls, may be connected to the IT enterprise. They may be vulnerable to attack, yet the constituency may depend on their availability.

In fact, OT is now often connected to IT networks, including those accessed by mobile devices through wireless and traditional network web interfaces, or in cloud environments. Because OT is intertwined with IT, it has become subject to the same attacks as traditional IT, including spear phishing, ransomware, and other categories of malware [284].

7.7.1 What's Different in Monitoring OT?

For one, the priority of security in OT is different. The goal of SOC monitoring is to align with other aspects of OT security, which usually means that availability is often the highest priority more so than confidentiality or integrity.

In addition to basic security goals, the devices themselves are usually different than traditional IT endpoints. The challenge with instrumenting and monitoring OT compared to IT is that there are different protocols used, vast number of device vendors, restrictions on where devices can be placed due to, and not as many security products available for monitoring.

Also, some if not most OT is managed outside IT operations, and the SOC, and even the CIO or IT directors may be unaware of the interconnectedness of the OT. Further, the deep skillsets for managing OT may not reside in the SOC or elsewhere in the cybersecurity apparatus. Challenges to the technologies include:

- Different, OT domain-specific protocols
- Connection or communication medium (serial, ethernet, wireless, etc.)

- Existing SOC correlation, such as SIEM, may not support OT protocols or log formats
- Limited security monitoring options for OT
- OT often relied on air-gapped technology where security was not considered
- Limited computing, and ability to place security agents on devices
- Damage to devices can be more severe and cause physical harm to device or people.

Despite these challenges, bringing OT into the SOC and correlating with other IT sources can augment security on both IT and OT. For OT, attacks might be detected sooner, and even prevented before rendered unavailable. For IT, OT alerts can tip off adversaries that might be hiding in the IT infrastructure, through highlighting data not previously available.

7.7.2 Detection Strategies

When considering what to monitor in OT, considerations include addressing the following questions:

- What are the OT systems, and how are they connected to the IT infrastructure?
- Is there security built into the OT? How effective is it?
- What happens if the OT system is damaged or unavailable?
- Can the current IT infrastructure meet the OT availability needs?
- Are vendors or other third parties accessing the OT remotely?
- What are the interfaces between OT and IT and where are they?
- If an OT alert or cyber attack is identified, who needs to be included in the response?

Potential data sources to include:

- Energy Management Systems (control electrical building loads, HVAC, lighting, etc.)
- Physical access control systems (e.g., badge readers)
- Security camera feeds (e.g., CCTV)

When combining OT into a SIEM for correlation or monitoring, the SOC analysts need to learn what is normal for the OT systems under their watch. In best case, the OT that is connected are configured and maintained by OT staff, who can assist the SOC analysts on what is normal or a baseline.

One of the easiest, most familiar ways for the SOC to get started in monitoring OT is potentially in the following places:

- Connection points between OT and IT networks
- Systems connected to the OT network that use commodity operating systems, in particular Windows- and Linux-based HMIs
- IoT devices that utilize commodity operating systems AND will accept ordinary IT monitoring packages, OR feature cloud integration, such that monitoring can be accomplished through cloud IoT management/telemetry interface or API

Events to monitor:

- Operational events and alarms
- Failed login attempts on OT exceeding a set threshold

- Direct Connections to OT systems (VPN to OT, suspicious IP addresses, network traffic, etc.)
- Unexpected software/firmware updates
- Events and operations specific to OT protocols such as Modbus TCP, DNP3, EthernetIP, BacNet or OPC-UA
- Unexpected network traffic (that might suggest unauthorized command and control or data exfil)

Interestingly, the SOC might partner with physical security forces. For example, the physical access control systems (PACS) data regarding authorized employee credentials can be correlated with security camera surveillance footage and SOC SIEM data to identify malicious activity of an authorized insider, or someone who has entered a facility without being authorized.

7.7.3 OT-Compatible Tools

Another consideration is what technologies might be used in the SOC to monitor OT. Some tools and technologies available for IT monitoring and analysis are also compatible with OT. Notably, when considering the tools used for OT, consider if the tool is passive or actively engaging the physical OT devices. Because of the damage that can be caused, SOCs should consider passive means of discovery and monitoring in most cases. Some examples of tools used for IT and OT include [284]:

- Wireshark: passive network packet analyzer, works for both IT and OT
- ARP Tables: built into network and systems, ARP tables, which link IP addresses with the MAC addresses are useful to link to physical devices
- NetworkMiner [285]: passive network mapping that includes OT
- Shodan [286]: discovers OT that is Internet accessible, including webcams, smart-TVs, and more
- Grassmarlin [287]: a free tool developed by NSA for passive network mapping tools for ICS devices
- Dragos Cyberlens, Sophia [288]: free, passive packet capture for networks with ICS devices, using OT protocols, and customizable for specific port-related profiles.
- Microsoft Azure Defender IoT (formerly Cyber-X) [289]: uses auto-discovery of IoT and ICS to provide network maps and behavioral profiles of devices

CTI for OT, including malware and incident information specific to OT is available through the following sources, as well as others:

- Industrial Control Systems Cyber Emergency Response Team (ICS-CERT) [290]
- Electric Sector Information Sharing and Analysis Center (E-ISAC) [291]
- Dragos Threat Activity Group Tracking [188]

7.8 Two Instrumentation Strategies

Building on the lessons and considerations discussed in this strategy, below are two common network and host sensor placement scenarios. These scenarios provide a starting point for sensor placement but certainly do not begin to cover all the possible variations. And emerging architectural principals such as zero trust will necessitate continued evolution of sensor design.

In summary, the SOC should consider the following when choosing where to place sensors, and which log feeds to gather:

- What is the mission of the systems being considered for monitoring and what is its criticality (monetary, lives, etc.)?
- How much trust is placed in users of the system(s) and hosted services?
- What is the assessed or perceived level of integrity, confidentiality, or availability of the system(s), data, and services?
- What is the perceived and assessed threat environment? How exposed are systems to likely adversaries?
- Are the systems (or their audit data) under legal, regulatory, or statutory scrutiny, outside of those directly related to the SOC, meaning is the SOC the right organization to support that audit data collection, retention, and review?
- Where in the architecture do assets sit? Instrumenting a DMZ may look different than instrumenting a payroll system only available internally.
- What is the number of sensors and volume of data the SOC can reasonably manage? Whereas NIDS/NIPS deployments can comprise dozens or hundreds of sensors, a SOC may have EDR deployed on every end host. In the case of a large enterprise, this could comprise hundreds of thousands of sensors.

7.8.1 Example #1: On-Prem Enterprise Network

In the first example, the main portion of enterprise network is instrumented. This is where users perform their regular business computing and access services from the intranet and the Internet (See Figure 25). The instrumentation of this network begins with the Internet gateway at the top left. Here there is a passive sensor or set of sensors that gather network traffic metadata and IDS/IPS alerts coming from the next generation inline firewall. Alerts coming from a sensor on the external side of a firewall, if used, should be triaged with an understanding that a large portion of scanning and exploit activity was probably mitigated by the firewall and is not great concern.[10]

Every enterprise border DMZ is shaped differently. This example has on-prem email and Web proxies hanging off legs of the firewall. In the gateway DMZ, email and Web content detonation is used to detect zero-day attacks from Web pages and email attachments. The

---

[10] As an aside, there is a philosophical debate amongst sensor architects as to whether network sensors go on the inside or outside of the firewall. Here, the network traffic metadata sensor is placed on all sides of the firewall, to cope with NATing performed by the firewall.

architecture pictured allows blocking of malicious email and Web content by placing the content detonation devices in-line with their respective proxies.

Moving to the internal LAN at the top right, all user systems are instrumented with a host EDR suite. Most notably, if there is anywhere in the enterprise where the EDR suite is turned to active prevention mode, this would be the best and usually least-risky place to do so.

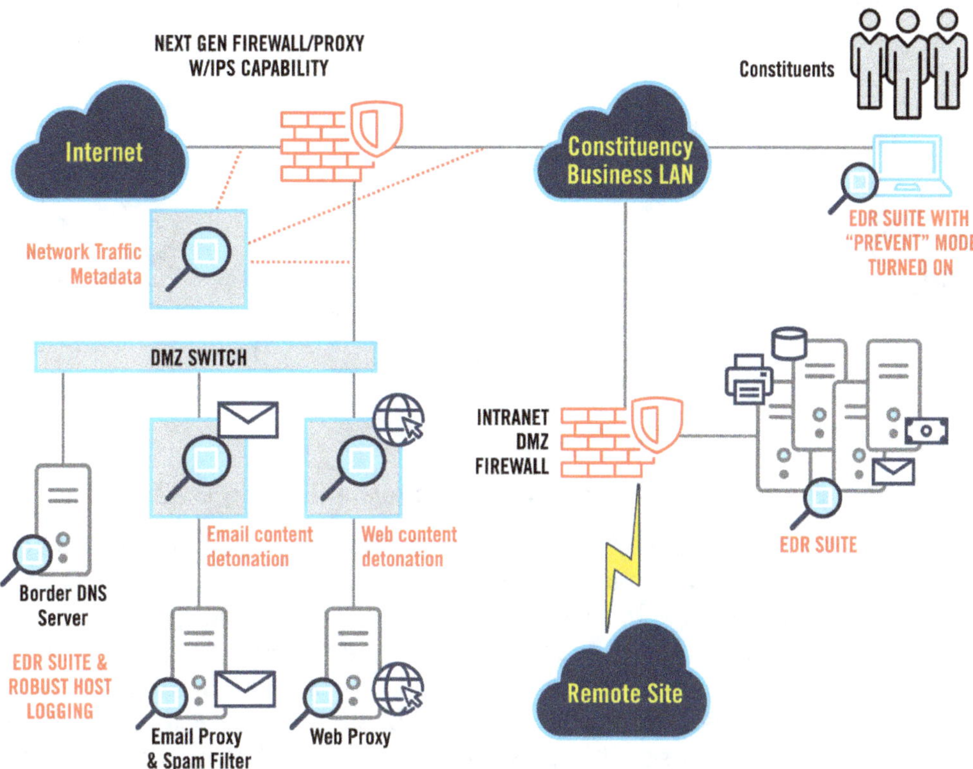

Figure 25. Instrumenting an On-Prem Enterprise Network

This constituency has an internal-facing set of Intranet services behind a firewall. This is an excellent point in the network to look for adversaries traversing the internal network and insider threat. In this case, the firewall is being used by itself and other network sensing may not be necessary. Rather, all servers are performing robust log collection along with EDR. This is an example of making a tradeoff between sensing capabilities to minimize duplicative data and maximize collection at the most effective point in the architecture.

Finally, a remote site (or sites) is hanging off the bottom of the diagram. These are good places to use an all-in-one affordable security appliance such as one that contains a firewall, reputation-based content filtering, NetFlow-style traffic logs, IPS/IDS, and content detonation.

7.8.2 Example #2: On-Prem-Based or Cloud-Based DMZ

In this example, the constituency has chosen to offer access to sensitive data to external parties such as the public or another institution such as a business or a government agency (See Figure 26). The data is probably being provided through an interactive Web portal and/ or machine-to-machine Web services. In either case, there are common elements to the monitoring architecture for both those approaches.

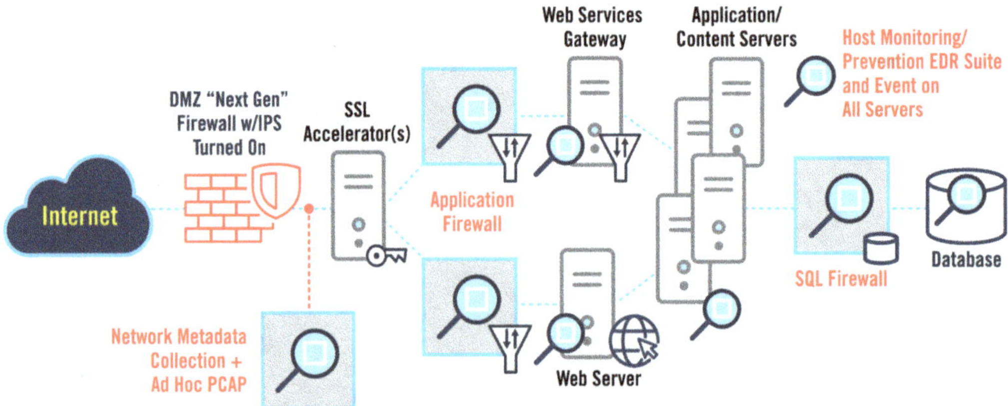

Figure 26. Instrumenting an External-Facing DMZ

On the left, is network traffic metadata collection at the border of the DMZ. There are several key points to note here. First, this construct does not include dedicated NIPS because there are so few protocols flowing in and out of this DMZ that a dedicated general-purpose NIPS sensor will provide limited benefit. Second, although the network link in between the SSL accelerators and firewall is being monitored, this traffic is encrypted. Therefore, monitoring is best limited to perhaps inspection of SSL handshakes and connection metadata; PCAP collection here is of almost no use. Third, the firewall is expected to provide commodity IDS/ IPS capability, which again is limited because it is only inspecting encrypted traffic. In this case the SOC is relying heavily on Web server logs, WAF logs, and Web application logs, and dispenses with pure-play NIDS/NIPS altogether.

Inside the SSL accelerators, the SOC can leverage an application-specific firewall, such as Web services or XML firewall, to inspect highly structured data flowing in and out of the Web services interface. Similarly, the SOC may also choose to use an SQL firewall in front of a relational database on the backend. Some implementations may feature both capabilities, though this may be seen as overkill. Finally, there is a host monitoring and prevention package on each server.

Perhaps most interesting, at this level of abstraction, a nearly identical approach for monitoring would take place if this DMZ was based in a cloud. There would, however, be some differences:

- The web server, SQL server, and possibly application servers would not be IaaS, but instead their analogous PaaS offering. In this case, an EDR would be irrelevant and instead the SOC would rely on security monitoring telemetry and protections available specifically for those platforms as a built-in feature that can be activated.

- For example, in a SaaS SQL, Postgres, MySQL, etc. situation, there may not be an independent SQL firewall device. Rather, SQL vulnerability scanning, and attack detection/filtering is simply a transparent feature that can be turned on as part of the PaaS resource (and usually at an additional expense).
- At the time of this book's writing, the throughput available for third-party firewalls in a cloud setting is sometimes limited; it may be necessary to utilize the cloud vendor's own firewall or network ACLing capability if a virtual firewall from a company like Cisco or Palo Alto is not available.
- Any kind of network sensing will require the use of a cloud-native "virtual tap" capability which may be specific to the cloud platform in use.

7.9 Summary – Strategy 7: Select and Collect the Right Data

7.1. Selecting and collecting the right data is a balancing act between having too little data and therefore not having the relevant information available and too much data such that tools and analysts become overwhelmed.

   - A starting point for selecting the right data is to identify mission drivers and develop use cases around the top threats. There are many common data sources available to the SOC. Consider the quality, availability, and coverage of the data when selecting what to collect.
   - In conjunction with identifying available data sources, the SOC needs to ensure it has the proper authorities and agreements in place for the data it is going to collect.
   - As the SOC is developing its data collection plan it needs to consider how it will pay for the transport and storage of the data.
   - There are two classic approaches that SOCs may take in selecting and tuning data sources: tune up from zero or tune down from everything. There is also a third, which is a hybrid of the first two, and focuses on considering what events are processed locally vs transported back to the SOC for central processing.
   - Data feeds require continual maintenance and should be routinely checked on to ensure data is being collected as expected.
   - The length of time the SOC needs to retain data is driven by a combination of legal and regulatory requirements, the risk profile of the constituency, and financial constraints.

7.2. Sensing and detection technologies can be described most succinctly along three dimensions: behavior-based vs knowledge-based detections; network vs host sources; and active vs passive response modes.

7.3. Data and instrumentation from endpoints are generally considered more informative and provide more clarity than from network traffic for both detecting and confirming intrusions.

   - Host observables include mounted file systems and other storage; system memory and processors; attached devices and system I/O.

- Key features of EDR capabilities include strong detections across multiple observables; rich telemetry relevant to incident investigation; interactive command and response; and integration with managed services.
- Application deny listing is a technique whereby an OS module or protection agent blocks specified unwanted processes running on the end host. Application allow listing policy uses a default deny approach. Both require additional management overhead to implement.
- Traditional host security protections such as host-based firewalls and antivirus/antispyware still provide some value but should be part of a layered approach to host defense and monitoring.
- Data loss prevention and user activity monitoring are two additional capabilities that SOCs will want to consider based on missions needs and use cases.
- Not every constituency will be able to instrument every constituency host. Consider the criticality of the system, trust relationships, user privileges on the system, vulnerability and attack surface, and the ability to instrument the system based on type.

7.4. Although network-based monitoring had greater emphasis than in the early days of security operations, network-based monitoring is still both prolific and provides a complementary approach. However, as encryption of network traffic increases, the value of network monitoring tends to diminish.

- Network-based monitoring technologies can sometimes be the most cost-efficient and simplest means by which SOCs can gain visibility and attack detection coverage for a given enclave or network, especially in cases where they have no other visibility.
- Considerations for deploying network sensors include false positives; signature details and context, response choices, response actions; presence; latency and bandwidth; cost of decoding; failure modes; integration with network operations; and price.
- The SOC has a number of capabilities it can use to generate records of network traffic from the least detailed (NetFlow), to layer 7 traffic metadata, to full PCAP. Sustained full PCAP collection on busy links is usually unnecessary, especially in the presence of other monitoring techniques like traffic metadata collection.
- Malware detonation chambers can accept files uploaded in an offline manner, but they are also used to scan network traffic, usually web or email, in real time, extract files, and detonate them fast enough to block malicious content from reaching its intended target.
- WIPS assists the SOC in detecting: rogue access points; wireless LAN attacks; unauthorized endpoint access and hotspots, unauthorized network connections, and some provide denial of service protection.
- There are several approaches to directing traffic to network monitoring devices. These include using layer 2 switches to SPAN traffic, and passive or active network taps to redirect traffic to a passive sensor. An active sensor can also be placed in-line with the traffic.

7.5. With the increase in use of cloud technologies, the SOC's visibility should extend into constituency cloud systems, services, and resources.
  - Many concepts from on-prem monitoring, intrusion detection, and incident response carries forward to the cloud.
  - Cloud asset topology is generally more distributed, meaning a further shift away from network monitoring and toward monitoring endpoints and services.
  - One big challenge for a SOC supporting a constituency moving to the cloud is harnessing telemetry types and detection capabilities that are new, different, or arguably more important than on-prem. These include cloud provider control plane, cloud identity providers and cloud to on-prem identity federation; "Serverless" code execution/ FaaS (e.g., Azure functions and AWS Lambda); and cloud email and productivity like G Suite and Office 365.

7.6. In zero trust deployments and scenarios, several security products can be combined for monitoring. For end points, EDR provides monitoring of devices. For data assurances, DLP monitoring can be used. For application portions of zero-trust monitoring, a combination of monitoring user access, such as Single Sign On (SSO), plus monitoring isolated application resources can be used.

7.7. OT systems, or the hardware/software for controlling and monitoring physical systems, is increasingly a security monitoring consideration for the SOC.
  - Availability may be a higher priority for OT systems vs confidentiality or integrity.
  - The challenge with instrumenting and monitoring OT compared to IT is that there are different protocols used, vast number of device vendors, restrictions on where devices can be placed due to and not as many security products available for monitoring.
  - To get started with OT monitoring, the SOC can look for: connection points between OT and IT networks: systems connected to the OT network that use commodity operating systems, in particular Windows- and Linux-based HMIs; or IoT devices that utilize commodity operating systems AND will accept ordinary IT monitoring packages, OR feature cloud integration, such that monitoring can be accomplished through cloud IoT management/telemetry interface or API.

7.8. There are many possible instrumentation strategies. Two common ones include monitoring the on-prem enterprise network and monitoring an on-prem or cloud-based DMZ.

Strategy 8: Leverage Tools to Support Analyst Workflow

Previous strategies discussed the sources of data and threat intelligence the SOC has at its disposal, including sensing technologies and log feeds. Each piece of data is valuable on its own, but its only when combined that the true power of the data becomes available. To achieve this goal the SOC needs to bring all this data together into an architecture that can help turn the data into information, and information into knowledge. And this architecture must support the analyst workflow within the SOC. As with many aspects of the SOC there is not a one-size-fits-all answer about how to do this. Different SOCs will put different tools at the focal point for their workflow: SIEM, SOAR, case management, EDR, threat intel management, and so forth. And it seems like every new set of technology promises centralization and a "single pane of glass" for SOCs. However, experience tells us this is rarely the reality. Rather, reducing the number of panes of glass, and providing integration between them is the best strategy with an emphasis on automation and integration for repeated tasks, escalation, and incident handling.

8.1   Tool Integration Overview

Figure 27 shows a common data flow and tool integration story for a large, sophisticated SOC. While not all SOCs harness all these technologies, this approach can serve as the starting point for discussing the possibilities. To illustrate, when an analyst starts their workday, they will typically start in one of three places:

- Incoming fresh alert triage in the SIEM or EDR tool
- New threat intel and news found in the intel/indicator management platform
- New or updated cases in the case and workflow management platform

From there, they will have several options on what to do next:

- The analyst can pivot based on entity to drill down in EDR platform, looking at detailed process execution trees and file changes on the end host
- The analyst can pivot to network metadata capture to see what is known about that hosts' communications
- Using either, the analyst may take information about processes or files and look them up in the threat intel management platform
- If available, a file captured from the host or unencrypted traffic can be moved to a malware detonation chamber to see if the file is malicious
- Queries can be executed using analytic notebooks, executing various predefined, curated queries in a couple minutes
- Information on involved users and hosts may be retrieved from asset and entity knowledgebases; in the presence of a UEBA product they may be even further enriched

- Finally, the analyst can record this information captured in a case management tool for further action

This kind of workflow and quick pivoting is common place for mature SOCs. It allows the SOC to perform more data gathering and reasoning in a half hour than other SOCs might achieve in a week. This kind of quick, accurate analysis is both essential to proper alert triage and incident response, but it is also essential to analyst quality of life. This does not have to be a mere dream for smaller or less mature SOCs, either. With just a little planning and integration, many variations on this are possible, as discussed in this strategy.

Each SOC brings to bear a different set of tools, capabilities, and integrations in support of their analysts. This strategy shows multiple configurations of popular tools. Figure 27 shows one example architecture, which also serves as an overview for the chapter. Other example architectures, variations, and integrations are offered later in this strategy.

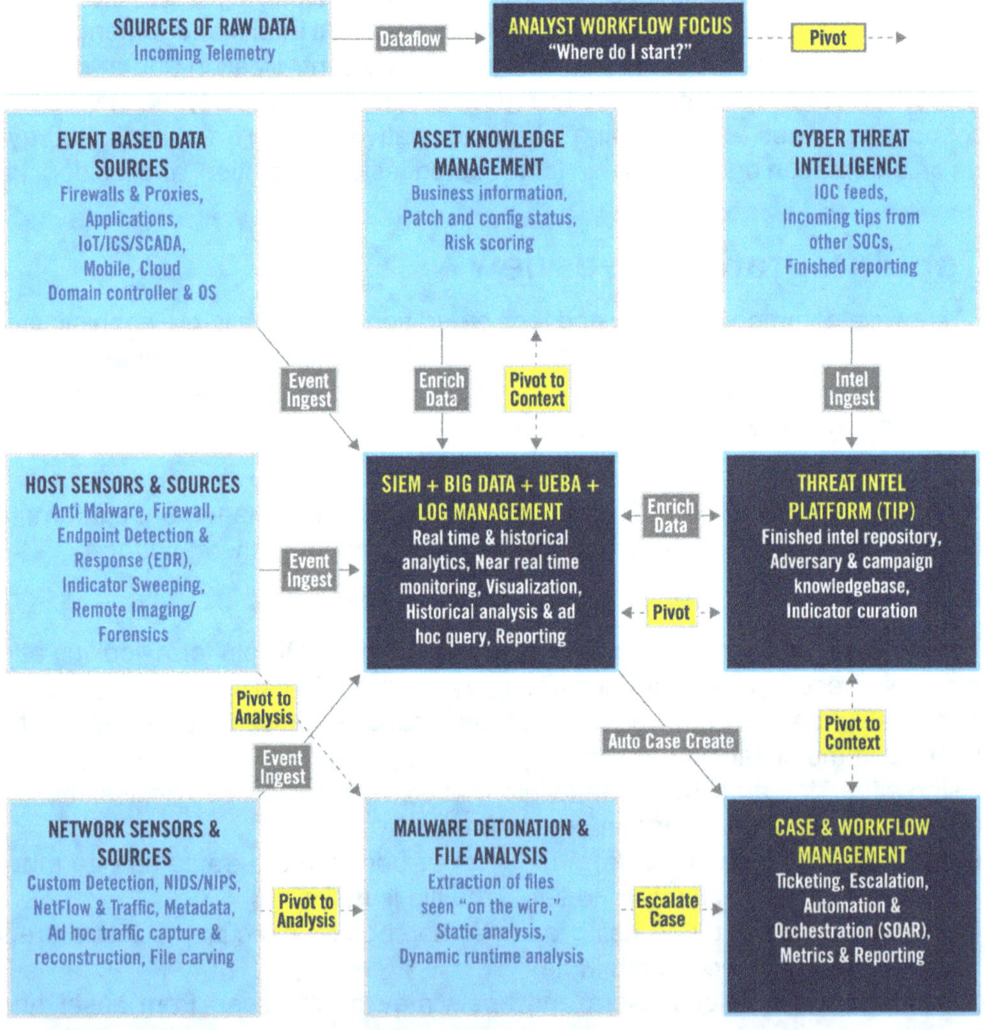

Figure 27. SOC Tool Integration and Pivoting

8.2 Security Information and Event Management

SIEMs promise the ability for the SOC to not only cope with, but also maximize the value of the billions of events collected every day. There is a large corpus of best practices and available material offering various judgements and best practices around their use. SIEMs can be very expensive both to acquire and to use; like any other SOC technology, the value found from a tool is largely proportional to the effort put into that tool. SIEMs are no different.

This section explores:

- The common traits and value factors of SIEM
- Their architecture
- Best practices for their acquisition and operationalization, including how to keep SIEM "healthy"
- Their relationship and overlap with similar and related technologies, including log management, SOAR, and UEBA
- General expectations, best practices, and lessons learned
- Alternative ways to achieving the same outcomes as off-the-shelf SIEM

Due to their combined cost and operational significance, this strategy dwells longer on SIEM than any other technology.

8.2.1 SIEM: the 10,000 Foot View

SIEM collects, aggregates, filters, stores, triages, correlates, and displays security-relevant data, supporting both real-time and historical review and analysis. SIEM workflow is targeted for the SOC, ranging from the ad hoc security team model to a hybrid centralized/distributed model. Figure 28 shows some of the many different types of data and processes that can come together in a SIEM.

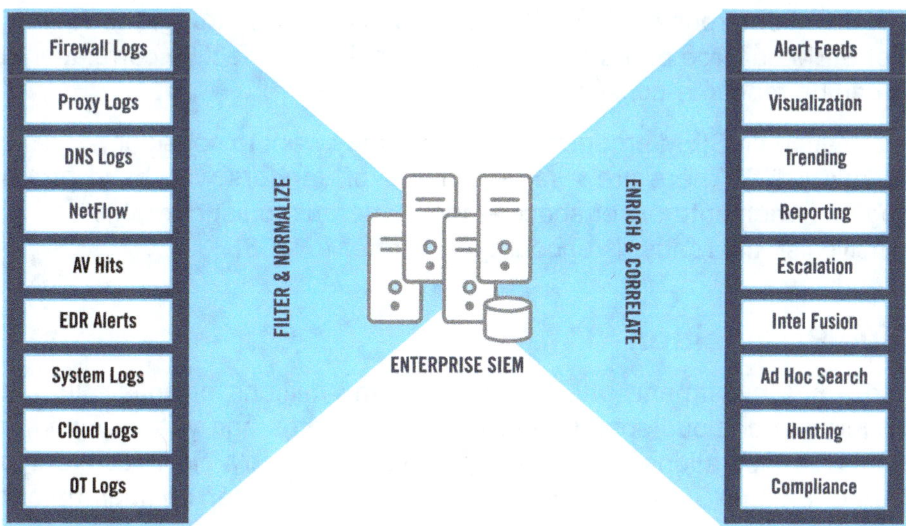

Figure 28. Notional SIEM Architecture

Best-of-breed SIEM acts as a force multiplier, enabling a modest team of skilled analysts to extract cyber observables from large collections of data, including virtually all those listed in "Strategy 8: Leverage Tools to Support Analyst Workflow." Though the same outcomes can be achieved by composing a SIEM of disparate software components, an off-the-shelf SIEM solution typically makes this job easier to achieve and in less time. The purpose of SIEM is to enable the analyst to turn information collected by the SOC into knowledge that can be acted upon in a timely fashion. Modern best-of-breed SIEMs can support many compelling use cases:

- **APT detection:** Including piecing together disparate data indicating lateral movement, remote access, command and control, and data exfiltration
- **Incident analysis and log forensics:** Including retention and investigation of historical log data
- **Workflow and escalation:** Tracking an event and incident from cradle to grave, including ticketing/case management, prioritization, and resolution
- **CTI fusion:** Integration of tippers and signatures from CTI feeds
- **Trending and threat hunting:** For analysis of long-term patterns and changes in system or network behavior
- **Perimeter network monitoring:** Classic monitoring of the constituency for malware and external threats
- **Insider threat and audit:** Data collection and correlation that allow for detection and monitoring for profiles of suspicious internal threat activity
- **Configuration monitoring:** Alerting on changes to the configuration of enterprise servers and systems, from password changes to critical Windows registry modifications
- **Cyber SA:** Enterprise-wide understanding of threat posture
- Policy compliance: Built-in and customizable content and reporting that satisfy elements of various regulatory compliance, such as PCI, SOX, and FISMA.

SIEM products have been on the market since the very early 2000s. They have proven their value in many enterprise SOCs. That said, some SOCs struggle to realize the value proposition of SIEM, in large part due to their complexity, as effective correlation rule writing and upkeep can be resource-consuming.

For more information on SIEM products, the reader may want to consider [292], [293], [294], [295], [296], and [297]. There are a large number of vendors who have products in the SIEM and log management market space— too numerous to list here. Instead of providing a comprehensive list, the reader is encouraged to refer to [298].

8.2.2 Value Proposition

SIEM can be a big investment, often involving many millions of dollars in software and hardware acquisition or cloud spend, along with the months and years of work required to integrate it into SOC operations and keep it updated. With such a big investment, one should expect a big return. Some SOCs recognize SIEM as little more than an aggregator of massive quantities of log data; this is only the beginning. To realize the full potential of SIEM, the SOC must leverage it throughout the event life cycle, as shown in Figure 3.

Figure 29 follows the basic SOC workflow: alert enrichment, prioritization, triage, investigation, escalation, and response. From left to right in the diagram, the event lifecycle "inverted funnel" goes from billions of events to a handful of potential cases. In this process, SIEM moves from automation on the left, through correlation and triage, to workflow support and enabling features such as event drill-down, case management, and event escalation on the right.

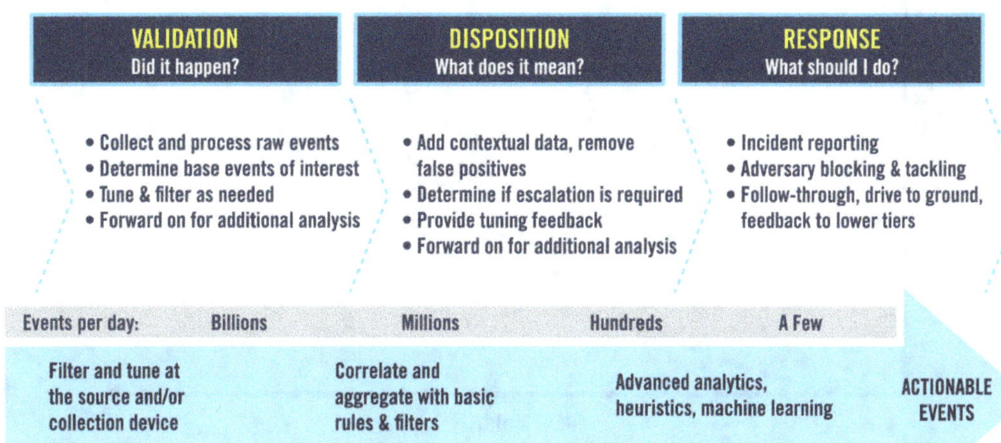

Figure 29. SIEM through the Lifecycle

Because SIEM acts as a force multiplier, fewer analysts can get more done in their workday or shift, assuming SIEM has been outfitted with the right data feeds and good content. In fact, one might conclude that a very mature SIEM implementation would reduce the number of analysts a SOC needs. This is not necessarily true. There is no replacement for the human analyst. In practice, as a SOC implements SIEM, it follows the classic S curve [299]: allowing the SOC to move through cycles increased effectiveness and efficiency. In practice, this means the SOC begins to recognize all the activity in its logs that previously went unnoticed. So, instead of staffing levels going down, they might instead go up, because the workload typically increases, as shown below in Figure 33.

In fact, some industry experts [300], vendors [301], and SOCs have claimed "we no longer need tier 1 because our SIEM's analytics and SOAR automation are so good." In reality, what they have done is shift staffing requirements away from low-efficiency, low-effectiveness alert triage, and toward increasingly better detections, data enrichment, and automation. Put another way: they have shrunk, but not eliminated, the staffing pool responsible for triaging alerts, because those alerts are likely fewer in number, higher quality, and more enriched.

The best place for the SOC to be is on the right-hand portion of Figure 30 where a mature SIEM implementation enables a modest team of analysts to achieve what a team of a thousand unaided analysts could not. Even though the SOC may have invested a few FTEs in maintaining and writing content for SIEM, it has more than made up for that with the capability and efficiencies gained. This theme returns during the discussion on SOAR, below.

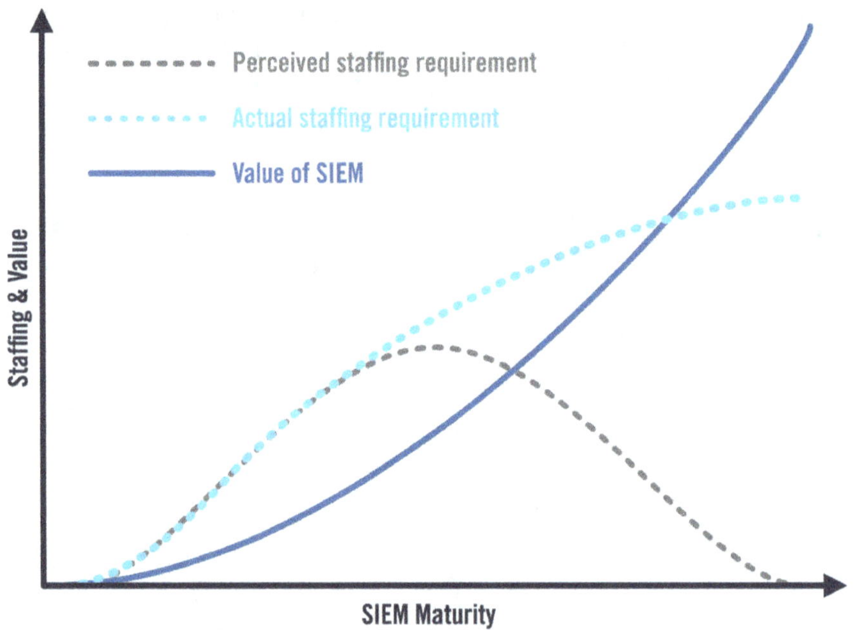

Figure 30. SOC Staffing Requirements as a Function of SIEM Maturity

8.2.3   SIEM Architecture, Common Features, and Expectations

Each SIEM vendor brings its own approach to bear in providing its blend of functionality. With that said, there are some common functions and components in modern SIEMs.

SIEM Data Acquisition and Collection
This component is often known as an agent or collector. It will reside on either:

- **The monitored host:** Where it has direct access to logs such as through local APIs, or files accessible from a filesystem seen by the host.
- **Remotely:** Where it either interrogates one or more devices for data (pull) or accepts data sent to it (push); the agent can gather this data through various native protocols such syslog, RESTful APIs, and Java Database Connectivity (JDBC).

Consequently, there is no SIEM architecture that is fully agentless, meaning there is always a piece of software that must ingest the data; what is in question is the location of that agent–on the host being monitored, running as software on a nearby system, running on an appliance, an API on the SIEM, or in the cloud as a SaaS capability; best-of-breed SIEMs should offer most or all these scenarios.

Software agents should have several configuration options, allowing administrators to optimize their CPU, memory, disk footprint, and multithreading. Administrators should be able also to optimize the load agents place on systems they pull data from, such as their data polling interval and retrieved batch size.

The agent should provide various facilities for ensuring timely, lossless receipt of data, such as filtering, batching, deduplication, caching, congestion/flow control, and prioritization. Additionally, the agent should provide authenticated and encrypted SIEM data transfer, typically using TLS. Accordingly, the data collector is often the first and best place to tune data, thereby reducing network bandwidth and storage costs downstream. See Figure 31.

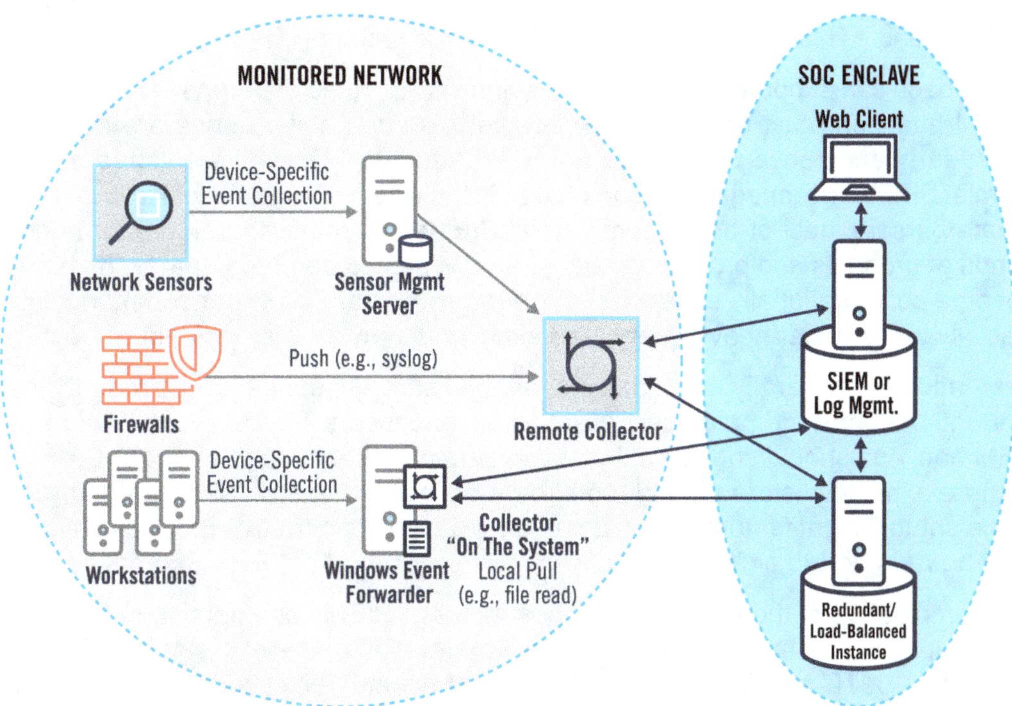

Figure 31. Example SIEM and Log Management Data Collection

SIEM data normalization and persistence

In many SIEMs, data is collected at a central location. Data is typically stored in a backend that supports high-speed queries and condenses on-disk storage. Most SIEMs offer a distributed, horizontally-scalable architecture that uses NoSQL [218] techniques such as MapReduce [302], document stores [303], Key Value Stores [304], or columnar stores [305] to fragment or "shard" data across many nodes [218]. Traditional SQL RDBMS as a bulk event persistence and retrieval technology has been mostly eclipsed in the log management and SIEM space, due to its comparative complexity, scalability, and performance. Consequently, many SIEMs, both commercial and FOSS, can be viewed as a special application of big data technology, thus blurring the line between both.

Data collection and correlation "master" nodes historically are offered as hardware appliances or software, but increasingly are offered in virtual appliance, pre-built cloud IaaS images, and cloud SaaS offerings.

Some SIEMs take the approach of data normalization, meaning semantic feature extraction, parsing, and schematization at the agent/collector. This approach is most associated with SIEMs that enforce a comparatively more rigid approach to data handling and are commonly typically used in concert with columnar or RDBMS data stores. This is associated with a traditional approach of Extract, Transform, and Load (ETL), sometimes referred to as "schema on write."

Other SIEMs take the approach of comparatively less data transformation at ingest and persistence. Instead, data transformation is favored at query time, using the approach of "schema on read." This approach is generally favored with document stores.

Either approach can support data query achieving search rates of billions of rows per second, but careful attention must be paid to expectations around how queries are constructed in relation to the persistence technology and data structure. One differentiated value that a commercial SIEM/log management can deliver is the size and maturity of its library of "off the shelf" data parsers; best-of-breed commercial SIEMs will ship with parsers for hundreds or thousands of products and product versions. Also, it should be noted, that "schema on read" can become computationally expensive when the same data is referenced many times; this is especially acute under heavy query load such as during a major incident or hunt.

Data retention is generally limited by available storage space and product licensing, consequently supporting multi-year retention as necessary. To keep costs under control, the SIEM backend should support at least two tiers of storage, such as "hot" recent data on faster, more expensive storage, and "cool" data on less expensive, slower storage. Further, very large setups might feature a third tier of un-indexed "cold" data exported in a format that can later be re-imported or "thawed."

Using a combination of the above techniques, modern SIEMs can persist, index, and query data measured in the multi-petabyte range. In situations where a single SIEM exceeds roughly 20TB-100TB per day or 100,000-500,000 events per second, the SOC may wish to pursue "cluster of clusters" approaches whereby disparate SIEM cluster instances share persistence, query, and analytic load. These techniques are often known as federated query [306], [307], federated search, cross cluster join [308], or when geographically distributed, geo sharding [309].

Federated query is an extremely powerful technique whereby a single user, running a single query, can interrogate multiple disparate SIEM or log management systems in parallel, with the results collated in a manner that makes the experience seamless for the user. This approach can address many technical and political hurdles for the SOC and others in IT. Simply put, rather than getting in a protracted debate about who is responsible for and has access to what data, disparate teams and organizations can query each other's data in place. That said, it is not a panacea for scalability; in particular, attention should be paid to how correlation and analytics will execute in a distributed fashion. See Figure 32.

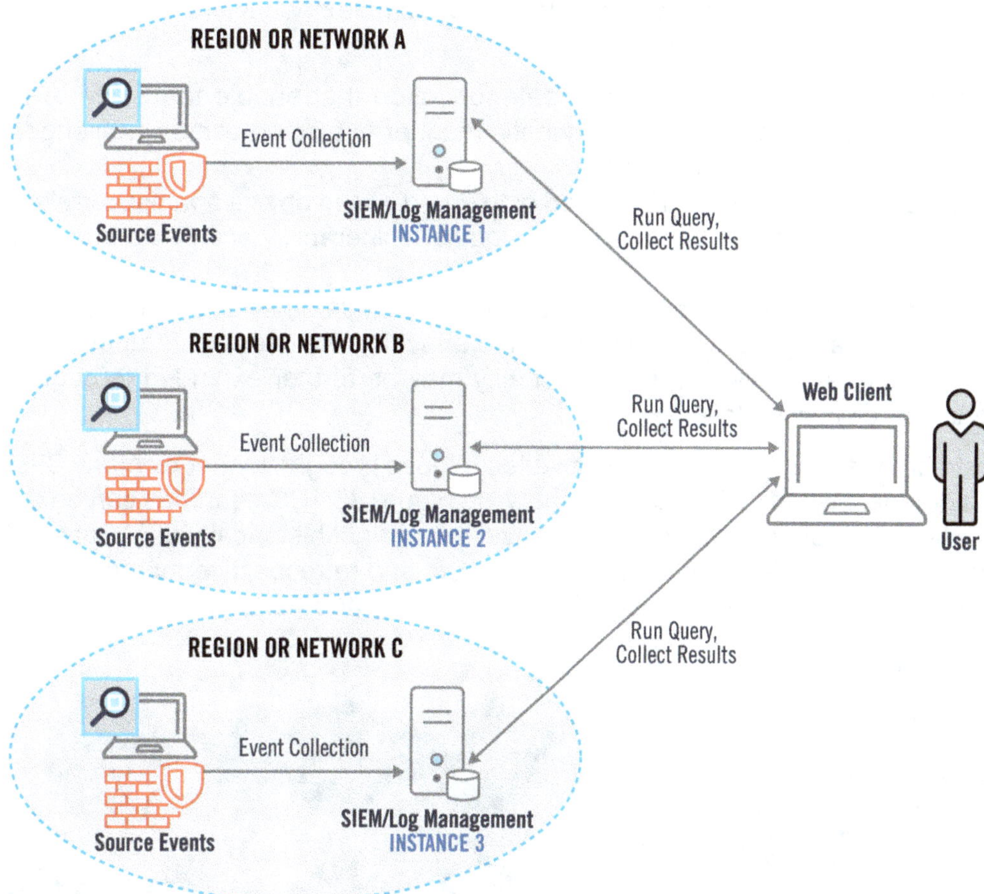

Figure 32. Federated Query

SIEM data analytics

The SIEM will generally support two or three approaches to analytics and detections:

- **A near-real-time alerting and correlation engine:** Supports alerting on single event matches (known as atomic rules) and sets of events, potentially utilizing a state machine (e.g., "true" multi-event correlation)
- **An analytic engine that executes analytics against data persistent on disk:** Sometimes referred to as "query on a timer" that executes on a schedule defined in each analytic
- **A machine learning (ML) engine:** For more advanced SIEMs (and SOCs), the ML engine can run on data in-memory as it streams in and/or against data persisted on disk

The SIEM analytic engine is its most complex and defining feature in contrast to ordinary log management systems. It will ship with "stock" rules targeting various cyber defense, insider threat, compliance, and other use cases to detect complex behaviors or pick out potential incident tip-offs. These may include:

- Normalization, prioritization, and categorization that enable the SIEM to leverage various data feeds in a device-agnostic manner but is sometimes challenged due to the large diversity of data [310].
- Events can have their priority raised or lowered based on hits against correlation rules and enrichment such as comparison against vulnerability scan data and various ML techniques.
- Alerted fired by various analytics can trigger various other user-configurable actions such as creating a case within SIEM and attaching the event to it, running a script, or emailing to an analyst. This functionality may be further extended through a SOAR product, discussed later.

To illustrate both real-time and retrospective techniques brought to bear by many SIEMs, one may consider the case of IOC matching, shown in Figure 33. IOC searches can be done both against data as it streams into the SIEM, as well as against historical value. Some SOCs do both, enabling both near-real-time alerting on IOCs, and retrospective matching on recently ingested IOCs matched against events ingested a week ago (for example).

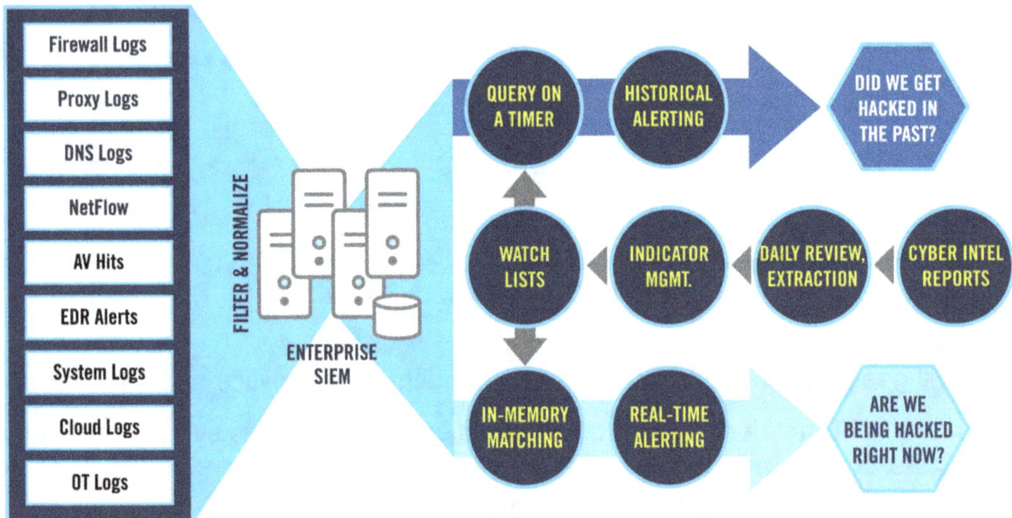

Figure 33. Using In-Memory and Query on Timer Techniques to Perform IOC Matching

Data query, interaction, and workflow

Using a Web-based UI, analysts interact with the data through various output channels, either through real-time scrolling alerts, event visualizations (bar chart, pie chart, tree maps, event graphs, etc.), through ad hoc query, and through curated analytic notebooks [311]. By curating, saving, and reusing queries and visualizations into a series of notebooks, efficiencies and consistency can be gained through repeated actions, and results more easily shared.

The SIEM should provide a robust, performant query language enabling complex questions to be asked of the data, ideally using the same syntax as with its analytics above.

In operation, using queries that have been written according to the best practices of the SIEM, should allow the analyst to ask questions of several days' data, and get results back in a matter of seconds or a few minutes. Failure to do so usually means a) the data was not properly parsed, b) the query was not written correctly, c) the SIEM was sized incorrectly for its data and user load or d) other concurrent users are presenting unreasonable load to the system. In some situations, SOCs may choose to curate summary datasets, in the spirit of OLAP cubes [312], that can answer frequently asked questions more efficiently, with a comparatively lower on-disk data footprint and cost. Creating curated data sets is particularly important if the same question is being asked of data frequently, e.g., "when on which days did Bob log into the HR portal," and are often a key step in executing structured hunt.

The SIEM should be written and deployed to support the SOC analysts and other partners. It is not unusual for a SIEM or log management solution to serve multiple concurrent queries and 20+ active users at any given time. Some SIEM products will offer additional capacity scale-out specifically for hosting more users or concurrently running queries.

The system provides some level of alert tagging/acknowledgement, incident ticketing or case tracking, allowing users to acknowledge and escalate both alerts and cases.

SIEM users can view content created by other SIEM users whose access is controlled through groups and permissions, leveraging both vendor-supplied "stock" content as well as customized reports, rules, filters, dashboards, and other content.

Flexible integration

Most best-of-breed SIEMs provide methods for users to move SIEM content in and out of the SIEM. This functionality may be further enhanced through online SIEM user communities or marketplaces where users can share and collaborate on content.

Externally facing APIs and message busses should allow external systems to query the SIEM, export data out of the system in near-real-time and perform bulk export.[11]

Best-of-breed SIEMs support multitiered, peered/clustered, or redundant deployment scenarios:

- With tiering, one SIEM can forward some or all its alert data to a parent SIEM, perhaps leveraging the same agent or collector framework that gathers data from end hosts.
- With peering or clustering, as is the case with federated search or cross cluster joins
- In redundant scenarios, multiple SIEM instances can ingest the same data, potentially through dual reporting from one agent to multiple SIEMs, or through synchronization between disparate SIEM nodes.

---

[11] For big data, log management, or SIEM solution the SOC uses, it is wise to evaluate its after-the-fact log export capability before fully committing to the platform. Many security teams have used such a tool for months or years, gotten to a critical incident, litigation hold, administrative proceeding, or lawsuit, and been stymied at challenges with bulk data export. It is best to not leave this process to a moment of crisis; the SOC should exercise this feature early in the tool's use to ensure it meets expectations.

8.2.4 Log Management

Collecting and querying events from a disparate set of systems or applications does not always necessitate the features and cost associated with a full-blown SIEM. Oftentimes a less-expensive log management system, which is usually simpler to set up and use, is a better choice. Log management systems incorporate some of the aggregation, storage, and reporting capabilities found in SIEM, but with a comparatively smaller feature-set.

Some SIEM and log management systems perform "dual duty" meaning they can serve both general IT use cases and the SOC. Two very good examples of this are Elasticsearch and Splunk. However, it is possible to differentiate a log management system from a SIEM as follows:

- SIEMs are generally built with the SOC as the primary or sole user, whereas log management fits a more general IT use case.
- Log management systems usually do not strongly support differentiation of different SOC roles or personas, such as through workflow or automation targeting SOC use cases.
- Log management systems' analytic engine is often focused more on "query on a timer" and metric-based alerting, and *usually* less on sophisticated ML, though this area is evolving rapidly at the time of this book's writing.
- A SOC should expect that a SIEM comes with a library of "stock" detections and analytics, tailored to SOC use cases, that it can leverage with comparatively little effort. Log management tools are less likely to contain this feature.
- Log management systems *may* offer less parsing of specific fields, instead favoring ease of data collection.
- Due to lack or parsers or domain-specific data schema, a SIEM *may* not be suited to handling log data that is not typically regarded as security-relevant.
- Either product family:
  - Can typically be found as software, appliance, or cloud-based offering.
  - Should be equally strong in storing and searching massive amounts of log data.
  - Frequently offers off the shelf features supporting standards compliance, including pre-built reports that match specific regulatory controls and requirements.
- Log management systems are *usually* cheaper than a SIEM sized for the same volume, velocity, and variety of data, though either product *may* cost $millions in acquisition, annual subscription, and maintenance costs, given a multi-TB/day ingestion scenario.

Today, many vendors describe their log management systems as having features found in a full-blown SIEM. Therefore, understanding where a given product falls in the SIEM, and log-management spectrum can sometimes be challenging. When considering acquisition of either, it is important for the SOC to compare specific capabilities of either tools against their own business requirements and future roadmap.

As discussed at the end of this strategy, it is possible for a SOC to achieve the same outcomes as a SIEM by using log management tools in concert with other technologies.

One way to look at the difference between full-blown SIEM and log management systems is that SIEM is meant to serve as the cornerstone of SOC workflow, whereas a log management system typically has a more generalized set of use cases and supported personas. Recall the discussion of SOC organizational models from "Strategy 3: Build a SOC Structure to Match Your Organizational Needs" which detailed that there are some constituencies where security operations are performed in an ad hoc manner (e.g., with a security team). These organizations have few resources and do not devote many (if any) full-time staff to alert triage and incident response. Discussed below in SIEM Alternatives, their needs are likely satisfied by log-management system in concert with an EDR product. Full-fledged SIEMs requires care and feeding that small SOCs, and security teams are historically challenged to provide, although cloud based SIEM "as a service" are changing this equation.

8.2.5 Acquisition

SIEM is often the largest single purchase a SOC will make. Before investing in a SIEM tool, the SOC should consider whether the following baseline conditions are true:

- The SOC's needs exceed what a log aggregation tool typically offers, particularly around workflow or security analytics
- The SOC performs a substantial portion of its analysis duties on near-real-time data
- The SOC has identified multiple data feeds beyond network and host sensors that it intends to consume in a sustained, near-real-time fashion
- The SOC is prepared to dedicate resources to SIEM content management and SIEM tool performance, capacity, and health management

There are certainly exceptions to each of these conditions, but these hold true in most cases where a SOC is ready for SIEM.

The next logical consideration in acquisition is requirements. Requirements are what drive acquisitions—business, functional, performance, system, user, operational, and the like. What does the SOC want to get out of the capability? What major use cases will it serve? Does the SIEM's architecture and feature set support where the SOC will be in three years? These are just the beginning.

As with any major acquisition, the SOC may want to consider a bake-off (e.g., proof of concept or pilot) of best-of-breed products. In this case, one or a few tools can be compared side by side against well-defined, repeatable, measurable requirements. SOC engineers may wish to leverage a scored, weighted, repeatable requirements comparison chart such as Kepner-Tregoe [313]. Vendors will typically grant 60- or 90-day temporary licenses to SOCs performing an in-depth trial. In fact, if the SOC is considering an on-prem installation of a SIEM, using a cloud-based instance may be a much easier way to perform an evaluation or pilot, assuming the cloud version meets the SOC's security and privacy requirements; cloud-based SIEMs pay-as-you-go licensing can be especially helpful here.

SIEMs can sometimes have complex licensing schemes. Each vendor will likely base its product cost on one or more of the following:

- Number of "central" or "indexer" nodes (e.g., hosts that hold the "brains" of the SIEM, such as the correlation engine, or index and persist data); this may be measured in the number of CPU cores belonging to one or more central processing nodes
- The amount of data ingested, persisted, or indexed by the system, often measured in gigabytes per day, events per day, or sustained events per second
- Number of users accessing the system (e.g., the number of "seats")
- Number of appliances purchased (in the case of virtual software or hardware appliances)
- Number of devices sending data to the system and, possibly, the number of device types (Windows, UNIX syslog, database, application, etc.)
- Additional features or add-ons such as content packs

For on-prem and cloud IaaS installs, some SIEM vendors institute hard ingestion, query, or indexing limits through the license file applied to the product instance. Operationally, this can be frustrating when a SOC hits a hard limit on event ingest rate, the number of devices it is collecting from, or the amount of data retained by the system. Sometimes these are hard predictions to make when first buying a SIEM, so careful planning is key. The SOC should also plan for out-year costs as part of its TCO. Annual maintenance and support fees frequently measure between 20 and 30 percent of the initial acquisition cost.

For cloud-based SaaS SIEMs, costing is less fixed and is generally capacity-based. While this places fewer constraints on the SOC, it should be watched to ensure costs do not spiral out of control, perhaps due an overzealous data ingestion, lax tuning policy, or unexpected increases in source data volume.

8.2.6 Operations

The perennial challenge for SIEM owners is in the months and years after initial acquisition. Any analytic platform, SIEM or otherwise, is a long-term investment, and many have opined at these challenges; this has led some to say "SIEM is dead" [314], [315], [316], with their replacements being UEBA and SOAR, which are discussed below.

Custom analytic creation and post-installation tuning enables the SIEM, and thus the SOC, to reach maximum effectiveness.

Regardless of the platform—SIEM, log management, UEBA, SOAR, big data, or machine learning—the SOC needs to dedicate resources to several areas for their investment to yield promised value [317]. Conversely, a SIEM, SOAR, UEBA or big data platform without custom content, use case, workflow, and other tuning, usually offer little value to its users; this is where most such installs go wrong.

Keeping data feeds healthy and data quality high

A mature SIEM is likely to have dozens or perhaps hundreds of distinct data feeds. While any one data feed may be reliable, at any given time there may be multiple degraded or downed feeds. While agents or collectors may be the culprit, often this is an upstream problem: changed network routes, changed firewalls, deactivated accounts, updated device configurations, decommissioned equipment, and so forth.

To solve this, the SOC should have daily or weekly routine for alerting on drops in data across multiple dimensions, such as: collector, enclave, Windows domain, customer, or geographic region. While monitoring collector up/down status is appropriate, the SOC will also need to place two key measures on the data collected: data volume (such as events per second), and coverage (such as devices "seen" per collector). In addition to this, the SOC may wish to measure other aspects of data quality, including statistically significant changes in data parsing and normalization, field cardinality, data enrichment success/frequency, and event size [318].

Creating, curating, updating, and tuning content

The star of the SIEM show is the content crafted by its users. That includes detections, analytics, reports, dashboards, queries, notebooks, and the like. With even a handful of users, there can very quickly become a plethora of custom content found in a SIEM instance.

The SOC should dedicate resources to not only creating this content but managing it. That typically manifests as a routine cadence for content review and vetting, along with one or more analysts designated as "content managers." Some SOCs may implement agile scrum processes [319] around work planning and work execution for creating and maintaining custom analytics and detections, just like it would for ordinary code development.

Optimizing queries

Every query executed occupies system resources. Too many bad queries, and the system will grind to a halt. The SIEM should provide statistics for query resource utilization, such as duration and CPU time. The SOC should in turn a) write some simple automation to alert when heavy or unoptimized queries are seen repeatedly from a given user, and b) take these opportunities to coach users on how to write more efficient queries.

Knowledge management

Some SIEM products include a robust knowledge management capability aside from its case and incident tracking. In such a situation, such as with content, the SOC may designate a steward for this to keep articles about users, computers, enclaves, services, etc., fresh and orderly.

Workflow management and case closeout

The SIEM typically provides basic alert and case management. Users will escalate alerts to cases (and to their peers), and if the case management feature is used, record details in hundreds or thousands of cases per year. It is therefore necessary for the SOC to designate analysts or leads to ensuring cases and alerts are driven to closure in a timely and orderly fashion, and they are not left in an untidy, orphaned, incomplete, or lost state. This can further be driven by routine metrics and reporting on alerts and case handling, typically native to any major SIEM. With that said, SOCs wishing to pursue custom and complex ticketing and workflow scenarios should look carefully at the pros and cons of what they can do with the SIEMs they have or are considering, and what is available in SOAR and case management products. For more on case management, see Section 8.4.

8.2.7 Observations and Tips for Success

As with any other tool discussed used by a SOC, an entire book could be written on SIEM. Our emphasis here is on sound decision making from an architectural and operational perspective. As a result, this section ends with some lessons learned. These takeaways are written as they apply to SIEM but also have strong bearing on log management and SOAR.

Security and network management tools are not interchangeable
SIEM and network management systems have many similar architectural features, such as the ability to aggregate lots of log and alert data into one place, process it, and visualize it. The confidence that can be placed in logs and alerts for security products tends to be less than in other areas of IT such as networking. When a router says its fans are spinning slowly and a switch says a port is down, that is almost certainly the case. When a network sensor says, "Something unusual occurred," it will require a lot more investigation to determine what is happening. SIEM has a rich feature set that supports the workflow necessary to drive events to ground, evaluating whether a given alarm is a true or false positive.

Network management systems lack many of these security-specific features; enterprises seeking to maximize value for network and security management are, therefore, advised against trying to combine network management and SOC workflows into one system.

The best SIEMs were built from the ground up as SIEMs
Some log management and SIEM products in today's marketplace were first architected and coded with a very narrow scope of features and use cases, compared to what they currently advertise. Just as it is important to lay a strong foundation for a tall skyscraper, a good SIEM product is built with a scalable backend, robust correlation engine, and extensible data-ingest capabilities.

Some SIEM products lack a strong foundation and, as a result, have run into problems as developers bolt on more and more features. A poor foundational architecture can manifest itself through reduced ingest rates, slow query speed, fragmented workflow, lack of key use cases, limited real-time visualization, lackluster user interface capabilities, and clunky reporting.

When a SOC is acquiring a SIEM, it is important to look for features that suggest the product was built as an enterprise-grade SIEM from the start, not as a one-off or homegrown project that later turned into a commercial offering.

Consider the whole package
When contemplating a log management or SIEM purchase, many SOCs are narrowly focused on one criterion— "Can the product ingest data type X?" This is probably most analogous to buying a car solely based on what tires it comes with or what color paint it has.

These are certainly important features, but (1) they can be changed by the owner at modest expense, and (2) there are vastly more important considerations such as speed, reliability, and operator experience.

In general, SIEMs can accept almost any possible security-relevant data feed, either out of the box or through an agent software development kit (SDK) or API.

What is not important is, "Can the SIEM ingest my data?" That is a given; what is more important is, "What can I do with the data once my SIEM has ingested it?" This means everything from real-time dashboards to correlation to reporting and escalation.

A day to install; a year to operationalize
The initial setup of an on-prem SIEM is largely straightforward and can usually be accomplished in a day or two; cloud-based SIEMs are deployed in minutes. In a few weeks, the first few data feeds can be hooked up and tuned, with content created that provides quick wins.

However, outfitting the system with the right data, tuning it, writing content for the constituency, training users, and integrating it into operations can take several months. In many cases, getting data owners to provide reliable, sustained log feeds can be a politically arduous process. As a rule of thumb, the more robust the SIEM, the steeper the learning curve: not because the interface may be clunky, but because it takes even the smartest analysts time to understand the fundamentals of the tool and what it is capable of. A college Computer Science student can learn the fundamentals of a new language in several weeks; becoming truly proficient in a language takes much longer. The same is true of SIEM. One must also consider the time commitment for integrating the features of SIEM into SOC operations.

Most successful SIEM implementations worked because they had a champion within the SOC who understood the technology, invested the time necessary to get data feeds working, and adapted content to the constituency environment. This is the case for every security product used by the SOC, though SIEM especially.

Out-of-the-box content serves as a good start for most SOCs, but the best content is customized for the constituency.

Each part of the SOC will Use SIEM differently
Each work center in the SOC has different uses for the data and features SIEM can provide. Triage analysts will be interested in efficiencies and accuracy of alert enrichment, prioritization, and annotation. Incident investigators will likely be interested in gathering as many details about a potential event or incident as possible, meaning they will run free-form queries over long periods of time. Those responsible for hunting and trending activities will likely fuse various sources of cyber intel and tippers into the tool, repeatedly running long, computationally expensive queries. Managers will be interested in case management and metrics from the tool, validating that their respective part(s) of the SOC are following procedures, following up on anomalous activity when the system catches it, and not letting cases languish in the system.

Each of these parties has an equally valid need for training on the tool, well-written content to fulfill their mission role, and performance that meets their ops tempo.

Again, many SOCs meet these needs by designating a SIEM champion or content manager. Moreover, each SOC section will have overlapping needs for the tool, and it is important that the SOC designates one or more people to ensuring content and that queries are effectively consolidated and deduplicated.

New tools mean new processes
When a new SIEM is introduced to a SOC, some analysts may cling to old, familiar habits formed with old tools. Using a SIEM to triage and view data in the same way as the previous tool (another SIEM or the native console perhaps) may add little value. When a SIEM is brought in, updated SOPs and training are essential tools to enable analysts to take advantage of new, unfamiliar features and approaches.

Once again, this is an opportunity for the SIEM champion and content manager(s) to open minds to new ideas. Just a few examples crafted for each SOC "persona" can get them thinking, collaborating, and using new approaches, and seeing new value from the SIEM.

A SIEM is only as good as the data you feed it
The old saying, "garbage in, garbage out," applies perfectly to SIEMs. "Strategy 7: Select and Collect the Right Data" discussed how the value of even the most relevant, detailed security logs can be completely diluted if the SOC is not discriminating about what data it brings in. It is of utmost importance to select and tune data feeds according to the constituency environment, threats, vulnerabilities, and mission.

One of the by-products of a healthy selection of data sources is that a SOC's own sensors are essentially put on the same footing as any other source of data (e.g., Web proxy records or application logs). From the perspective of the analysts, network sensor telemetry and EDR alerts are just more data feeds among many they can choose from when tailoring a report, correlation rule, or dashboard to a given threat.

Lack of a single common data standard can be overcome
The history of audit data aggregation and security data management is paved with industry standards, none of which have had comprehensive adoption: Common Intrusion Detection Framework (CIDF) [320], Incident Object Description and Exchange Format (IODEF) [321], Common Event Infrastructure/Common Base Event (CEI/CBE) [322], Common Information Model (CIM) [323], Common Event Format (CEF) [324], and Common Event Expression (CEE) [325]. A common theme among them all is the desire to provide a vendor-agnostic format for recording and ingesting security-relevant data. CEF for example does not cover all event and log sources [326].

Some may find a single standard overly limiting as a single standard does not support the full variety of data of interest, and consequently, the meaning and value of some data fields are lost when they are forced into a single standard. Moreover, the increasing use of key/value pair data formats, notably JSON, lessens the need for a rigid data schema. For example, Windows standard security audit events [327] and Zeek's various log types [328] are both extremely rich. One can only use their imagination as to what will happen if both are put in the same table with the same schema. Instead, many SOCs find value in preserving these raw formats because they are both understandable and give the analyst the full fidelity of what was originally recorded.

Modern SIEMs feature hundreds of data parsers, which vendors expend a lot of resources to keep up to date. With that said, no SIEM ships with parsers for every data feed in existence. It is common for SOCs to tweak existing stock parsers and fashion their own from scratch. SIEM vendors will support multiple disparate event schemas, each tailored to a comparatively finite set of use cases, as is the case with CIM.

SOCs should architect their collection and retention to support criminal, civil, and administrative proceedings
The electronic evidence a SOC collects *may* be used by law enforcement, legal counsel, and various investigative bodies, depending on the role of the SOC. Just as with any artifact collection procedure, the SOC should ensure that the way it gathers, stores, and analyzes security-relevant data supports these activities if required. Moreover, applicable privacy laws may impact how the SOC collects and retains certain log types of content, such as with PII and based on the geographic region of the users being monitored. While computer forensics and specific regulatory compliance is out of the scope of this book, the SOC may wish to discuss this with legal counsel and examine applicable laws and regulations for further guidance on this matter [329]. Also, it should be noted that the interpretation of such laws varies widely and can have a profound impact on the cost of the SOC's log collection and storage architecture. Ensuring that common sense is integrated into system design, along with a well-informed understanding of the law's impact, is critical.

SIEMs are not "dead"
As has been discussed, the SOC has an enduring need to collect, normalize, store, retrieve, analyze, and share security-relevant telemetry. This need is not going anywhere, though the way to achieve it has more options than it did ten years ago. SIEMs have been (perhaps rightfully) criticized as expensive to buy and time consuming to maintain. As will be articulated in the next section, sometimes a full-fledged SIEM is too big for a SOC; maybe a cloud-based log management solution is a better choice. For some of the largest SOCs, they may take still other, more sophisticated approaches.

8.2.8 SIEM Alternatives

By now, the myriad use cases, success factors, and rationale for a SIEM should be clear. But not all SOCs that want to achieve these outcomes run a SIEM. How and why?

The best way to answer this question is by providing two archetype SOCs that do not operate a SIEM, and how they achieved the same or similar outcomes through alternative approaches.

8.2.8.1 Small and New SOCs: EDR Plus Log Management

Many small and young SOCs do not have the resources or expertise to stand up a SIEM. Specifically:

- Not enough resources to operate and maintain a SIEM.
- Not enough data sources to justify running a SIEM.
- They have an incumbent or shared log-management solution that gives them access to most of the log data they need and standing "queries on a timer" in the log-management solution fill the handful of detection requirements they have.
- Based on their mix of desktop/laptop endpoints and cloud-based services, they choose to satisfy their most important monitoring and analytic use cases with a combination of EDR and log management.

This is a very pragmatic approach for monitoring a small enterprise. In combination with some investments in managed security services, this SOC has satisfied its needs on a very modest budget. An example of this approach is shown in Figure 34.

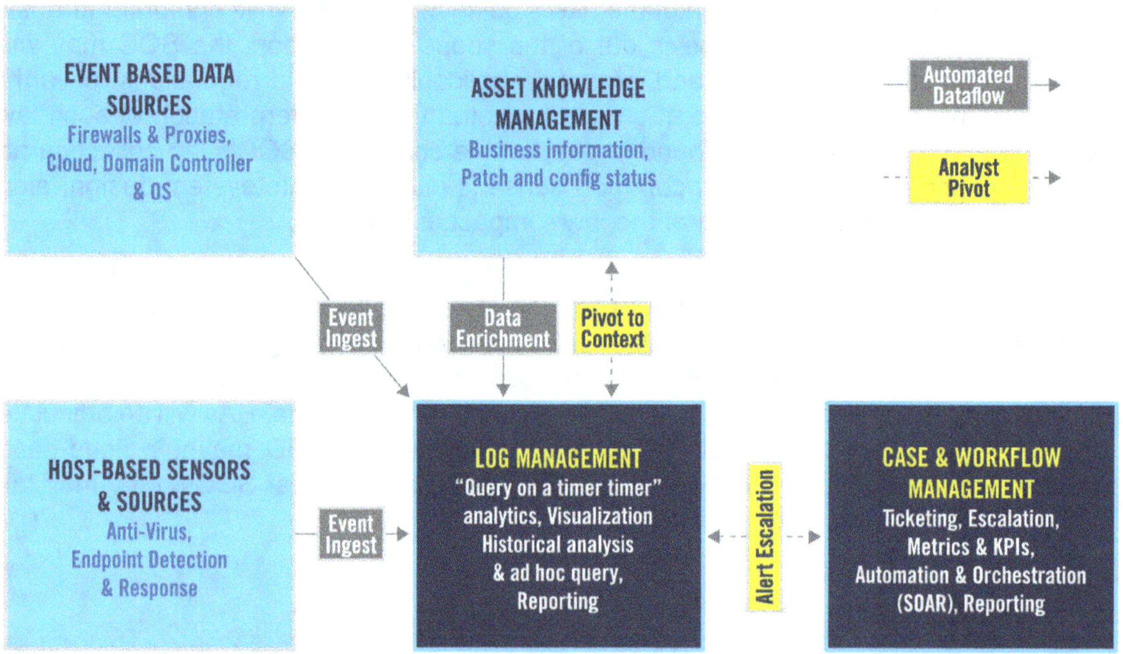

Figure 34. Small SOC Tool Architecture

8.2.9 High End: Building a SIEM from Parts

Some very large, mature SOCs feel that they have "outgrown" SIEM in part or in totality. They have:

- A large shop of over a dozen tool admins and engineers savvy in development and big data
- Dozens of disparate data feed types, tens of thousands of nodes to monitor, and well over 10TB/day of data ingestion
- Complex analytic and detection requirements, with dedicated resources for daily detection authoring and tuning, including specialists in intel fusion, hunting, and maybe one or two data scientists
- A large user base of SOC analysts and partner system/service owners that have been deputized by the SOC to participate in analytic creation
- Experience with commercial SIEMs such that they understand its inner workings and requirements well, and have felt constrained by its limitations, or burdened by its cost model

In this case, this SOC may be justified in shifting its focus toward away from SIEM, either relegating a SIEM to specific use cases with smaller data volume and thus reduced cost[12], or eliminating the SIEM entirely, and possibly eschewing a commercial EDR as well. It is possible to achieve this with a blend of technologies (See Figure 35):

- Rich system instrumentation using OS Query [60], GRR [330], Sysmon [59], WEC/WEF [331] and auditd [332]
- Data ingestion, parsing, and transformation using tools like Logstash [333], Fluentd [334], and Beats [335]
- Data ingestion and movement using a publish/subscribe model with a message bus like Apache Kafka [336], perhaps aided by Apache NiFi [337]
- Ad hoc query, investigation with a data store like Elasticsearch [61], Graylog [338], Azure Data Explorer [339], or Cassandra [340], along with Jupyter [341] notebook and Kibana [342]
- Near-real-time and batched analytics using several of the following: Flink [343], Samza [344], Spark [345], or Storm [346], running against data in Kafka or stored in HDFS [347], perhaps with jobs and notebooks run out of a cloud-based framework like Databricks [348] in a prevailing language like R [349] and Python
- Alert triage, escalation, workflow, case management, and automation using a case management or SOAR platform, discussed in the next section

---

[12] The practice of running a traditional SIEM side-by-side with a big data or ML platform is common enough that is often referred to as running "dual stack" monitoring platforms

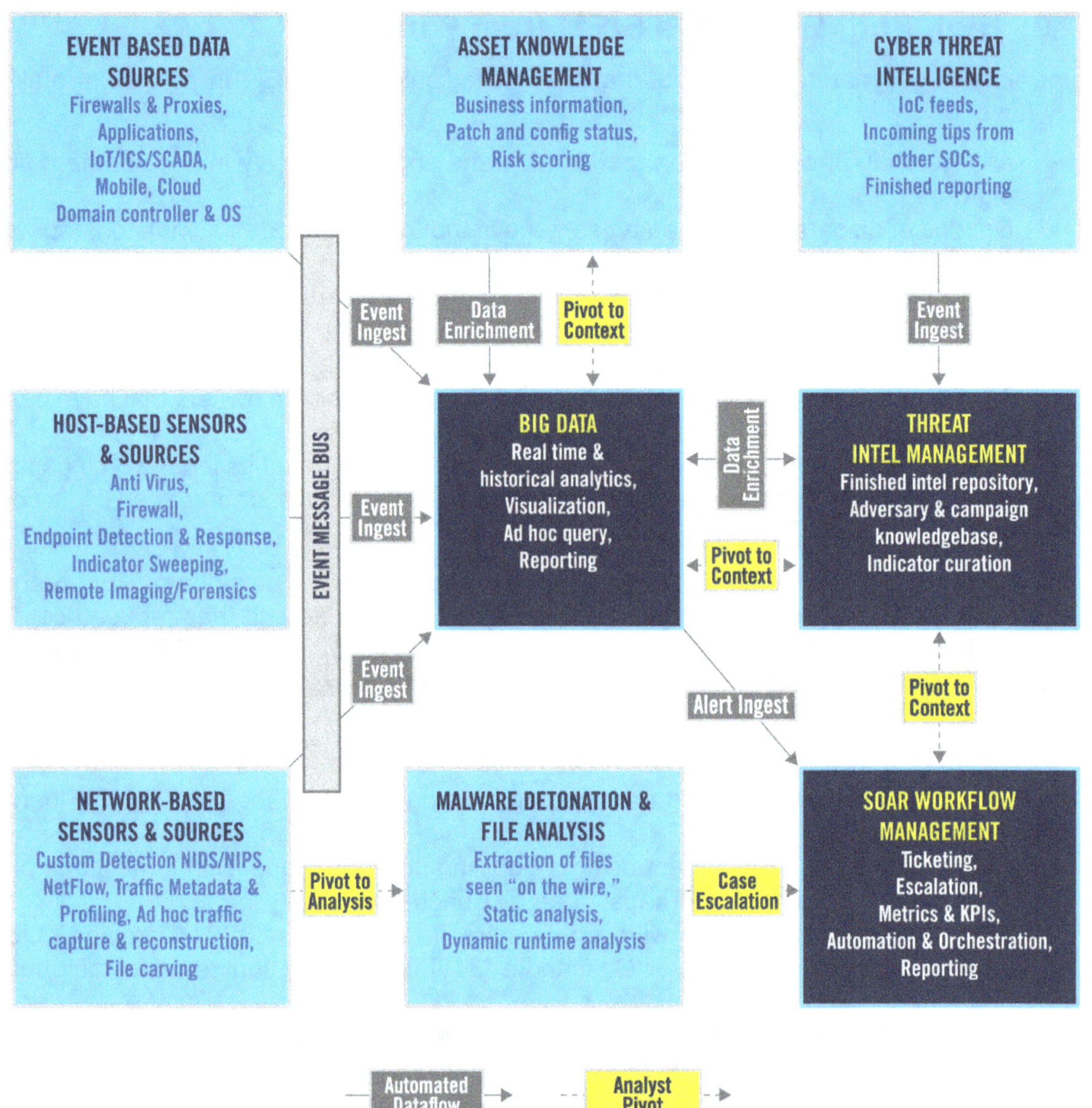

Figure 35. "SIEMless" SOC Architecture

At a *very* high architectural level, little has changed from an architecture that includes SIEM. Implementation and sustainment compared to pure-play SIEM is substantially different, however. Here, the SOC is:

- Getting less "off the shelf" and consequently the amount of integration and sustainment work has gone up considerably, in particular: data parsing and writing analytics.
- Compared to a SIEM-centric design, is more reliant on SOAR to "stitch the pieces together," SOAR or case management to support alert triage, and the big data analytics platform to perform correlation.

- Ideally, using a message bus as the hub for data transfer, given that multiple data platforms will likely publish, enrich, and consume event data.

On the other hand, given the economies of scale involved, some SOCs will end up saving money by taking a "build not buy" approach. When considering this approach, it is important for the SOC to consider the full cost involved, including labor for integration and sustainment. The success of this approach is highly dependent upon strong documentation, sound system design, and comparatively low employee turnover. If, for example, this SOC's engineering team was handled by in-house contract staff, a contract changeover may be devastating to the health and wellbeing of the monitoring architecture. This approach is certainly not for everyone, but for very well resourced SOCs, they may feel this is the best way to get the scalability and flexibility they desire.

8.3 User Entity Behavior Analytics

User entity and behavior analytics (UEBA) refers to the set of products and product functionality associated with uncovering users' and other entities' deviations from normal patterns, which in turn is likely to reveal malicious activity. In contrast to traditional SIEM, UEBA tends to have an increased focus on internal users and threats, and the right-hand portion of the cyber kill chain. UEBA is typically found as a free-standing product, a product that augments a SIEM, functionality within a SIEM, or functionality in other security software. Because of the variety of functionality delivery, UEBA may be [350] defined [351] by its attributes, described in three pillars:

UEBA Pillar 1: Use cases
- Detecting users that are acting in potentially or clearly malicious fashion, or simply outside their normal "pattern of life"
- Compromised insiders (such as due to stolen credentials or lateral movement)
- Data loss detection and data exfiltration, such as across enterprise or enclave boundaries
- Changing alert priority and enriching alerts based on knowledge of entities involved in the alert (such as based on risk scoring)
- Supporting analytical aids and techniques
- Peer grouping: measurements of similarity clustering across multiple dimensions for groups of entities, e.g., what users or computers act in similar manners
- Tools to stitch together disparate entities, e.g., 1.2.3.4 == 11:22:EE:23:8C:36 == evilhost.company.com
- Leveraging entity relationships and connections to spot lateral movement
- Data visualization that enables analysts to view significant entity behavior in a chronological order, with a richer visual representation than simple "flat" generic events

UEBA Pillar 2: Data sources [352]
- User authentication and access control systems like Windows domain controller logs and VPN systems
- Configuration management
- Employee data, such as from human resources databases
- Next gen firewall, NIDS/NIPS, and deep packet inspection
- Host sensors, including anti-virus and EDR

UEBA Pillar 3: Analytics
- Supervised machine learning, which is to say statistical methods of finding bad behavior that are reinforced with analyst feedback (sometimes referred to as "labeling")
- Unsupervised machine learning, meaning statistical models that look for deviations from statistically normal patterns, without the need for analyst input
- Rule-based detections, similar in fashion to those in a legacy SIEM
- Various techniques fitting the above categories, which are beyond the scope of this book: Bayesian belief networks, social network analysis, classic statistical clustering, neural networks, and regression models

8.3.1   UEBA Function Selection Considerations

When selecting a UEBA product, or add-on feature set, one can use the three pillars of UEBA to weigh value and cost factors. Consider the following questions:

- Has the SOC reached the state of maturity where it really needs (and can sustain) a UEBA tool [353]?
- How easy is data ingestion for the UEBA, given the candidate datasets the SOC intends to use?
- What integration considerations apply with existing tools and systems, SIEM or otherwise?
- What is the applicability of the UEBA's models and analytics to the data sources and analytic objectives of the SOC, its customers and partners?
- How is the quality, effectiveness, and "pedigree" or the statistical and ML models?
- What is the overall true and false positive rate?
- How much time is needed for the UEBA feature to train, and the amount of analyst or administrator intervention it takes to make the tool work as intended
- What is the overall "gut" feeling from the analysts: would the UEBA provide insights about users, hosts, etc. in the enterprise that would not otherwise known and proven accurate based on follow-up human investigation?
  - When the UEBA said a user or host was acting "weird," is it clear why that conclusion was made?
  - What proportion of entities are judged as being abnormal?
  - To what extent is the UEBA product or functionality is "connecting the dots" correctly?

One of the best conditions of success for a UEBA initiative is when it can be applied to several large, trusted user populations whose pattern of life behaviors can be defined and are relatively consistent. Under such a circumstance, it is more likely a UEBA product will be able to correctly and effectively spot outliers. A user population that does not demonstrate a strong sense of "normal" may be more challenging to achieve a successful outcome using traditional UEBA techniques.

Finally, UEBA success or failure is subject to nearly all the same factors and guidelines as ordinary SIEM, including:

- Selecting the right product and technology for business-driven requirements
- Correct implementation at scale with the right data feeds
- Champions in the SOC who are accountable to UEBA success
- Content authors and detection tuners that understand the UEBA analytical approaches well enough to effectively harness the tool and integrate its findings into the alert handling lifecycle
- Analyst resourcing to triage, analyze, and escalate the UEBA's alerting and leverage its UI/UX

8.4 Case Management

Every SOC needs a way to track incidents. Generally, the more mature the SOC, the more sophisticated and customized that incident tracking capability needs to be. Case and workflow management, perhaps more than any other SOC tool, is intimately involved in SOC operations. Consequently, there is no one size fits all.

This section covers requirements, architectural options, and success factors for incident tracking. It also discusses areas in which an incident tracking system (by itself) is insufficient, indicating the SOC should seek out additional forms of knowledge management.

8.4.1 Case Management Requirements

The SOC has a handful of prominent requirements and use cases for case management that should be considered:

- Allows consistent and complete information capture across incidents for each state of the incident life cycle—alert triage, in-depth analysis, response, closure, and reporting
- Can record both structured information from analysts (incident category, time reported), semi-structured data (impacted users, impacted systems) and unstructured information (analyst narrative), along with time-stamped notes
- Has necessary interfaces with constituents, usually including one or more of the following:
    - Direct mediated access to cases a constituent or constituent's service/systems are involved in
    - Communication with one or more constituency ticketing and case management systems

- ◦ Email messaging and interface with constituents, such as notifying constituents on case updates, automated escalation, tasking, and email replies that will update the case
- Is insulated from adversary access, particularly in the case of a large-scale breach
- Supports escalation and role-based access control for different sections within the SOC
  - ◦ Supports access control around specific sensitive or confidential cases, thereby supporting insider threat case tracking for SOCs that perform that function
- Supports trending, metrics, and feedback:
  - ◦ Including mean/median time to acknowledge, respond, eradicate, and close
  - ◦ As a feedback loop to inform detections and analytics tuning, and for advanced SOCs, labeling for supervised ML models
  - ◦ The SOC should expect to routinely gather and report on metrics and workflow
  - ◦ Supports measures discussed in "Strategy 10: Measure Performance to Improve Performance"
- Can incorporate artifacts or pointers to artifacts, such as events or malware samples.
- Allows analysts to capture information about specific entities (particularly users and hosts) that can be referenced and correlated with other cases, thereby providing continuity across disparate analysts and cases, and potentially avoiding redundant work
- Allows linking and parent/child relationship between cases, including the ability to spawn child cases and tasks to other parties, such as request malware analysis or follow up by another analyst
- Some ticketing systems will overlap with related product segments, in particular:
  - ◦ Information collection, data enrichment, and response automation, discussed in the SOAR section below
  - ◦ Cyber threat integration and enrichment, such as threat indicator matching

Constituents should be able to submit security case incident reports from a constituent-facing Web portal. Usually, this form is either a) a tailored/locked down portal to the SOC case management system itself, or b) a set of web automation that will take user-submitted form input and post it to the case management system API, thereby creating a case. Although this information might automatically populate a case, the information visible to the constituent after submission should be tailored. For example, the constituent should not be able to read all the internal analyst notes.

8.4.2 Implementation Options

When contemplating how it will support case management and case tracking; SOCs have several disparate approaches, each with their own pros and cons, shown in Table 21.

Table 21. SOC IT Case Management Options

Approach	Pros	Cons
Existing IT case management system used by the constituency IT operations organization	• Usually comes with polished feature set, documented setup, and central administration. • Economy of scale; acquisition and O&M free or less expensive. • Larger pool of case management system experts. • Robust reporting and metrics. • Seamless integration with IT help desk and IT operations	• Less likely to be flexible to SOC needs. • SOC may need to adjudicate customizations through other groups and approval boards. • Sensitive data is comingled with general IT help tickets. • Admins can see SOC's cases which increased the likelihood of compromise of internal threat cases. • If general constituency systems are compromised, adversary may be able to see or manipulate SOC cases. • Customization may be lengthy or expensive, especially if vendor does not provide "out of the box" SOC extensions, use cases, templates, and workflows. • Depending on customization options, may feel equivalent to a "walled garden."
SOC instance of COTS IT case management system [354], [355]	• Comes with polished feature set, documented setup, and central administration. • Usually has out of the box customization specific to SOC use cases • Robust reporting and metrics. • Case details available only to parties designated by the SOC. • Usually, the most flexible and powerful approach amongst all ticketing options.	• Can be very expensive. • Customization may be lengthy or expensive, especially if vendor does not provide "out of the box" SOC extensions, use cases, templates, and workflows. • Depending on customization options, may feel equivalent to a "walled garden."
SOC instance of FOSS security case management system	• Depending on tool chosen, may come with polished feature set, documented setup, and central administration. • Many (such as TheHive [356], SCOT [357] and RTIR [358]) are designed specifically for use by SOCs or have incident handling modules or plug-ins. • Free to acquire. • Reporting and metrics possible. • Case details available only to parties designated by SOC.	• Unless there is a commercial vendor that offers professional services, setup and sustainment may be a challenge with community-only support. • O&M & customization may require staff with experience in scripting, programming, or databases.

Approach	Pros	Cons
SIEM or SOAR case tracking system [359], [360]	• "Free" and usually easy to leverage if SOC owns a SIEM or SOAR. • Likely very strong if part of a SOAR • Specifically built for tracking security incidents. • Leverages user groups, permissions, and escalation paths shared with other SIEM/SOAR tasks and functions. • Users can attach events and some artifacts to tickets.	• Not appropriate if SOC workflow is not focused on the SIEM. • Typically, more limited flexibility, depending on SIEM product. • Usually less robust than purpose-built case management (especially with some legacy SIEM products) • If SIEM/SOAR goes down, nearly all aspects of SOC operations (triage, analysis, case tracking) are also down. • Optimized for alert handling from SIEM, usually less-so for other platforms (e.g., email or EDR).
Custom-built ticketing system	• Extremely flexible • Reporting and metrics possible • Case details available only to parties designated by SOC	• Expensive to build and maintain. • SOC must build system from scratch, requiring staff with extensive experience with programming and databases. • Development of system may take a while, since SOC must start from nothing. • Less and less justified due to availability of robust COTS and FOSS SOC ticketing capabilities.
Cloud-based SaaS ticketing system	• Has pros associated with SOC tools and data in the cloud (See Section 8.7) • Turnkey capability which can be up in a matter of hours or days. • Usually no upfront costs. • Sustainment, O&M, upgrades very simple. • Generally, will require less staff to support.	• Has cons associated with SOC tools and data in the cloud (See Section 8.7). • Heavily dependent upon security features and internal protections of the vendor. • Could be compromised if vendor or underlying cloud is breached. • Single errors in configuration and management can easily open system to outside exposure or compromise. • Not appropriate for air gapped SOCs and enclaves. • Depending on customization options, may feel equivalent to a "walled garden."

8.4.3 Success Factors

The case management system will receive daily, even hourly attention and use by most analysts in the SOC. It can greatly enable efficiency and repeatability in its standard operating procedures. The SOC's investment in this system tends to be deep, proportional to the size and maturity of the SOC. Replacing this tool can be very disruptive to operations, so making a wise choice with several years' growth in mind is wise. Here are some observations and suggestions:

- New, less mature, and small SOCs may wish to consider a ticketing system that is cloud-based, the same solution as the IT helpdesk, or built into one of its other tools like a SIEM or EDR, to minimize timelines and acquisition and sustainment costs.

- SOCs that strongly focus on their SIEM tool will sometimes use the SIEM's ticketing capability and live within whatever limitations that tool presents. The SOC should maintain awareness of this unintended bias as it may exert undue influence on operations.
- It is very common to see case management and automation comingled in the same system, particularly in the case of SOAR; traditional case management vendors will offer SOAR capability and SOAR vendors will often offer case management capability, making the distinction blurry.
- Complex, free-standing ticket management systems can be a strong choice for large, mature SOCs and SOCs that are hierarchical, federated, or tiered with multiple operations floors.
- The SOC may find that it will naturally gather information not only on constituent systems and networks, but about adversary actors and campaigns.
  - When making a choice regarding case management, the SOC should evaluate whether it expects to capture this information principally inside its case management system, or in a disparate threat intelligence management platform. See "Strategy 6: Illuminate Adversaries with Cyber Threat Intelligence" for more information on TIPs.
- For SOCs unsure where to get started in how to structure incident tracking and metrics, they may wish to consult the VERIS framework [361].

After choosing ticket management platform, the SOC should observe the following keys to success, many of which will mirror SIEM and SOAR:

- For on-prem installations, assign staff responsible for installation and system administration.
- Designate staff who are responsible for use case development, workflow curation, and examining analyst routine to find new opportunities for efficiencies and automation.
- As necessary and appropriate, consider configuration management control commensurate with the complexity of the case management platform and the size of the user base
- Evaluate the SOC's needs for incident data capture, paying careful attention to:
  - Collaboration needed between different SOC roles and constituents.
  - Building a knowledgebase over time about frequently seen activities, systems, networks, and enclaves.
- When tailoring the case management capability for analysts performing case and alert triage, the following actions may help:
  - Minimizing the number of clicks per investigation, such as through making frequently used fields and information visible on the first page, frame, or tab of an incident view.
  - Ensure alert deduplication/aggregation and throttling is put in place to avoid a phenomenon known as ticket storms.
  - Enabling triage analysts to "squelch" or temporarily pause a busy detection that would otherwise overwhelm them with unwanted cases.
  - Automate alert enrichment and frequent emails sent to constituents, such as through email templating.

- Drive a metrics program, with careful attention not just to repeatability and efficiency, but supporting and quality measures and positive behaviors, and detecting when team members are manipulating case handling simply to make metrics look good. For more, see Strategy 10.
- Leverage data export mechanisms and APIs to:
  - Fuse case data with other systems, supporting alert enrichment, detection tuning, and reporting.
  - Enable COOP and failover scenarios as necessary.

This is just the beginning for what a good ticket and case management system can do. This conversation quickly shifts in focus from information capture to workflow automation. The next section addresses the latter.

8.5 Security Automation, Orchestration, and Response

SOAR are a set of products and features that, as their name implies, enable the security operations user to quickly and efficiently design and leverage repeatable processes common to the SOC. Although SOAR is regarded as its own market segment [362], [363], [364], both case management systems and SIEMs achieve many of the same outcomes. In this book, the term "SOAR" is used to refer to any product or tool principally used to achieve automation and orchestration for the SOC. Leveraging SOAR, the SOC can:

- Gather incidents from disparate systems, presenting a single pane of glass view for alert triage and alert management.
- Enrich and prioritize alerts, integrating threat intelligence and knowledge of entities involved in an alert.
- Execute automated queries or other information gathering activities when an alert fires, like sending a file to malware detonation chamber, gathering vulnerability scanner results, or looking up a user's HR data.
- Run a series of frequently used queries against a log repository.
- Perform routine constituent interactions, such as sending alert details to a constituent, asking for confirmation or repudiation, "was this expected" or "was this really you?"
- Automate response actions like terminating network connections or disabling user accounts.

There are many reasons for the SOC to harness SOAR capabilities:

- Too many alerts and not enough time to manually analyze all of them
- Bring better, more prioritized, and enriched data to the analyst
- More repeatability and consistency in triage and investigation
- Enabling junior analysts to "snap to" practices and procedures codified in workflows by more senior and experienced staff, and at the same time allowing those junior analysts to improve their TTPs more quickly
- Not enough staff with both experience and time to manually implement the same integration and automation that SOAR can support "out of the box"
- Improvement in quality of life for the analyst, meaning fewer manual tasks to accomplish

- Faster triage time (mean/median time to acknowledge and investigate)
- Faster response time (mean/median time to contain, respond, and eradicate)
- As an overall force multiplier to all other analytic and detective capabilities

8.5.1 When to Consider SOAR

Generally, the SOC is ready to harness new or additional SOAR capabilities when it has met several of the following conditions:

- It has established a routine incident handling process, codified that process in writing, and practiced following that process across its pool of analysts in a sustained and consistent manner.
- It has used indigenous and existing scripting and automation capabilities and finds that either it is increasingly difficult to achieve its business outcomes with those tools and/or is spending an inordinate amount of time keeping those automations up to date and evolved with its overall processes.
- It is finding it increasingly difficult or necessary to push down TTPs from senior analysts to junior ones and/or junior analysts are not consistently following guidance for handling certain routine incidents.
- Analysts and engineers are finding it increasingly difficult or time consuming to understand, share, and co-contribute to the same automation and workflows.
- Analysts simply find that they have reduced certain repetitive incident investigation steps to routine practice, and they want to devote more time to more activities like hunting.
- The tools the SOC wishes to integrate with a prospective SOAR product or functionality have documented or compatible APIs.

Regardless of the specific tool, the idea of automatically closing out incidents and responding to the adversary can be very compelling. However, for a SOC that does not already have firmly established processes in place, this can also add risk. Extending the conversation from IDS/IPS around risks and rewards for putting an IDS into prevention mode, the SOC should examine the balance of benefit and regret in each scenario [365].

Here are some tips for making the most out of a SOAR investment:

- Make investments in key tools with well-documented APIs, notably log management, network and host sensors, SIEM, case management, and cyber threat intelligence management.
- Perform good project management, as with any other tool investment:
  - Designate members of the SOC staff who will act as champions for making the SOAR successful and ensure they have the time and resources to meet that commitment.
  - Establish requirements, success criteria (like analyst hours saved), and a definition of done.
  - Allow time for buildout, testing, and operational integration of SOAR use cases.

- Identify and start with high-reward, low-risk items that are frequently repeated in the SOC's daily and weekly incident handling process; the best initial use cases are typically information gathering and alert enrichment.
- For each IT systems and business unit that the SOAR product will integrate with:
  - Identify stakeholders and engage early to harmonize objectives.
  - Integrate understanding of the targeted systems' behaviors, and device playbooks in case response automation does not go as planned.
- Avoid high-risk workflows until both the SOC has reached strong maturity in its incident handling, and executed low-risk integrations; high risk workflows are those that satisfy several of the following criteria:
  - Interface with systems the SOC does not control.
  - Implement a change that cannot be reversed or undone.
  - Enact changes that may (or are intended to) interrupt services, accounts, network traffic, or system activity.
  - Enact changes across a wide set of systems or services.
  - Have a potential to cause hardship, disruption for many users or customers or endangers revenue generation.
  - Influence or impact systems responsible for life or safety.
- When implementing high-risk workflows:
  - Avoid active blocking actions against firewall, VPN, or identity solutions in the first three to six months of using a SOAR product.
  - Ensure reversibility is built in. For example, if a SOAR product can add or enable firewall rules, the SOC and firewall owners should have a playbook, which allows them to roll back that change quickly.
  - Whenever possible, attempt these integrations in a controlled manner, with non-production assets, or during non-critical hours of operation.

8.5.2 SIEM and SOAR Comparison

Having discussed SIEM and case management, this should sound very familiar. Truthfully, SIEMs have advertised automated response capabilities since the early days (meaning the early 2000s) In contrast to related products like SIEM, SOAR emphasizes:

- Gathering alerts and cases from disparate systems, such as incoming constituent emails, SIEM correlation alerts, and EDR detections, thereby forming a single pane of glass for alert triage, including some interesting enhancements, like:
  - Suggesting related cases
  - Performing machine learning that link information and activities seen from other cases
  - Suggesting next steps based on the way past cases were handled
- Graphical interaction with complex workflow and automation development through browser-based drag and drop flowcharting
  - This allows users to achieve similar outcomes with comparatively less custom coding and in less time.

- Most good SOAR products will also allow the workflow author to uncover and edit underlying code and/or script portions of workflows manually, when the need arises.
- Giving the automation author a framework to write the automation and the analyst using that automation strong visibility into the workflow being followed through the GUI
- Providing a curated list of APIs and integrations to *existing tools*, in the same way a commercial SIEM typically ships with a library of data collectors and parsers
- Providing a set of APIs, allowing a workflow author to integrate and actuate new and custom tools

Conversely, SOAR deemphasizes high-scale data collection and log management, and consequently the development and support resources to achieve it; and case management for areas of IT outside cybersecurity, thereby catering to a tight market segment.

8.5.3 SOC Automation and Orchestration Without SOAR

There are many ways to achieve strong automation and orchestration for SOC alert triage and incident response. So many, in fact, that an entire page could be filled with links to scripting languages and IT automation platforms. In achieving the same outcomes advertised by SOAR products, the SOC may leverage multiple approaches:

- Simple scripting using a popular language like Python and documented product APIs. While this may be cheaper than buying a SOAR for some SOCs, it runs into challenges with staff turnover and scaling "up" with analysts who are not strong scripters and coders.
- Alert enrichment can be done by SIEMs and through popular tools like Logstash.
- If the focus is purely on structuring and automating routine alert investigation, a user notebook like Jupyter may be the best fit.
- Any good SIEM can integrate alerts from other SIEMs and EDR products and are likely to have email integration. If the objective is to have a single pane of glass for incident triage, an incumbent SIEM may be the best fit.
- If the SOC has or is considering a robust case management tool, the SOC may want to see how much automation that tool can achieve on its own.
- If the SOC or its parent organization has an indigenous automation platform like if-this-then-that (IFTTT) [366], Zapier [367], or Microsoft Power Automate (formerly known as Microsoft Flow) [368], this may achieve some of the same outcomes, though careful attention should be paid to the security requirements and implications of using shared platforms for security tasks, and the suitability of that platform for the SOC's needs.

In summary, SOAR is an indispensable capability for the modern SOC. Like any other tool, the SOC will generally get out what it puts into SOAR. The outcomes offered in this market segment can be found in many tools, and the SOC consequently has many options for orchestration and automation. In general, the SOC should focus its early SOAR investments on automating repetitive steps in alert triage and investigation.

8.6 Protecting SOC Tools and Data

The SOC is a source of tremendous insights and situational awareness across the constituency, not to mention the raw data it collects. The SOC also has the responsibility to protect these tools and data from misuse. For example, even the best SOCs have gaps in their visibility, which is useful to an adversary. Knowledge of what monitoring tools are in use might allow the adversary to mount direct attacks against them or, more often, shape its attacks to avoid detection. Yet, sharing insights with others in the constituency is also important for defense. This causes a natural tension, and every SOC must decide the mix of protections and sharing it should put in place.

> *A SOC can execute its mission in part because the adversary does not know where or how monitoring and response capabilities operate.*

Best-of-breed SOCs operate most of their systems in an out-of-band fashion that isolates them from the rest of the constituency. Yet, at the same time, they need to share and collaborate with constituents. The SOC must find a balance, as shown in Figure 36.

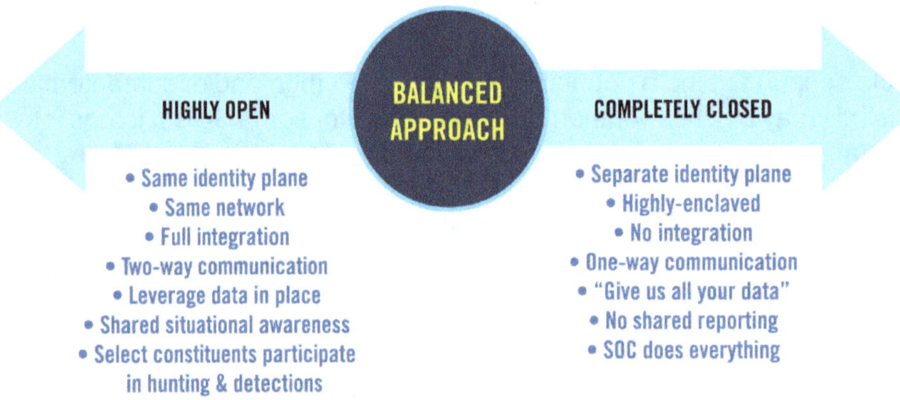

Figure 36. Open Versus Closed SOC

8.6.1 SOC Enclave

Operating a SOC presumes that, at some point, the constituency will be compromised. Moreover, actors of concern include individuals with legitimate access to constituency IT resources. Following this logic, the SOC must be able to function without complete trust in the integrity, confidentiality, and availability of constituency assets and networks. While the SOC must have strong integration with constituency IT systems, it must be insulated from compromise.

The assets defended by the SOC are usually bound through one or more identity planes with transitive trust relationships, such as membership in a Windows domain for an AD forest. One of the most common goals for the APT is to gain privileged domain rights, allowing the attacker to move laterally across any system that is a member of the domain or administered from the domain.

Consequently, any aspect of the SOC's mission operating on these domains is at risk, and many SOCs are adapting an "assume breach" approach. Consider a SOC that operates a fleet of network sensors and SIEM. It is fair to assume that the APT with domain administration privileges can install keyloggers and RATs on any system on the constituency domain. Even if sensors and log management systems are not joined to the domain, they may still be compromised if analysis and maintenance are performed from desktops that are connected to the domain. As a result, a key recommendation is:

With an assume breach mindset, any aspect of the SOC mission or its data that flows across systems that are joined to the constituency Windows domain could be considered compromised. The SOC is considered the "inner keep" of the constituency castle and should be the least likely asset to be compromised. As a result, the SOC must be even more vigilant in securing its systems against compromise. This often means vigilant attention to and shortened patch cycles and a robust security architecture.

With this said, it may not be advisable for SOC to horde all its data and systems in a tightly insulated enclave. Network taps can isolate some systems, such as passive sensors, even though they are still on the enterprise network. The SOC must deploy monitoring systems where constituents are; this might be down the hall, but it might also be around the world. Some network and domain owners may stipulate that SOC systems, particularly commodity Linux or Windows, must join their domain and be maintained, at least in part, through their automation and not the SOC's. Moreover, some host sensors can more easily be deployed and updated if the management backend servers are joined to the same domain as the monitored endpoints. An absolutist approach to enclaving the SOC is not entirely feasible.

The SOC should consider several questions when planning the approach and degree to which the SOC should insulate itself against constituency breach:

- **What is the cybersecurity posture of the constituency?**
  - Is malware frequently found (or likely rampant)?
  - How well maintained are general user systems and networks? If the SOC depends on a general pool of IT resources and sysadmins, will it be vulnerable?
  - What proportion of the user population has local administrator on their desktops, thus making privilege escalation and lateral movement by the adversary a further-elevated risk?
  - Must the SOC be able to continue functioning and monitoring when the main constituency network is unavailable?
- **In how many disparate geographic locations will the SOC place monitoring or data collection equipment?**
  - For one or two places, isolating monitoring equipment from the enterprise may be simple.

- For many places (e.g., 15 different countries), isolation will necessitate a highly scalable and more cost sensitive approach.
  - Aspects of constituency computing environment, such as use of cloud infrastructure, influence how data is generated and collected. For example, placement of virtual taps, and distributed data collection buses.
- **What kinds of logical separations are already in use, and can they be trusted?**
  - What WAN technologies (Multiprotocol Label Switching [MPLS], VPN, VRF/GRE, etc.), that allow logical separation of assets based on trust domain, are core competencies of the constituency? How well enclaved (or flat) is the constituency to begin with?
  - What is the security hygiene of network management, meaning can it be trusted to offer insulation from the general user population?
  - Are existing firewalls trustworthy enough to hang the SOC network off one leg without further protections?
  - Is the SOC co-located with network operations so that there's increased trust and communication in their ability to manage SOC network infrastructure?
- **What equipment comprises most of the SOC's remote monitoring capabilities?**
  - Do those capabilities have built-in firewall and point-to-point or end-to-end encryption capabilities?
- **What proportion of the SOC's monitoring assets must operate in band?**
  - What is the deployment, trust, and certificate management model for its host sensor and EDR capabilities? Will deployment of that EDR host agent be made much easier by joining EDR management systems to the same identity plane as monitored systems?
  - To what extent will its log collection systems need to join monitored Windows domains, such as to perform WEC/WEF?
  - If the SOC chooses not to join its log collection systems in the monitored domains, how well equipped and resourced is it to perform inter-trust boundary certificate and key management?
  - How does the SOC expect to interface with constituents? Does the SOC need to serve provide reporting, situational awareness, and ticketing in the same identity plane as the constituents? The answer to this question is almost always yes.
- **What can the SOC afford?**
  - Can it afford to purchase (and manage) networking infrastructures such as switches, routers, VPN concentrators and firewalls? Can the SOC trust network operations outside the SOC to administer them?
  - Is the SOC postured to operate its own domain controllers, SharePoint servers, SAN, virtualization infrastructure, patch management systems, vulnerability scanners, health and welfare monitoring, and other backend resources?
  - If the SOC must cope with multiple independent zones of trust, what are the costs associated with the placement of monitoring equipment and the SOC enclave? Can the SOC afford to maintain a presence on each?

Modern IT enterprises leverage centralized system administration and user authentication for a reason; so should the SOC. Many SOCs operate their own enclave, with own identity plane (such as a Windows domain). Centralized user management can also be extended to some monitoring technologies in other physically disparate locations. Taking this a step further, if the SOC domain went down or was compromised, would analysts still be able to log into monitoring consoles? Having emergency use-only local user accounts with different usernames and passwords than the Windows domain is always a good idea.

Typically, each analyst will have at least two desktops: one desktop for SOC monitoring and analysis, and one standard enterprise desktop/laptop for email, Web browsing, and business functions. These may not be two separate desktops at all:

- The SOC "desktop" could in fact be a laptop, with a local VM image or thin client providing standard enterprise productivity/email suite.
- The standard enterprise desktop/laptop could be used to access a VM, virtual desktop, or thin client with SOC tools and data.
- Some hybrid approach: perhaps "casual" users only have thin clients or SSL/TLS VPNs providing a web interface, whereas shift analysts get separate systems.

Maintaining this separation introduces some inconvenience for the SOC analyst, but this is usually outweighed by maintaining the highest level of integrity. When mixing and matching thin clients or virtual desktops, it is wise to use device management capabilities to measure and enforce basic IT hygiene on the host computer, such as patch compliance. One way or another, the analyst must have access to monitoring tools, constituency network(s), and the Internet (See Appendix D).

In summary, large and mature SOCs should consider leveraging the following best practices when constructing their monitoring network:

- The SOC has its own workstations used for general monitoring and analysis, usually bound together as a Windows domain that has no trust relationship with the rest of the constituency.
- The SOC supports remote, telework, and work at home scenarios, such as through remote VPN, virtual desktop/thin client, or cloud-based virtual desktop.
- In situations where the SOC has a cloud presence, that presence is in a separate cloud instance container and identity plane than the main constituency; in Azure terms, this might be a separate AAD tenant and set subscriptions; in AWS terms, this might be a separate AWS account.
- Malware detonation, reverse engineering, forensics, or other high-risk activities are performed on isolated/stand-alone systems, virtualized or cloud environments.
- The SOC's local systems are protected from the rest of the constituency by a modestly sized COTS or FOSS firewall.
- The SOC may use a combination of network taps and out of band management to isolate network sensors.
- Remote sensors and log-management systems are accessed via a software-based network encryption technology based on IPSEC or TLS.

- If there are several remote sites with a dozen or more pieces of monitoring equipment (sensors, log collection servers, etc.), the SOC sometimes may choose to put them behind a hardware VPN concentrator. This establishes a VPN tunnel back to a concentrator near the main SOC enclave.
- Host instrumentation, such as EDR, is managed by a central management server. This server may reside within the SOC enclave if appropriate restrictions are put in place.
- Digital artifacts (such as log snippets) and case data reside on NAS/SAN or cloud storage account devoted to the SOC, not shared with the constituency, not accessible from the general Internet, and heavily audited.

Even if SOC analysis systems are placed out-of-band, other systems that are frequently used for analyst-to-analyst collaboration (e.g., webcams, persistent chat, wikis, or VoIP phones) may still be on the enterprise network. Imagine, for instance, an APT that is listening in to the SOC's VoIP calls and learns that it has been detected. This is something the SOC should consider when designing guidelines for how remote analysts collaborate.

In addition, it may be tempting for the SOC to connect its equipment directly to general constituency network switches, with only VLAN separation. This can be risky, resulting in accidental misconfiguration, compromise of the network topology, or the lesser threat of VLAN hopping [369]. Once again, resourcing will drive the right choices here.

These best practices are depicted in Figure 37.

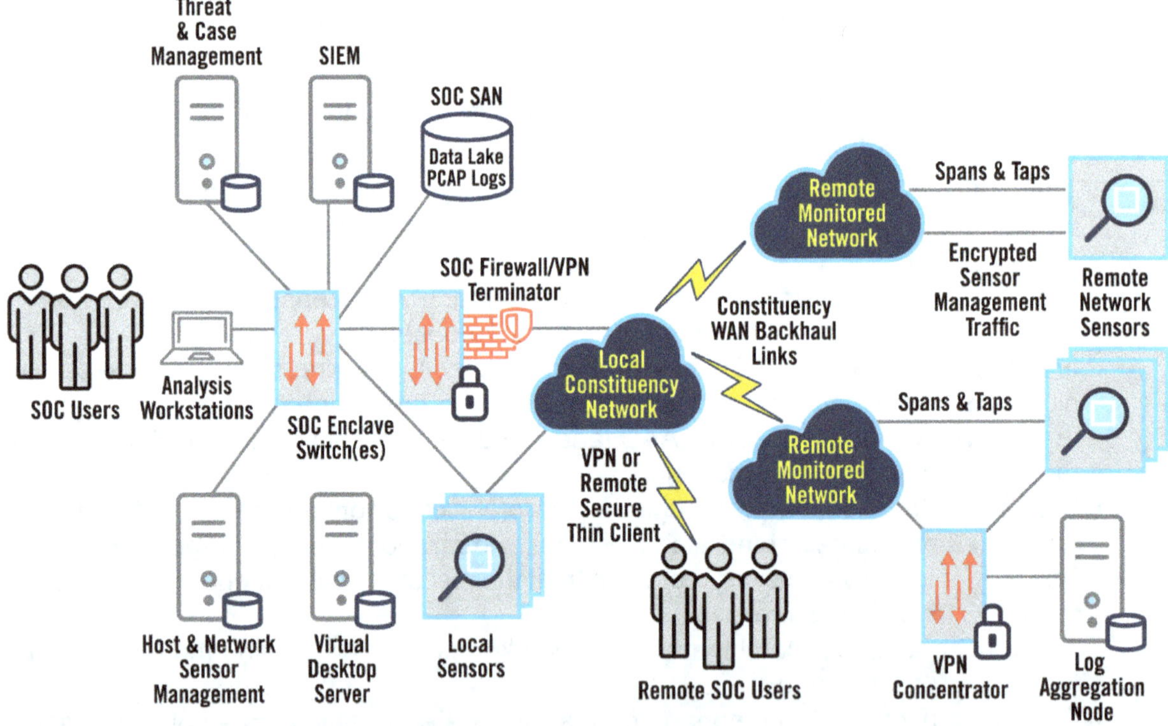

Figure 37: SOC Enclave

This architecture is optimal for large SOCs that have a dedicated administration team. Smaller SOCs may have to take shortcuts such as co-mingling their assets on general constituency networks. Hierarchical, federated, and distributed SOCs can also leverage this architecture by adding combination VPN concentrators/firewall devices at remote sites where small teams of analysts reside. In protecting the SOC enclave, some additional controls that SOC sysadmins should observe include:

- Maintain top-notch vigilance with patching SOC systems, updating open-source software and dependent libraries, and updating anti-malware definitions.
- Use widespread and consistent EDR on analyst endpoints, which ensure 100% coverage, including all form factors: Windows, Linux, macOS, virtual desktop, mobile, etc. The SOC may choose to use an additional or different vendor or agent not seen elsewhere in the constituency.
- Enforce code signing [370] and/or OS lockdown [371] on as many systems as possible (understanding that malware analysis systems are excepted from this rule).
- The SOC's analysis environment is on a platform that is more resistant to malware infections than general constituency workstations. "Sandbox" routine analysis functions through virtualization or on a non-Windows OS may help.
- Use multi-factor authentication where appropriate, particularly with any sort of remote login or remote management capabilities and using authentication infrastructure separate from that of the constituency.
- Ensure SOC sysadmins and analysts avoid using shared user accounts, such as with designated administrator accounts, use of sudo, and just-in-time privilege/identity management [372], understanding the following:
  - The goal is to support attribution of privileged actions to a specific user in the unlikely event of configuration errors, compromise, or sabotage.
  - It is very hard to remember a different password for each disparate system.
  - Tying authentication to the domain for every type of device usually is not possible.
  - Some general monitoring systems such as those projected on big screens on the ops floor (if one exists) will require generic accounts.
  - It may be best to strictly limit use of generic root or administrator accounts to emergency "break glass" situations.
- Keep on-prem SOC equipment under tight physical control.
  - The SOC has its own physical space with electronic badge access.
  - Local SOC equipment is in a SOC-controlled server room, server cage within the local server room, or at least in racks that close and lock with non-generic keys.
  - Physical and logical access control is commensurate with the sensitivity of information being stored, such as with PII, PCI, HIPAA, and other related data, as would be the case with storage of data gathered from SSL/TLS break and inspect.
- Subject cloud based SOC assets to strong cloud best practices, including:
  - Locate them in their own separate management place, such as a separate Azure tenant and subscriptions, or AWS Organization.
  - Use multi-factor authentication.

- Minimize using permanent administrative rights and traditional guidelines of least privilege, including lockdown at the control plan of the cloud resource management layer.
  - Pay extra attention to ensuring neither cloud services nor their underlying data are exposed to the general Internet.
- Ensue there is robust (but not overzealous) logging for all SOC systems, such as sensor management servers, domain controller(s), storage, and firewall(s).
- Ensue both SOC sysadmins and a third party regularly review SOC logs for any evidence of external compromise, sabotage, or infection, thereby addressing the question, "Who watches the watchers?"
- Build appropriate levels of resiliency into key systems such as sensor management servers, network switches, and log aggregation servers.

8.7   Cloud Considerations for SOC Tools & Data

Virtually every major SOC technology investment area: log management, EDR, SOAR, case management, intel management, storage, virtual desktop either leverages some aspect of cloud technology, is available in a cloud-based form factor, or both. When considering use of cloud technologies, the SOC should take the following into consideration:

- Managed infrastructure service, code, and version updates have become seamless and sometimes even invisible, but with them there is an inherent risk from tacit acceptance of code updates that the SOC itself may not have vetted.
- Critical security data is stored on systems and in identity planes the SOC may or may not control, meaning the risks of the cloud provider transfer to risks borne by the SOC.
- Data may be stored in a geographic region which has certain legal implications, such as due to GDPR and other data sovereignty laws.
- Costing and cost planning will change. There may not be any up-front cost commitment, however the SOC must pay attention to week-to-week and month-to-month fluctuations in cloud costs.
- There may be opportunities to try out different tool options without the commitment of purchasing and deploying on-prem.
- The barrier to entry may be substantially lower when deploying at scale.
- Scale up needed resources generally becomes much more flexible.

8.7.1   Cloud Dependency Considerations

It is less and less likely for any SOC today to say, "we don't use the cloud." Rather, the question is the extent to which the SOC is choosing to place dependencies on cloud technologies. Different tools or products can leverage cloud technologies to different degrees. Starting with the lowest cloud dependency and moving to the greatest and most overt dependency:

- New cyber threat intelligence, such as with signature, detections, and analytics refresh automatically from the cloud with no user intervention, i.e., most modern anti-virus products

- On-prem product components receive executable code updates from the cloud (and consequently must go through a periodic service restart), i.e., some EDR and agent-based monitoring tools
- The tool in question is available as a PaaS/SaaS cloud form factor, as a COOP or BC/DR backup instance to an on-prem instance, i.e., some ticketing systems, SIEMs, and log management tools
- The tool or product exfiltrates signals, events, and other signals to backend analytic engine running entirely in the vendor's cloud infrastructure, with no on-prem alternative, i.e., some recent EDR and SIEM products
- The product, in its entity, is available only as a PaaS or SaaS service, i.e., cloud-based automation platforms
- The SOC enclave itself, and all its tools, is partially or entirely in the cloud, i.e., a mix of cloud-based storage, IaaS, SaaS, PaaS, analytics, and automation services native the cloud provider(s) used

When making the choice about how much "cloud" the SOC should leverage, there are four main questions to answer:

- What are the cost implications of moving a given solution or tool to the cloud, e.g., how much more or less will a cloud-based log management solution cost versus on-prem, given the SOC's event volume and velocity? This can be prevalent with other technologies that require high-performance, high-volume storage, such as PCAP collection.
- How much more can it achieve in each period of time versus on-prem approaches, e.g., the barrier to entry and deployment time in cloud is typically much smaller?
- To what extent are the first two offset by the risk calculus for placing confidentiality, integrity, and availability dependencies on its potential cloud providers?
- Can the SOC achieve the same or similar outcomes as it has in the past; for example, a "lift and shift" from on-prem to the cloud may be substantially more costly but leveraging a different approach to achieve a similar outcome in the cloud may yield a very different outcome. For example: would use of WAF log collection deprecate an otherwise huge investment in PCAP collection in front of a Web server farm?

8.7.2 Transitioning to Cloud

If the SOC and/or the constituency does not have a lot of experience moving to the cloud, the SOC should consider a more gradual transition until the SOC gains expertise in operating in the new environment. Actions the SOC, or a supporting engineering team, should take before transitioning to the cloud include:

- Ensure identity plane integration and identity management.
- Consider networking, such as how to connect cloud networks into on-prem enclaves
- Determine how to enable and secure user access to cloud-based systems and infrastructure
- Review data management and determine adjustments needed

- Assess capacity and performance management along with asset health and welfare
- Leverage the capabilities integrated into the cloud platform such as automation, systems management, asset provisioning, and issue management
- Establish strategies for inter-region and cross-region data and service redundancy and failover

Finally, the SOC should ensure the following questions are answered before it procures cloud-based services:

- What is the nature of the service being offered? Is it truly an elastic PaaS or SaaS service, or is it the same technology as on-prem, simply deployed on IaaS?
- Where geographically will the assets and data physically reside?
- What security measures does the vendor employ to protect the SOC's data and compute?
- How does identity & access federation work with the SOC's existing enclave and systems?
- What is the forensic quality & integrity of data stored in that cloud?
- What protections & regulatory compliance does the cloud service provider offer?
- How will analysts integrate and pivot between systems and tools in the cloud vs on-prem?
- What costs are associated with moving bulk data back out of the cloud provider's hosting service, and technically how easily is this achieved?

Each SOC will make its own choices about how and where to leverage cloud-based technologies in its architecture and tool suite. Increasingly, though, it enables its users to do much more in much less time.

8.8 Summary – Strategy 8: Leverage Tools to Support Analyst Workflow

8.1. Consolidating and harmonizing the number of views into tools and data and providing integration between them is important to enhancing the SOC workflow.

8.2. SIEMs collect, aggregate, filter, store, triage, correlate, and display security-relevant data, supporting both real-time and historical review and analysis.

- SIEMs act as a force multiplier allowing SOCs to shift staffing requirements from low-efficiency, low-effectiveness alert triage towards increasingly better detections, data enrichment, and automation.
- Common features and expectations for SIEMs include data acquisition and data normalization and persistence; data analytics, data query, interaction, and workflow; and flexible integration with other capabilities.
- SIEM and Log Management capabilities may overlap. In general, SIEMs offer enhanced capabilities specific to a SOC while log management tools can often support multiple IT use cases.
- A SOC should consider purchasing a SIEM when the SOC's needs exceed what a log aggregation tool typically offers, the SOC performs a substantial portion of

its analysis on real-time data, the SOC has identified multiple data feeds beyond network and host sensors it needs to integrate, and the SOC has the resources to dedicate to SIEM management.

- Maintaining a SIEM requires dedicated resource over time, not just during installation. Ongoing maintenance tasks include keeping data feeds flowing smoothly; updating and tuning content; optimizing queries; knowledge management; and workflow management.
- Small SOCs or new SOCs may want to start with EDR and Log Management before investing in a SIEM. Very large SOCs with dedicated engineering resources may want to build a more custom solution than they can get from a single commercial SIEM, such as with big data technologies.

8.3. UEBA refers to the set of products and product functionality associated with uncovering users' and other entities' deviations from normal patterns, which in turn is likely to reveal malicious activity.

- Data to support UEBA will come from several sources including user authentication and access control systems, employee data, network sensors, and host sensors.
- UEBA lends itself to rule based detections as well as supervised and unsupervised ML analytics.

8.4. Every SOC needs a way to track incidents. Generally, the more mature the SOC, the more sophisticated and customized that incident tracking capability needs to be.

- Implementing a case management system for the SOC can take multiple forms: tailoring an existing IT case management system; purchasing and tailoring either a commercial or an open-source standalone case management system just for the SOC; leveraging the case management capabilities within a SIEM or SOAR system if the SOC already owns one of them; building a custom system; or utilizing a cloud-based SaaS ticketing system.
- New, less mature, and small SOCs may wish to consider a ticketing system that is cloud-based and/or the same solution as the IT helpdesk to minimize timelines and acquisition and sustainment costs.
- It is very common to see case management and automation comingled in the same system, particularly in the case of SOAR; traditional case management vendors will offer SOAR capability and SOAR vendors will often offer case management capability, making the distinction almost non-existent.

8.5. SOAR capabilities are a set of products and features that, as their name implies, enables the security operations user to automate repeatable processes quickly and efficiently.

- SOAR emphasizes gathering alerts and cases from disparate systems; graphical interaction with complex workflow and automation; and curated APIs and integrations to existing tools.
- SOAR de-emphasizes high-scale data collection and log management, and consequently the development and support resources to achieve it; and case management for areas of IT outside cybersecurity, thereby catering to a tight market segment.

- The SOC is generally ready to harness new or additional SOAR capabilities when its integration and workflow needs exceed existing SIEM or case management tools provide, and it has the staffing to support customization and maintenance of a SOAR capability.
- When first implementing a SOAR capability, start with high-reward, low risk items that are frequently repeated in the SOC's daily and weekly incident handling process. Gradually implement higher risk workflows, such as enacting changes that interrupt services, accounts, network traffic, or system activity, as confidence and maturity grow.

8.6. The SOC is a source of tremendous insights and situational awareness across the constituency and the data it collects can be valuable to other parts of the constituency. Yet the SOC also has the responsibility to protect these tools and data from misuse.

- Best-of-breed SOCs operate some or all of their systems in an out-of-band fashion that isolates them from the rest of the constituency. Yet at the same time, they need to share and collaborate with constituents. The SOC must find a balance.
- The SOC and thus its constituents can both benefit and prosper greatly if the SOC is able to not only engage constituents but democratize aspects of the SOC mission by directly involving security-minded stakeholders.
- The SOC should not give all stakeholders access to all information. What is made available should be specific to the needs of the stakeholder and consistent with a risk management strategy.

8.7. When considering a move to the cloud, the SOC should recognize both the benefits such as the flexibility of scaling up as well as some of the drawbacks like having less control over where data is stored and what updates are pushed to systems (if using SaaS).

- It is less and less likely for any SOC today to say, "we don't use the cloud." Rather, the question is the extent to which the SOC is choosing to place dependencies on cloud technologies. This can range from simply receiving data from the cloud (such as with AV products) up to fully moving the SOC enclave itself into the cloud.
- When making the choice about how much "cloud" the SOC should leverage, there are four main questions to answer: what the cost implications are; how much more can be accomplished; what the risk calculus is of using a cloud provider; and can the SOC achieve the same or similar outcomes.

Strategy 9: Communicate Clearly, Collaborate Often, Share Generously

Our ninth strategy addresses the importance of engaging with others to exponentially improve the SOCs effectiveness. This includes engaging within the SOC itself, within the constituency, and with other external organizations. As discussed throughout this book, a successful SOC will be part of the larger business ecosystem of its constituency, it does not function as an island unto itself. This means it needs to be able to clearly communicate and provide risk and value context to its constituency and stakeholders. It also must collaborate with others to improve operations and incorporate business priorities back into its activities. Participating as an active member of the larger cybersecurity community is vital to maximizing the SOC's capabilities.

9.1 Getting Started

No matter how well-funded or well-staffed a SOC is, it can never know everything about the cyber threat and vulnerabilities the constituency faces. Yet at the same time, it may know things of value to other SOCs. By building skills and capacity to engage with others, the SOC can improve its own capabilities while contributing to the overall security posture of the broader community. Helping to improve the community through partnerships helps the SOC to best address larger systemic issues. These actions should be built into plans and policies from the beginning and structures should be in place to encourage these engagements by all SOC personnel.

Partnering and sharing with others creates a stronger
cyber defense community for everyone

Some interactions will have immediate or easily visible benefits. For example, providing senior leadership risk recommendations may result in an increase in funding for a new defensive capability. Others will not be as easy to quantify or will take more time to come to fruition; however, they are still an important part of being a good citizen of the community. Imagine a member of your SOC presenting at a conference; outcomes could include brand recognition for the SOC, thereby improving its ability to hire hard-to-find staff, or the building long-term relationships that will make acting during a major incident easier. These benefits may be hard to quantify at the time of the event or even after the event is over, but they both have clear positive outcomes for the SOC.

When thinking about how to improve engagement, the SOC should consider what they have to offer others as well as what they want to get out of the engagement for themselves. First, the SOC should think about how it will present and receive information from others as part of its routine practice as well as during a major incident. To do this, it needs to build its skills in presenting information, ensure others are receiving it clearly and correctly, and taking feedback or outputs from those sessions and incorporating them back into SOC activities. The SOC should also think about how collaboration can lead to a better joint outcome such as a mutually developed decision or product. Usually, having stakeholders be part of the process rather than just receiving and end decision or product will result in a more successful outcome. Additionally, the SOC should think about how it shares information in a way that raises the collective knowledge and capability of many.

When thinking about these different approaches it is important that all the SOCs constituency groups are considered. Table 22 gives examples of how the three engagement activities described above could manifest themselves across different groups:

**Table 22. Examples of Communicating, Collaborating,
and Sharing with Different Groups**

	Inform and be informed	Collaborate	Share
Within the SOC	Pass information from one shift to another.	Bring together incident responders and the CTI team to create a new analytic.	Mentor a colleague.
With Stakeholder and Constituents	Provide risk summaries and recommendations to stakeholders and executives.	Pre-plan with constituents how to respond to incidents and jointly publish guidance.	Hold a lunch and learn about the latest cyber threats and how they might impact the business.
With the Broader Cyber Community	Provide incident TTPs, IOCs, detection tactics to other SOCs, and receive some back.	Compare best practices, chosen joint activities such as hunt.	Hold cross training with other SOCs; incorporate and hold lessons learned sessions.

The following sections are organized by the same groups as Table 22 (within the SOC; with stakeholders and constituents, and with the broader cyber community) and dive more deeply into how the SOC can engage with each of those groups.

9.2 Within the SOC

Improving communication, collaboration, and sharing should begin within the SOC itself. Some roles, such as team leads or managers, will intrinsically have more opportunities to exercise these skills. However, everyone in the SOC should be given the opportunity to be an active participant in these areas as only then can the SOC fully leverage its most valuable resource, its people. There are many ways a SOC can approach improving its performance regarding these activities. Internal to the SOC these actions can manifest themselves as:

- Ensure SOC staff know how to perform their duties.
- Pass operational information among SOC team members.

- Create, plan, and share information about activities the SOC is performing.
- Solve problems collaboratively.
- Improve the feedback loop for tools, data, intelligence, or other functions to maximize their value.

9.2.1 Ensuring SOC Staff Know How to Perform Their Duties

One of the main ways the SOC ensures clear and consistent communications among its team members is through the use of written processes and documentation. This can include everything from shift schedules and pass-down logs to COOP plans. Appendix C includes a listing of some common documents to consider. Additionally, Section 5.2.2 provides an overview of planning and documenting that plan in items such as SOPs and Playbooks.

9.2.2 Passing Operational and Workflow Information among SOC Team Members

When most think of information sharing between members of the SOC, they often think of incident case tracking. This is just the beginning, however. Modern SOCs will structure formal and informal communications through both structured and semi-structured means, including:

- Cyber threat intelligence, IOC, and adversary tracking
- Detection and analytics work status
- Engineering and development activities
- Structured hunt planning and execution
- Agile sprint planning, scrum, and task execution
- Major case and incident updates
- Daily standup and pass-down

Most mature SOCs host routine standups where team members report operational status to the rest of the team; more effective standups encourage a culture where each individual contributor speaks plainly and informally about what they are working on yesterday, today, and tomorrow. Many larger SOCs will host a standup with representatives from each section or team, and then separately, each team run its own sync. Some of these activities are purposefully overlapping. For example, a SOC may choose to organize structured hunt using the same agile scrum/DevOps methodology it uses for any other project work such as tool engineering and development.

A healthy SOC should also sponsor and drive collaboration in the following additional forums and channels:

- Monthly or quarterly all-hands
- Team building events
- Bilateral 1:1 sharing among different team members, particularly amongst those separated by geography
- Deliberate norms that address shift structure, work habits, and local time zones SOC team members, such as synchronous vs asynchronous meetings.

SOCs need to consider how they will engage those SOC staff members that are not normally working during scheduled interactions, such as with follow-the-sun operating models (see Section 3.7.8). One way to do this is to plan these meetings around shift change times and have staff stay late or arrive early so that shifts overlap. Another option is to record sessions and make them available for playback later. These should be planned to ensure maximum inclusive behaviors, which consider cultural norms, geography, job roles, and experience for team members to be included and encouraged to contribute.

9.2.3 Create, Plan, and Share Information about Activities the SOC is Performing, and Current Status

The SOC should be transparent about its how it is expending resources, both how they are currently fulfilling the mission and how they are planning for the future. Transparency of planning to all SOC members, and inclusion in the process boosts chances of plan success. This can be accomplished through routine work and mission improvement planning, agile scrum planning, and mission readiness scoring and tracking; for example, a this could occur through a detailed "scorecard" that decomposes the SOC's major functions, capabilities, and services, and shows their current status: mission capable, partially mission capable, not mission capable, or some variation.

9.2.4 Collaborative Problem Solving

Employee engagement and quality assurance are enhanced when analysts work together on tough problems, including both reactive incident handling and proactive project work. This usually revolves around analysis, code, and data: the queries, scripts, and tools used to achieve some analytic outcome, and the input and output of that activity. Fostering this collaboration goes to many other issues discussed in this chapter, as well as Chapter 4. In light of hybrid and remote/virtual teaming, there are several ways to achieve this technologically, keeping in mind balance with securing communications (see 8.6.1 Security Enclave for more information):

- Persistent chat technology such as Slack or Teams
- Real-time screen sharing (in lieu of presence in a physical operations floor)
- Ordinary case management
- Cyber threat intel management
- Knowledge repository such as SharePoint or wiki
- Agile work tracking (JIRA, Azure Dev Ops, etc.)
- Any other technology that supports knowledge storage and retrieval with attribution back to who contributed what, and when

It bears repeating: one of the things SOC management and leads need to contribute toward is fostering diversity and inclusion, such that each person can bring their best ideas to the table, feel heard, feel supported, and are able to support others.

9.2.5 Improving the Feedback Loop for Tools, Data, Intelligence, or Other Functions to Maximize Their Value

Beyond identifying the right mode of communication, the SOC should think holistically about what to share among its teams. These teams have much to offer one another. "This incident went really well," "here's what we found in that malware sample last week," "APT3 was seen doing X Y Z last week," "hey, when you send me tickets of type X, I need you to fill in section Y better," etc. When SOC functions are broken apart or artificially siloed organizationally or physically from one another, they tend to communicate and share less. Building on the SOC models show in "Strategy 3: Build a SOC Structure to Match Your Organizational Needs," SOC functions must be set up to support coordination within the SOC itself. Different elements of the SOC coordinate and support each other in myriad ways, as depicted in Figure 38.

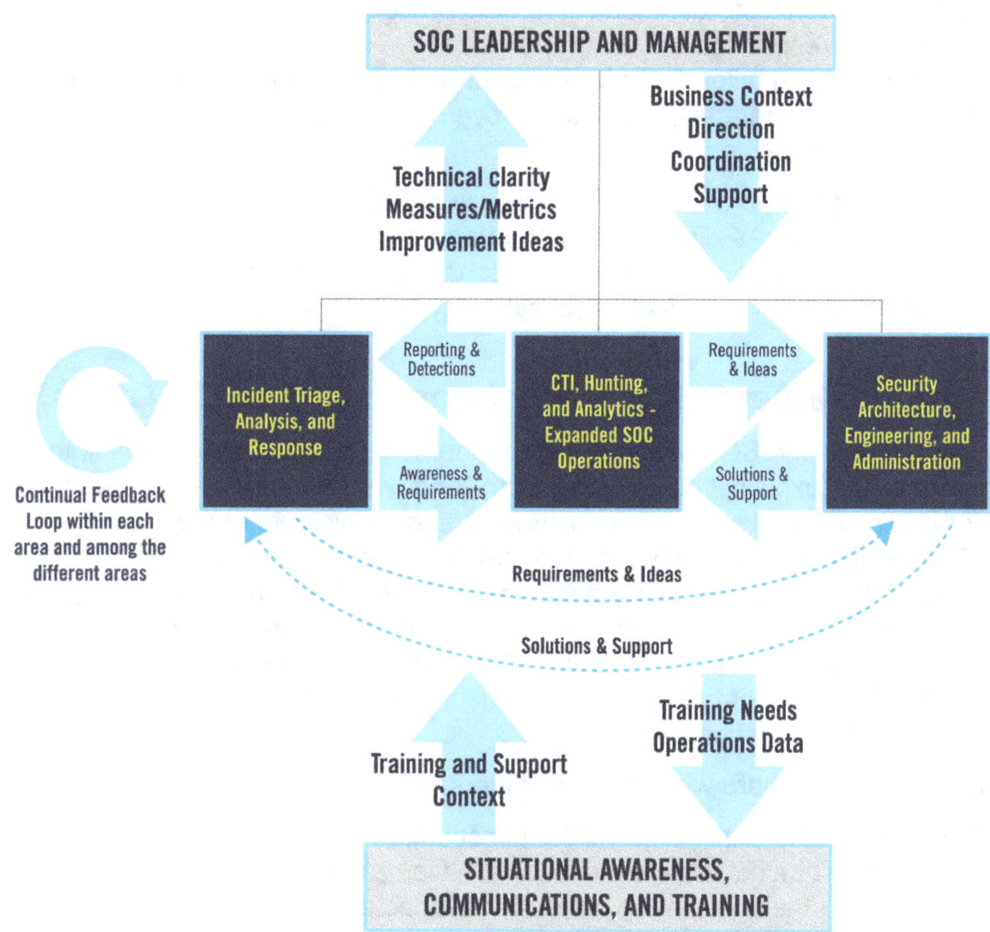

Figure 38. All Functions of Security Operations in the SOC

9.3 With Stakeholders and Constituents

Engaging with constituents, including IT leaders and executives, and gaining their support does not have to be difficult, but is incredibly important. Being able to express clearly and succinctly what the SOC is doing and what the SOC needs will go a long way towards building strong relationships with the SOC constituency. Additionally, engaging at a more tactical level with individual stakeholders will create the momentum necessary to get the SOC the visibility and partnership it needs further up the chain. This section addresses various forms of engaging with constituents, while keeping SOC processes and data secure, including:

- Let constituents know what services the SOC offers
- Understand the business and communicate change
- Convey risk in business and mission terms
- Share SOC architectures, tools, data, and processes
- Receive and respond to incident reports
- Jointly create and tailor detections and analytics, monitor, and respond to incidents
- Seek and incorporate feedback to improve operations
- Educate the constituency

As always, SOC engagement with constituents is usually done in partnership with the larger cybersecurity apparatus. As discussed in "Strategy 1: Know What You Are Protecting and Why," this is often facilitated through customer champions in the security organization, and/ or cybersecurity champions across the constituency.

9.3.1 Let Constituents Know What Services the SOC Offers

This is usually where the SOC will start its engagement journey. It should maintain a basic set of materials that explain "this is who we are and what we do." These materials should include both online content such as a constituency-facing Web presence, even if just through a wiki or SharePoint. The SOC should also maintain a standard slide deck that SOC leads and management can use for routine briefs to others. As the SOC matures, it may expand its offerings to include items such as special purpose materials for specific capabilities or deep dives into past incidents and outcomes that can help the constituency understand the value of the SOC. No matter the form of the artifact, it should be routinely updated to ensure it reflects the latest information about the SOC.

Areas the SOC should share with their constituents include:

- How and why the SOC supports the constituents
- How the constituents enable the SOC to accomplish the mission
- How to reach the SOC in the case of a suspected incident, and what constitutes a suspected or confirmed incident
- The SOC's offered service catalog (see the SOC Services table in "Fundamentals")
- Supporting governance that formally sanctions the SOC's existence and business functions (See "Strategy 2: Give the SOC the Authority to do Its Job")
- Links to various resources that the SOC offers

The SOC may also wish to combine general mission briefing with the work it does toward general constituency education and awareness.

9.3.2 Understand the Business and Communicate Change

As discussed in "Strategy 1: Know What You Are Protecting and Why," four of the five dimensions of knowing what you are protecting and why pertain to the constituency itself. This knowledge is gained by working in partnership with the constituency and helps the SOC align investments to constituency mission, get the support and resourcing it needed, and help drive down risk.

What this looks like, concretely, is the cybersecurity apparatus (including the SOC) work together to interface with key stakeholders at the mid management, lead, and individual worker level on a routine basis. During these engagements, the SOC should coordinate with any other cyber functions to cover the following ground:

- Proposed investments and security requirements, such as new sensor packages and preventive measures
- Ongoing and completed projects and investments specific to that area of the business, such as sensor deployments and analytics
- Recent "wins" by the SOC relevant to that area of the business, such as incidents of note
- Threats and risk that can impact the constituent's mission
- A curated set of constituent-facing metrics, as discussed in "Strategy 10: Measure Performance to Improve Performance"

In turn, the SOC should proactively seek and hear back from its constituents:

- Recent and upcoming changes in their org that impact the SOC, such as new or upgraded services, and business reorganization
- Reality check on its perception on its own risks and concerns in terms of cyber
- Feedback on SOC investments, past, present, and future
- An evolving sense of the IT, OT, cloud, and mobile device landscape, that can be reconciled against the SOC's composite asset inventory (See Section 1.5.1)

This engagement should be harmonized with—but distinct from—basic security maintenance and change management cycles. These kinds of forums will give cybersecurity and IT/OT stakeholders the chance to speak plainly and in partnership, with a focus on being forward looking vs reactionary. In very large businesses, these sorts of engagements are likely to be done with different segments of the constituency; in smaller constituencies, the SOC may be able to achieve this interaction with a single meeting. In either event, a monthly, bimonthly, or quarterly engagement may make sense.

9.3.3 Convey Risk

The SOC and the red team, if separate, (see description in "Strategy 11: Turn Up the Volume by Expanding SOC Functionality") are best postured to understand risk from an in-depth, technical perspective. It is incredibly valuable if they can tell their story in context of business impact to get constituents to act in a meaningful and positive fashion. The SOC has the following avenues to report on risk (and resulting consequences/impact):

- As part of routine engagements such as periodic emails, status briefings, or online website/forum updates
- Through metrics gathered and reported in partnership with the larger cybersecurity apparatus, including cybersecurity executives (See Section 10.3)
- Through red team and purple team readouts and reporting
- Through incident reporting and post incident remediation (PIR) activities
- Through emergency patch and vulnerability notifications (if part of the SOC mission)

Patch and vulnerability notifications are some of the most frequent, high-priority communication and direction a SOC will give to constituents and will greatly influence its reputation. The tips below can apply to many types of communications but are particularly important for these types of highly visible notices:

- Prepare before the communication is even needed:
    - Curate a template and plan the distribution in advance. This will give the SOC time to pre-determine exactly what information it needs to share and how it will go about doing so.
    - Have email lists assembled in advance and have those lists fused against asset/ service data. Ideally, the SOC should be able to leverage its asset and service inventory (See Section 1.5.1) to maintain up to date lists. It should also consider how to best scope the communication to only impacted service owners. For example, if there is a database patch, only email database owners, not everyone, should receive the communication.
    - Pull in security champions and other stakeholders early, and especially those in charge of vulnerability patching and mitigation. Consider looping these parties before wide email distribution or actions are pushed to system owners; getting early feedback and support can minimize any pushback or friction resulting from the communication.
    - Make the content relevant and actionable.
- Get to the point… quickly:
    - What does the SOC need affected parties to do, on what assets, by what date, and why? Make the action clear. Go here, download that, upgrade this, turn off that.
    - Make it as data-driven as possible. Try to avoid asking "are you running X?" to a broad audience. This is why the SOC (or its close partner) hopefully has vulnerability scanners, asset databases, network mapping, and centralized IT purchasing. System owners are most likely to act if the notification they receive contains links to specifics on impacted systems, software, and versions relevant to them vs a generic message sent to all.
- Ensure quality content:
    - Have a documented process for review and distribution. Designate SOC reviewers such that editing and final checkoff can be handled cleanly and quickly.

- Ensure there is a reviewer for technical accuracy. The SOCs reputation is on the line with each communication it sends out, it is important to minimize any technically incorrect information. If any statements are based on analytical assessments make it clear what the confidence level is of the assessment and why.
- Ensure there is a reviewer for grammar and formatting. This is often missed, especially when information is developed by multiple parties; grammar and spelling errors can distract and detract from the content messaging. Additionally, take the time to place the content in the established template using the pre-determined structure, headers, fonts, etc. Details matter so do not let these issues undermine the SOC's messaging.

9.3.4 Receiving and Responding to Incident Reports

The SOC should have multiple avenues for constituents to report anomalous and suspicious activity, such as, a form on their website, an email address that is easy to remember, e.g., incident@company.com, and via the help desk. Using a standard e-mail alias for incidents is important for addressing staff turnover and minimizing single point of failure. For example, if Molly is an incident responder, and all reports go to her e-mail address, if she leaves, the other incident responders may not be receiving the reports to stay responsive. Reinforcing the use of these channels via annual security training is helpful. Also, the SOC should have standard templated email communication and ticket handling expectations to support responding to these reports and letting the reporters know their input was received and is being acted upon.

In addition, the SOC should be in the routine business of interfacing with constituents about its alerting and investigation. Mature SOCs serving medium to large constituencies are likely to follow up with constituents on a regular basis: "was this you?", "was this expected?", and "can you help me understand this activity in the context of your service?"

Also, as discussed in "Strategy 2: Give the SOC the Authority to do Its Job," the SOC should articulate its roles and responsibilities so that participation is enabled. And, as discussed in "Strategy 5: Prioritize Incident Response," the SOC should have a routine, structured means of capturing post incident bugs and risks to be worked by other cybersecurity and security stakeholders as appropriate.

9.3.5 Sharing SOC Architectures, Tools, Data, and Processes

To gain trust across the constituency, a certain level of transparency is necessary. Giving the right people a sense of how a SOC gets the job done breeds respect and acceptance. However, there are some pieces of information that should not be shared with anyone outside the SOC without a compelling need. And, sharing information in a contested (or assumed breach) environment can inadvertently tip off adversaries (See Section 8.6 for this discussion).

For instance, were an adversary to obtain a list of sensor types and locations, it would then understand where there is not coverage, helping it avoid detection. Some SOCs get requests from external stakeholders on a semi-regular basis for lists of their sensor tap points.

Blanket requests for such information can add unnecessary risk to security operations. SOC managers might discuss this with requestors to understand the goal of their inquiries, and work with them on how to meet the goal while minimizing the release of information outside the scope of the request. The more people the SOC shares this information with, the more likely it is to end up posted on an Internet website or found on a compromised server. Table 23 covers common pieces of information shared by the SOC and some likely circumstances under which it should and should not be released.

Table 23. Common Information Shared by the SOC

What	Who Gets Access and Why
Monitoring architecture. High-level depiction of how the SOC monitors the constituency	Anyone the SOC briefs on its mission: IT and cybersecurity stakeholders such as constituency executives, security personnel, partner SOCs, and others
Monitoring tap points. Exact locations of sensor taps and full details on how they are protected	No one outside the SOC except those who maintain or deploy sensors, if this function is separate
Monitoring hardware/software versions, patch level	No one outside the SOC except those who maintain or deploy sensors, if this function is separate
Network maps	Organizations with a need to understand the shape/nature of their networks, such as IT ops, network administration, or the offices of CIO/CISO
Vulnerability lists and patch levels (scan results)	System owners (limited to their scope), other constituency cybersecurity professionals as needed, such as those who calculate risk, configuration compliance, and vulnerability compliance
SOC system and monitoring outages	Those directly above the SOC in its management reporting chain, such as the CISO or head of IT operations
Observables, indicators, and TTPs, including analytics, playbooks, detections and SIEM content	Partner SOCs, particularly in federated or tiered scenarios, with some potential exclusions for extra-sensitive signatures or insider threat indicators
Major incidents (possibly in progress)	Those directly above the SOC in its management reporting chain, possibly the CISO, in accordance with legal or statutory reporting requirements, such as with a national SOC
Incident details, including PII	The appropriate investigative body, such as law enforcement or legal counsel
Incident roll-up metrics and lessons learned	Those directly above the SOC in its management reporting chain, possibly the CISO or CIO
SOC incident escalation CONOPS and flowchart	Any interested constituents
Audit logs for non-SOC assets	Individuals assigned the responsibility for monitoring IT asset audit records, such as sysadmins and security personnel
Raw security telemetry	Other supporting or supported parties, see next section

Sharing information about the types of techniques used—without giving away the "secret sauce" on exactly how it has done—will go a long way toward building trust with interested parties. The SOC is advised to share some details with select constituents about its TTPs for spotting external adversaries. This presents a lower risk than sharing details about its insider threat program. Even high-level architecture diagrams are okay to share on a limited basis, so long as device details (e.g., IP addresses, host names, and software revisions) are removed. Moreover, when the SOC demonstrates forward-leaning, robust capabilities, it informs users that their actions are indeed being monitored. This may potentially ward off some miscreant activity. The key, though, is not disclosing so much that a malicious user knows how to circumvent monitoring.

9.3.6 Jointly Create and Tailor Detections and Analytics, Monitor, and Respond to Incidents

Some SOCs with critical and high-needs customers may want to consider if the SOC and high-end customers would benefit from a more active partnership. Specifically, technical security partners, those constituents with technical skill and interest, can be given more robust access to SOC tools and be brought into detailed technical discussions with the SOC. If done right, this can be a huge win for everyone involved and can result in better tailored detections written against constituency systems and services, better integration of constituency networks and systems into alert triage processes, and faster and more accurate incident investigation and response. Key considerations for doing this include:

- The SOC should have core staff with the time and communication skills to engage with constituents as part-time hunters or detection authors.
- The SOC itself must have analytics and detection capabilities that can be easily extended to include parties outside the SOC.
- The expectations of both parties need to be made clear, e.g., what is the steady-state time commitment toward supporting the SOC.
- The technical security partner cannot share their tool and data access with others and must follow all other SOC protocols for proper information handling and dissemination
- The technical security partner must not escalate malicious or anomalous activity outside the normal SOC reporting chain; for example, if they find something, they cannot report to their own management chain and not the SOC.
- The technical security partner must not perform response actions except as directed by SOC incident response coordinators or leads.
- The technical security partner routinely syncs with the SOC and the deputy's management as appropriate.

9.3.7 Receive Feedback to Improve Operations

In addition to routine feedback as part of red teaming and purple teaming (described in more detail in "Strategy 11: Turn Up the Volume by Expanding SOC Functionality"), post incident reviews, and routine partner/stakeholder engagement, the SOC may wish to proactively augment these with more structured feedback. This can be part of ticket closure "how did we do" as well as periodic user satisfaction polling to constituents. As with any sort of form or feedback, the SOC may want to air on the side of brevity, to ensure constituents fill out the whole form. Finally, the SOC is well-advised to meet informally in small group or one on one settings with stakeholders to gain feedback and seek opportunities for improvement.

9.3.8 Educating the Constituency

The SOC has an important role to play in reinforcing the overall culture of cybersecurity across the constituency. This can and should be done in partnership with the greater cybersecurity organization. This means integration of the SOC's understanding of new adversary campaigns, major threats, incidents lessons learned, and lessons from other organizations in the same business or industry vertical. Ideally, routine cybersecurity training should reinforce some of the metrics and hygiene issues the SOC and other cybersecurity stakeholders are emphasizing.

As SOCs gain experience in incident handling, they should build up a set of "wins" that it may consider putting into its standard mission brief. These examples, successful hunts (see "Strategy 11: Turn Up the Volume by Expanding SOC Functionality" for more information on threat hunting), mitigated risks, expelled adversaries, insights about cyber threat, etc., should serve to illuminate the SOC as a set of uniquely abled professionals, but not grossly undermine its mission if shared with the wrong party.

9.4 With the Broader Cyber Community

The broader cybersecurity community is large and yet accessible. It is large because there are many sharing groups, conferences, collaborative projects, and other official and unofficial groups. It is accessible in that each SOC can likely find a subset of these groups with relevant benefits from engaging routinely; and where the SOC's members should find opportunities to both learn and share not otherwise possible. This is important as no SOC can know everything or have total situational awareness on its own, and SOC team members need to be continually learning and growing to keep up with changes in technology and adversary techniques. Additionally, the SOC may need to partner with other organizations as part of its response efforts and should identify and build those relationships before they are needed. Therefore, the two main thrusts for external engagement are to share with and learn from others and collaborate with various external entities for incident response.

9.4.1 Share with and Learn from Others

Participating in right sharing forums is one of the most unique ways SOCs can come together to tip the balance against the adversary. This is both give and take: share with others and you would want them to share with you. Strong engagement can make the SOC more informed and more agile, particularly in terms of TTPs, intel, and SOC tools. The SOC will know this is succeeding if it gets tips and intel through these relationships that are unique and not available through any other medium or feed.

Some SOCs are especially well-positioned to share outside the SOC in a more public way. Large SOCs may have a continual stream of new information and experience beneficial to others in the community and may have the personnel resources to dedicate to sharing information beyond their own constituency. Small SOCs likely also have some information and experience that would be helpful for others and should try to carve out opportunities for SOC team members to share their knowledge.

Sharing should be done thoughtfully and in concert with the constituency's broader legal, PR, and communications standards. When developing a sharing policy, the SOC should ensure:

- The sharing team members understand what they can and cannot discuss publicly. This includes general conversations along with each specific paper, presentation, blog posting, tweet, etc.
- There is an established content review policy.
- Legal counsel and SOC stakeholders (like the CISO) have correct buy off.
- The SOC has established a relationship with—and clearance from—the constituency's PR department, as applicable.
- The SOC management agree that the SOC employees have time to get involved, and the SOC has the capacity to support any resulting follow up from outside interested parties.

There are many different communities and venues with whom the SOC can engage. These communities can include interfacing with other SOCs directly, forums, networking groups, associations, alliances, and government/industry groups, and government agencies dedicated to cyber and supporting the public's cyber knowledge.

Other SOCs directly
These can sometimes start as personal relationships with staff, or as an incident that spans across disparate constituencies such as from a supplier. They are usually, but not always, based on SOCs that support similar industries and thus find common ground. These relationships can start completely organically and can stay that way; some move to formally named groups with programmatic support and structure, others do not. Engagement across SOCs gives staff a chance to compare their experience, successes, and challenges with others who do similar work. Analyst-to-analyst collaboration across SOCs can not only increase the skills of the staff but also build a sense of belonging and fraternity employees may not get through any other outlet.

Forums, networking groups, associations, alliances, and government/industry groups
These groups can take on many forms. They can be focused on a geographic region, an industry sector, or a particular type of technology (such as cloud or data analytics). They can be targeted towards everyone from students to seasoned security professionals. Some may focus on broad cybersecurity topics while others are exclusively focused on cybersecurity operations. The SOC should carefully review the goals and activities of the sharing group to ensure they align with the needs of the SOC. Also, most hold conferences and other events

(See section 4.2.3 for a list of conferences). While there are too many organizations to list them all, some examples include:

- **Forum of Incident Response and Security Teams (FIRST):** FIRST enables sharing as a seminal forum with conferences and on-line tools for incident response and security teams all over the world to collaborate toward the common goal of a safe Internet [95].
- **Information Sharing and Analysis Centers (ISACs):** ISACs share cyber threat intelligence and provide tools to critical infrastructure owners and operators by collecting, analyzing, and disseminating information to their members [373].
- **InfraGard:** InfraGard facilitates critical infrastructure protection through sharing partnerships among the Federal Bureau of Investigation and commercial and private companies and organizations [374].
- **Woman in Cybersecurity (WiCyS):** WiCyS is a global community of women and men creating opportunities to collaborate, share knowledge, network, and mentor through conferences, professional development, and career fairs [375].
- **CSA Cloud Security Alliance:** The Cloud Security Alliance (CSA) defines and shares secure cloud computing environment best practices [376].

Government agencies dedicated to cyber and supporting the public's cyber knowledge
Many countries or regions have an organization or organizations focused on cybersecurity. The organizations will often share alerts and best practices, and some may investigate significant cyber activity themselves. Some of these organizations run information presentations or conferences. Others may offer services to various constituencies or have the responsibility to develop cyber policy. These organizations include, but are by no means limited to:

- Cybersecurity and Infrastructure Security Agency (CISA) (US) [377]
- Federal Bureau of Investigations (FBI) (US) [378]
- European Union Agency for Cybersecurity (ENISA) (Europe) [379]
- Canadian Center for Cyber Security (Canada) Canadian Centre for Cyber Security [380]
- Cybersecurity Agency of Singapore (Singapore) [381]
- ACSC Australian Cybersecurity Centre (Australia) [382]

Much of a SOC's engagement with these government agencies may be limited to utilizing information that they publish. However, it is worth identifying if they are an organization that is set up to receive notification of cyber incidents as well as understanding what other services and capabilities they offer before the SOC has reason to contact them during an investigation.

9.4.2 Collaborate Across Organizations for Incident Response

Sometimes an incident does not stop at the boundaries of the constituency or is bigger than the SOC can handle on its own. For example, by sharing information about a botnet or ransomware incident with an organization that has the capability and authority to act on the infrastructure or actors behind the campaign, each individual SOC can be more effective than it would be on its own. Or, because of the interconnectivity between the SOCs constituency and a supplier constituency, there is the potential for an incident to span the boundaries between the two organizations.

When preparing for incidents that cross boundaries, the SOC should have up-to-date partner organizations' information stored in a central place. Partner organizations could include:

- Suppliers, especially those with direct connections
- Service providers such as Cloud, including SaaS offerings
- SOC tool vendors
- Law enforcement and/or government at the town/city, state/province, and national/federal level
- Federated or peer SOCs in a larger constituency
- Other SOCs in the same industry vertical

9.5 Summary – Strategy 9: Communicate Clearly, Collaborate Often, Share Generously

9.1. Sharing with others helps create a stronger cyber defense community for everyone. The SOC should consider how it engages within the SOC itself, within the constituency, and with other external organizations. The SOC should consider what they have to offer others as well as what they want to get out of the engagement for themselves.

9.2. Internal to the SOC, communication, collaboration, and sharing can manifest themselves as:
   - Ensuring SOC staff know how to perform their duties
   - Passing operational information between SOC team members
   - Create, plan, and share information about activities the SOC is performing
   - Collaborative problem solving
   - Improving the feedback loop for tools, data, intelligence, or other functions to maximize their value

9.3. With Stakeholders and Constituents, the SOC should plan for ways to:
   - Inform constituents what services the SOC offers
   - Understand the business and communicate change
   - Convey risk
   - Share SOC architectures, tools, data, and processes
   - Receive and respond to incident reports
   - Jointly create and tailor detections and analytics, monitor, and respond to incidents
   - Seek and incorporate feedback to improve operations
   - Educate the constituency

9.4. The two main thrusts for external engagement are to share with and learn from others and collaborate across organizations for major incident response. In particular, frequent collaboration with others SOCs, such as around CTI, can be a huge boon for all those involved.

Strategy 10: Measure Performance to Improve Performance

Measuring performance within the SOC enables the constituency to understand if the SOC is delivering on its mission. Without measurement, it is difficult to know what is working well and where improvements would be most beneficial. Yet, despite the power of identifying what is important and measuring it, only half of all SOCs recently polled by SANS have a formal metrics program, making this strategy particularly relevant to both new and existing SOCs [383].

This strategy covers the elements of a SOC metrics program, discusses trade-offs of enlisting an external organization to help measure SOC performance, provides some sample metrics, and discusses data trending as well as how to avoid negative outcomes that could result from poorly implemented metrics or manipulation of performance targets.

Throughout this strategy, the terms measure, metrics, key performance indicators (KPIs), and assessment are used. These are all related terms; here they are used to mean:

- **Measure/metric:** A measure is a single unit (e.g., number of incidents in a given month) while a metric may be made of multiple units (e.g., percentage increase or decrease in incidents year over year). These terms are often used interchangeably throughout the cybersecurity community.
- **KPIs:** those measures/metrics that will demonstrate how an organization is achieving key business objectives.
- **Assessment:** An approach, process, or way of evaluating something that results in measures/metrics.

SOC metrics are frequently caught up in two very related topics: how the SOC provides situational awareness to its customers and cybersecurity metrics in general. Because these topics are linked, this chapter will touch on these areas as well, but is not intended to exhaustively cover the broader landscape of cybersecurity measures and metrics.

10.1 Elements of a SOC Metrics Program

An internal metrics program, whereby the SOC defines, measures, and reports on KPIs of operational processes, the output of its operations, and/or its situational awareness of the constituency is often at the core of how a SOC measures itself. Such a program will identify the motivations that drive the metrics program along with the outcomes they wish to influence. A SOC metrics program should consist of five elements: business objectives, data sources and collection, data synthesis, reporting, and decision-making and action. These elements build upon each other to create a complete program as shown in Figure 39.

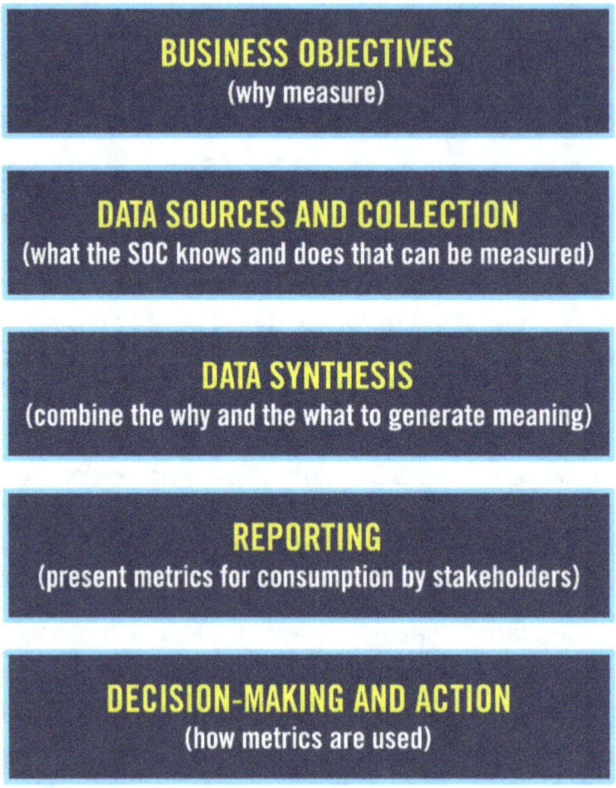

The figure shows five stacked boxes representing a process flow:

BUSINESS OBJECTIVES
(why measure)

DATA SOURCES AND COLLECTION
(what the SOC knows and does that can be measured)

DATA SYNTHESIS
(combine the why and the what to generate meaning)

REPORTING
(present metrics for consumption by stakeholders)

DECISION-MAKING AND ACTION
(how metrics are used)

Figure 39. SOC Metrics Program

10.1.1 Business Objectives

The first element of a SOC metrics program is clearly articulated business objectives including the rationale and outcomes for collecting, synthesizing, and reporting, and acting upon a given set of measures. These business objectives can be further broken into three categories: internal measures of SOC performance; externally-facing measures of SOC performance; and cybersecurity measures and risks that are not focused on the SOC itself. Specific examples of metrics for each of these three categories is included in Section 10.3.

It is worth noting that SOC performance metrics presume a strong understanding of the SOC's mission scope, services, and functions. If these have not been established, it is helpful but not strictly necessary to solidify them in writing first. Additionally, any kind of absolute thresholds should be defined with the business priorities, threat landscape, and resources available in mind. It is very easy to say "we should be 100% monitored" without recognizing how difficult it will be to get past 90%, or the fact that pouring resources into that target will detract from others (scanning coverage, variety and depth of monitoring capabilities used, etc.).

The audience for internal measures includes analysts, SOC section/team leads, and SOC management. These business objectives generally relate to quality and timely delivery of SOC services and are typically tactical in nature. Examples include:

- Improve effectiveness and quality of analyst efforts (detections, investigations, etc.).
- Ensure quality, stability, and service delivery of internal SOC tools and systems.
- Drive month-over-month, year-over-year improvement and optimization in SOC processes and output.
- Assess and close gaps in detection and prevention for adversary TTPs (such as with the MITRE ATT&CK framework).
- Understand and demonstrate readiness for certain objectives, services, mission areas the SOC is considering undertaking (e.g., is the SOC ready to perform hunting, does it need in-house malware analysis, should it consider purchasing a deception product).

Externally-facing measures of SOC performance

Externally facing measures are those consumed by the SOC's customers. These customers include executives above the SOC, the SOC steering committee (if it exists), IT executives, IT service owners, and anyone paying the SOC in a "fee for service" scenario. Business objectives here are typically focused on key performance indicators, impactful incidents, high level service delivery, and risk to the constituency. Examples include:

- Provide transparency and level set expectations on overall mission readiness as a synthesis of people, process, and technology.
- Ensure consistency of output and adherence to SLOs & SLAs and in so doing, deliver clarity and consistency in expectations for the SOC and its customers.
- Achieve compliance across customers, starting with routine hygiene such as scanning, monitoring, patching; as well as emerging and newer focus areas: compliant cloud services, enforcement of code signing and secure boot.
- Demonstrate cost and value factors for cybersecurity spend vs services performed.

Metrics that typically get the most attention are Service Level Objectives (SLOs): measures and indicators of SOC operational objectives but that the SOC is not necessarily held to, and SLAs: metrics and measures that the SOC is held to through formal agreement, contracts, and business relationships [384]. Put another way, an SLO is something a SOC may pick a measure that it, executives, or stakeholders want to achieve. In contrast, an SLA is a certain measure the SOC may be required to meet. Just because someone measures an operational KPI or strives to achieve a given threshold does not make that KPI an SLA, necessarily. If the SOC wants to understand their own KPIs in the context of how other SOCs perform the same, see: [47].

Cybersecurity measures and risks that are not focused on the SOC itself

The audience for these metrics extends beyond the SOC to the constituency at large. These metrics will often show to constituents, especially executives and IT service owners, how the SOC contributes to their understanding of cybersecurity risks and to an overall culture of cybersecurity awareness. Examples included:

- Progress in implementing a given set of cybersecurity controls
- Effectiveness in the output of a control or across several controls
- Impact on the constituency mission, such as cost avoidance or risk reduction including measuring and reporting on constituency impact of incidents handled by the SOC

- SOC contributions to overall risk/security metrics for constituency, such as a reporting, metrics and compliance reporting performed by the wider cybersecurity apparatus

It is very easy to emphasize one of the above categories above another however a balanced security metrics program should address all three.

10.1.2 Data Sources and Collection

The SOC can and often does gravitate toward metrics that can be calculated from data it already has on hand, especially when getting started with a metrics program. This can and should drive the SOC to be resourceful and leverage both the data that comes into its own systems as well as the data available from other parts of the organization. Data sources that are often readily available for SOC metrics may include:

- SIEM/analytics/log management platforms, operated by the SOC or elsewhere in the IT organization
- SOC ticketing/case management system and SOAR platforms
- SOC code repository and task management DevOps support systems, particularly if many analysts contribute to it, it is used for tracking project work at the analyst and dev level, and/or the SOC is using CI/CD for code, script, configuration, and detection management/deployment
- SOC budget and spend tracking
- Enterprise asset management, inclusive of cloud inventory (as applicable)
- Vulnerability management/scanning systems
- Automated attack frameworks, phishing as a service, and breach and attack simulation (BAS) platforms

An ongoing challenge for the SOC will be looking across data sets and trying to combine them to gain insight. In some case the data will be structured such that it can be easily queried across sources. In other cases, the SOC will have to create a bridge between different data types. The more the SOC has data in a structured and queryable format, the easier it will be to analyze for the purpose of metrics development. This is another situation where driving convergence across constituency datasets into one or a few persistence mechanisms that support federated query is immensely helpful and powerful and make the whole much greater than the sum of the parts.

As with other situations where the SOC leverages data from outside of the SOC, consideration will need to be given on how to best merge that data for analysis. If the data set is particularly large it probably does not make sense to try and duplicate that information within the SOC infrastructure. Depending on the data type, amount, frequency, and intended use for developing metrics, options for accessing and using external data beyond bringing the entire dataset to the SOC include summary or other types of aggregate reporting; exploring the data natively in the system it is stored in; or only bringing in small samples of the data to the SOC environment.

While using existing data can kickstart a metrics program, there are some cases where the SOC will want to generate different types of data to answer more generalized business questions. Or there may be a need for the SOC to measure itself under specific conditions. The choice of data collection or assessment approach for these needs will be influenced by the questions the SOC wants to answer as well as the resources and time the SOC has available to execute a given approach. Common ways the SOC might generate data to assess itself in these cases include:

- **Operational exercises and simulations:** Operations training, tabletop exercises (TTX) and simulated major incidents. This type of measurement allows the SOC to understand how they will perform under specific operational conditions and is often used to measure process, communications, and response actions. This is discussed in Section 11.8.
- **Focused technical measurement of the effectiveness:** Coverage and completeness of its own detection and investigation apparatus, such as through red teaming, purple teaming, and BAS products. These measurements are often tied to understanding the cyber threat as well as the constituency's key systems and data and using those inputs to scope the assessments. This is discussed in Strategy 11.
- **Use of an established cybersecurity framework or capability maturity model (CMM):** Used to examine the SOC holistically. This type of measurement tends to look at the full spectrum of people, process, and technology that makes up the SOC and is often used to assess the overall capability of the SOC.

Assessing the SOC against a cybersecurity framework or CMM can be a very powerful way to look holistically at SOC capabilities. Examples of these types of frameworks include the NIST cybersecurity framework [385] which can serve as inspiration for a wider set of cybersecurity metrics the SOC may wish to collect and the US government's Cybersecurity Maturity Model Certification [46]. Additionally, there are open-source methodologies such as the SOC CMM, which combines a set of measures and process specifically for measuring the SOC [45]. Some commercial cybersecurity companies also offer assessment services and use their own frameworks. See Section 10.2 for a discussion on using an external entity to assist with the SOC metrics program. When using a cybersecurity framework or CMM there are a few things to keep in mind:

- The SOC should look to implement other more basic aspects of a metrics program before turning to one of these types of assessments.
- These frameworks typically do not measure things like tradecraft or detection coverage. Other methodologies such as a BAS tool or the ATT&CK framework should be used if that is the primary goal of the assessment.
- The SOC should use the same framework or CMM over time so that it can measure progress between assessments. If a different framework is used each time, it will be difficult or impossible to normalize the findings.
- Be careful not to overly orient on the framework's scope. It may push the SOC to grow in a direction that is not beneficial or emphasize a set of services it does not need.

10.1.3 Data Synthesis and Measures

The third part of a metrics program describes how the SOC synthesizes the data it collects or accesses. Both new and established SOCs have been known to synthesize data reporting and metrics through spreadsheets and scripts. Whenever possible, however, the SOC is encouraged to leverage repeatable and automated processes. Most mature metrics efforts will leverage both batch (meaning periodic, routine), continuous transaction-based, or "at query time" on demand metrics calculations. These three approaches are not mutually exclusive and can be used in concert together.

When possible, the SOC should limit metrics that require human processing for parts of the process that could be automated. Instead, analysts and leaders should focus on making meaning out of data that is processed. A decade or two ago, it was not uncommon for a person to manually "stitch" together disparate vulnerability scanner output into reports that are published to constituents. This process could take hours and days and is, of course, subject to errors and variance in reporting month to month. SOCs should avoid this in sustained activities whenever possible. Today, data automation, robust dashboarding tools, serverless computing and federated query are powerful tools that can be used to fuse disparate datasets and bring processing times down from days to seconds.

As the SOC synthesizes various information, be aware of what data actually gets analyzed and considered for reporting vs. what data just stays in storage or is only used to make a pretty chart that cannot result in any action on new understanding. If data is not being used during this phase, consider reducing or eliminating that data source. Or look for ways to create and store aggregate information rather than full data feeds.

10.1.4 Reporting

The reporting aspect of a metrics program defines how the SOC packages and delivers its metrics output to the intended audience, and how it articulates the intent and objective for each metric. When thinking about how and what metrics to provide, there are several tips for success to keep in mind.

Scope the reporting to the audience and time

Before publishing metrics to any audience, the metric authors should carefully consider the reaction of the audience—that could be technical, budgetary, or political. It is very easy to get carried away and drive measures for dozens or even a hundred plus metrics. This is especially true when talking about measuring security controls. Consider presenting only a small number of the most relevant metrics to most audiences, rather than drowning them in 100 security controls. Some comprehensive security measurement programs can feel overwhelming; some new audiences might therefore consider them difficult to approach. By contrast, in both the daily/weekly SOC "ops standup" and a monthly or quarterly metrics rollup to executives, the SOC should be able to present and answer questions on KPIs in a fairly brief period of time: perhaps ten to thirty minutes, depending on the nature of the meeting and the audience.

Plan for the metrics to be shared

Expect that metrics and reporting are likely to be shared and that they will potentially be reused without context other than what is included in the reporting format. Therefore, when developing metrics, consider sharing not only the result, but also the methodology used to reach it. The SOC should also consider how it shares information with those who want to dig into the details. What this means, mechanically, is that in addition to sharing summarized results, the SOC should be prepared to share access to the data directly and the queries used to generate the output of concern with authorized personnel from outside the SOC. This approach applied here and more generally to sharing analytic results, helps breed an environment of trust, collaboration, and transparency.

Communicate in a format the constituency understands

Consider leveraging the prevailing reporting mechanisms used by the constituency. If the constituency leverages Amazon Web Services (AWS), that is likely Quicksight [387]; if the constituency is a customer of Azure, consider Power BI [388]. There are more possibilities, including tools like Tableau [389], Grafana [390] and Kibana [391] that are established, popular data visualization capabilities. To the greatest extent possible, reporting mechanisms should utilize formats that allow for rapid automated updates such as web-based dashboards, rather than static products that require manual updating.

Graphics are incredibly powerful

The SOC should also consider what types of visualizations will best convey their message. In some cases, visualization options will be built into the tool that is being used to collect and analyze the metrics. In other cases, the SOC will need to create and tailor their visualizations to their specific data and audiences. To get ideas and inspiration about visualization take a look at reports that are put out by various security organizations to see how they present the information. For example, the Verizon Data Breach Investigations Report (DBIR) [392] and FireEye M-Trends [393] both provide annual summary reporting on cyber threats and trends, feature several means of presenting high level metrics.

10.1.5 Decision Making and Action

The final part of a metrics program is using the information developed in the previous stages to decide and take action. Metrics should only be collected if they will be reviewed and lead to a decision, otherwise the time spent developing them could be better spent on other activities. That does not mean however that the decision needs to result in a change. The outcome from reviewing a given metric could result in a decision that things are going well, and changes should not be made. However, when there is a need to take action because of the knowledge gained through a metrics program, the SOC should make sure to have metrics in place that will help them identify if the changes are having the desired outcome.

10.2 Utilizing an External Organization to Measure the SOC

So far, this strategy has presented the elements of a SOC metrics program irrespective of who performs the various steps. However, there are elements of a program that could be performed in conjunction with an external entity. The phrase 'in conjunction' with is used specifically as certain elements of the program cannot be outsourced. The SOC needs to be closely involved in understanding the business objectives and at the end they, or other business leaders, have to make decisions about what actions to take based on the information presented.

There are several reasons why a SOC may want to use and external entity to assist with their metrics program, especially when it comes to collecting, analyzing, and reporting. These include:

- Needing an outside fresh perspective
- Having suspicions about issues but needing an independent advisor to validate it
- Needing a second opinion to strengthen the case for change
- Eliminating internal blind spots or bringing in skills that the SOC does not have internally
- To comply with regulatory or internal risk audit requirements

There are several options for bringing in an external entity. The first option is partnering with another SOC. This may be a viable choice if there are multiple SOCs with a close constituency, such as in a federated environment. The benefit to this approach is that the other SOC may be well positioned to understand the operational realities of the organization and be in a position to put finding in context more easily. Another option is to engage a National SOC or Coordinating SOC. Sometimes these organizations offer assessment services and for certain customers their support may even be free of charge. Additionally, if these organizations do offer such a service, they will most likely already have methodologies in place and will likely have the context of having worked with a number of other SOCs.

Alternatively, the SOC could turn to an external commercial consulting business. These organizations should have existing frameworks they can leverage and the SOC can tailor their request to exactly what they need.

If the SOC does decide to bring in an external entity, communication with SOC personnel and external stakeholders needs to be informed early and be part of the planning process to ensure a good result. Everyone should be made aware of the objectives and intended outcome, scope, and timetable. Staff should be made aware of their role in providing data to the external entity and informed on how to respond to questions in a clear and non-adversarial way. Using an outside entity does open the constituency up to having others know about weaknesses in their security operations program, so ensure appropriate non-disclosure agreements (NDAs) are in place before an assessment commences.

10.3 Example Metrics

This section provides a non-exhaustive set of metrics that serve as inspiration for the SOC getting started or looking to expand its metrics program. For each metric, the desired target and some tips on how to be successful are also included. Tables 4 and 5 represent the three areas discussed in Section 10.1.1 Business Objectives.

The following general guidelines and caveats should be observed:

Not all metrics are applicable in all situations
Each SOC will find a different mix of measures and outcomes that resonate for their own situation. A large coordinating SOC will have many different business outcomes, audiences, and behaviors it is trying to drive versus a five-person SOC serving a constituency of 10,000 users.

The metric targets are for illustrative purposes and may not be appropriate for some SOCs
For example, some SOCs cannot reasonably expect to move more than 90% of their alerts from their detection engine to their alerting console in less than 30 seconds. For others, this may be trivial.

Tailor your metrics to the audience
Some SOCs may wish to move certain measures around; for example: a given CISO may be very interested in how the SOC is making progress on ATT&CK coverage, while many other SOCs may consider this an internal-only KPI. Metrics have the power to distract audience members who do not carry the context the SOC does, so it is important to look for potential unintended consequences when reporting a new measure.

The SOC should put in compensating controls and metrics for measures that might be gamed
Time-based metrics and analyst output metrics should have controls and measures to ensure meeting the numbers does not come at the expense of quality.

Using median and percentile values are more resistant to outliers skewing the metric
For any metric where an average can be taken, consider taking a median value and/or percentiles such as 75th, 90th, etc. Using median and percentile values are more resistant to outliers skewing the metric. For example, a single missed incoming email from a constituent that went unaddressed for two days can demolish a mean time to respond metric, even if it is only a single instance in 100 other emails that were each addressed in less than 60 minutes from receipt. Median and percentiles often tell a more holistic, judicious story.

Consider targeting more than one level of service
Customers can be aligned by service level based on risk, criticality, and compliance targets. For example, most customers might receive 90% monitoring coverage, but the SOC could have a target of 95% for high criticality customers.

Different cybersecurity experts express varying viewpoints on various metrics; the reader is encouraged to consider different viewpoints before implementing a given measure. Additional resources include: [394], [395], [396], [397], [398], [399], [400], [401].

10.3.1 Metrics Primarily Targeted for Internal SOC Consumption

The metrics shown in Table 11. are typically gathered, measured, and reported on by the SOC with the SOC itself usually as the primary audience. If these are consumed outside the SOC, it is likely by the management chain above, such as the CISO, and in summarized form. Table 24 provides some examples.

Table 24. Internal SOC Metrics

Metric	Example Measure and Target	Remarks
SOC tool health and welfare	% Uptime: 99.5 % Events successfully processed: 99%	The SOC should keep internal service availability metrics, not just based on OS uptime but the % of time that the service is available and working.
Data feed health	% Of sensors up: 98% % Of data feeds/collectors up: 98%	Data feed and sensor health is a constant challenge for the SOC. Things are always breaking. While rarely always 100%, the SOC should establish targets that keep most feeds running as expected.
Data latency through pipeline	Minutes from source to ingest (median): 5 minutes Minutes from ingest to data persistence (median): 5 minutes	The SOC is well-advised to measure the propagation of events and alerts through its pipeline. This will reveal issues that need to be addressed like gaps, throttling, backpressure, and time synchronization issues.
MITRE ATT&CK framework coverage	% Of tiles of interest for which the SOC has a detection: 25%	The SOC is strongly encouraged to ensure its detective, investigative and protective capabilities match adversaries of concern. For more, see: [208].
True/false positive ratio for detections	Ratio of alerts tagged by an analyst as true positive vs false positive: 50%	Different SOCs overall have different thresholds for what it considers to be "good enough" detection accuracy. Measuring this over time and by tool can be very revealing, especially of the SOC writes its own detections.
Detection creation	Number of new detections moved to production: 2 per week	The SOC is encouraged to measure the velocity of how many detections it is producing, understanding this metric can be gamed.
Use of curated analytic notebooks	Ratio of number of alerts processed by each analyst to number of queries or notebooks run: 1:1	For a SOC trying to drive automation and consistency in incident analysis, this can be telling to see who is using what analytics and in what situations, vs analysts simply clicking on and closing alerts with no meaningful follow up.
Alert follow up ratio	Number of alerts escalated by a triage analyst to the next team that were closed as true positive: 25%	If the SOC separates out triage from in-depth analysis, this quality control can shed light on which analysts are producing quality leads and who is not performing the best cursory analysis. It can also tell the story of which analysts are performing throughout review vs acting as "pass through." Consequently, a high or low percentage here is not by itself necessarily good or bad but must be evaluated in this context.
Alerts not investigated	Number of alerts for which no investigation was performed: <25%	If analysts are being overwhelmed with alerts, or alerts are generated by known bad detections, they will get closed without any queries. In some SIEMs and SOARs, this can be easily measured.
Time consumed doing routine activities	Scanning, compiling scan results, investigations for a given alert time: varies, trending downward	Some SOCs can drive efficiencies with routine tasks month over month by measuring the resources dedicated to them. Caution is warranted here as asking analysts to "punch a clock" can be seen as very negative.

Metric	Example Measure and Target	Remarks
Time consumed by metrics program	Number of hours spent by SOC personnel doing metrics activities: target varies	The SOC may wish to ensure it is not devoting too much time to measuring itself.

10.3.2 Externally Facing SOC Metrics

In contrast to the previous set of metrics, the metrics in Table 25 are more likely to be reported on and consumed not just by the SOC itself, but parties near to the SOC such as other IT stakeholders, the SOC's executive management chain, and the SOC's steering committee (if one exists).

Table 25. External SOC Metrics

Metric	Example Measure and Target	Remarks
SOC Mission Readiness	Fully mission capable	It is often helpful for the SOC to break down each major component of its mission and report on readiness for that service or function: not mission capable, partially mission capable, or fully mission capable. While potentially un-scientific "anecdata" this can tell a succinct and powerful story about whether the SOC is able to fulfill its assumed mission. This is particularly powerful in cases where the SOC is not sufficiently resourced and/or is being driven too hard by its customers or executives.
Time to detect, aka "dwell time"	Mean/median time from when an adversary enters the constituency to when they are detected; easy to measure for some routine incidents (e.g., phishing transiting an email gateway) but more involved for serious breaches: trending downward for specific incident types	The intent of this metric is to assess how long the adversary is on the network before the SOC notices them. This varies widely by incident type and is separated out for routine threats vs serious cases. Used well, this can help the SOC lobby for better preventative and detective mechanisms, e.g., "we didn't have the tools we need." Conversely, this can be used as a weapon against the SOC, e.g., "why didn't you see the APT for three months?" Care should be taken with how this metric is used.
Time to engage	Mean/median time from constituent ticket submission to response from the SOC, usually from call logs, ticketing system or email, particularly those marked as high severity: static SLO/SLA threshold such as 30 minutes	This metric is an indication of how quickly the SOC is following up on customer tickets and critical severity referrals. This is a strong example of a metric that can have toxic effects if overdriven: triage analysts should not necessarily be glued to their inboxes, to the exclusion of meaningful replies or in-depth analysis.

Metric	Example Measure and Target	Remarks
Time to respond	Mean/median time from adversary detection to response action: downward trend over time, broken down by incident type	This indicates the duration from when an adversary is first "seen" to when the SOC starts taking action beyond analysis and investigation. Great care should be taken: strong, mature SOCs will handle routine threats swiftly and with automation (driving down this metric), but in some situations give the right time and study before engaging the APT. Overdriving this metric will turn the SOC into a "whack a mole" operation with no opportunity to learn the adversary's TTPs and motivations right of hack.
Time to contain	Mean/median time to assess extent of adversary presence and blocker their ability to spread further; usually comes from logs and will vary widely based on the incident type and degree of automation	As with several others, this metric will vary widely per incident type. Moreover, tools that provide automatic isolation of the adversary (some EDR, host app containerization, inline detonation) can push these figures down to sub-minute timespans. It is important to separate this out from containment timelines for major incidents.
Time to eradicate	Mean/median time to purge an adversary from the network; no specific target but measured and reported on for major incidents	See above.
Time to recover	Mean/median time to completion of incident remediation activities; no specific target but should be reported on for major incidents	See above. The main differentiator for this metric vice time to eradicate is that recovery is considered "done" not only once the adversary is removed, but other recovery actions are complete, in particular full service restoral.
Asset ownership	Percentage and absolute number of IT, OT, cloud, and mobile assets "seen" on the network that attributed to a mission/business owner: target of 100%	As discussed in "Strategy 1: Know What You are Protecting and Why," the SOC frequently has a strong role in network mapping and asset tracking. The SOC can help drive clarity on ensuring that all assets seen across the enterprise are assigned to a given owner and managed by that owner. Keeping both the absolute number and percentage visible for this metric as key, given that the SOC and its partners frequently discover previously unknown assets, that add to the denominator in this metric. In medium and large enterprises, it is likely the SOC will need to fuse disparate and overlapping sources of asset data to paint a comprehensive picture, particularly with hybrid on-prem + cloud scenarios.
Asset management	Percentage of known IT, OT, cloud, and mobile assets that have routine management and security hygiene applied such as CM and patching: target of 100%	

Metric	Example Measure and Target	Remarks
Scanning coverage	Percentage of systems by asset type, network or business owner that are vulnerability scanned: 95%	Scanning coverage is one of the most popular and critical metrics for the SOC to compile and report on. It also strongly ties into its role in asset management: to calculate monitoring coverage, it must have a strong understanding of the known asset landscape. Used wisely, this metric will help the SOC procure budget for monitoring and scanning capabilities and press business owners to participate in sensor deployment and vulnerability management. Differentiating coverage by asset type (Windows, Linux, cloud, IoT, etc.) will help drive investments with precision.
Monitoring coverage "breadth"	Percentage and absolute number of systems by asset type, network, or business owner that are monitored: 95%	
Monitoring "depth"	Percentage of systems by layer of compute stack (hardware/ firmware, OS, application) and completeness of telemetry (ATT&CK coverage, tool type)	This metric is intended to introduce additional dimensions to coverage. Low capability SOCs may consider an asset "monitored" if there is a single network sensor sitting on a nearby network segment. A more advanced SOC will likely have a suite of monitoring capabilities, and not consider a service "monitored" if it is not receiving telemetry across asset types, tools, and layer of the stack. This metric allows for both fidelity and differentiation in these regards.
Case volume by customer	Alerts or cases broken down by SOC customer or service: no target	Whether cases are going up, down or staying steady is not necessarily an indication of whether things are "better" or "worse," rather a useful observable to understand the SOC's focus and workload. This could drive further choices ranging from actions to address analyst confirmation bias to shifts in monitoring investments.
Case volume by adversary	Alerts or cases handled by named adversary: no target	

10.3.3 Cybersecurity Metrics Outside the Scope of Core SOC Services

Because of the tools it operates, and depending on the scope of its services, the SOC may be in a position to report on metrics further afield from core aspects if detection and response. Table 26 provides ideas of the reporting the SOC can do itself, or ideas for raw data that may be shared for others may leverage. With this said, the SOC should be careful not to let compiling and reporting on measures outside its core business functions become a drain on its resources.

Table 26. Other Cybersecurity Metrics

Metric	Example Measure and Target	Remarks
Patch compliance	Percentage of systems with 0 outstanding (critical) patches and/ or time to achieve x% patching, measured in days; target is likely 90% patched within ~7 days for routine patches	If the SOC is operating a vulnerability scanning solution, it will likely report those numbers to a select set of its constituents. Safeguards should put in place to reduce the likelihood that the adversary can get a hold of those results.

Metric	Example Measure and Target	Remarks
Outstanding patches	Patches available but not applied, down to individual system: target is 0 outstanding patches per system, or there must be a risk exception on file.	The SOC is strongly encouraged not to hoard vulnerability results. Instead, they should provide these to system owners and admins, using appropriate controls, so that admins can efficiently take action on the results. Even if the SOC does not operate vulnerability scanners, some of this information can be leveraged from process creation events, EDR telemetry, and most anything that includes OS/firmware patch/release (such as firewalls). Note: end of life software is considered a security risk when its creator/maintainer no longer provides security patches).
Presence of end-of-life software	End of life software and firmware present on end systems; target is 0 outstanding EOL software per system or there must be a risk exception on file	
IT utilization	Central processing unit (CPU), disk, memory, network utilization; unconventional targets like pages printed or website browsing habits by user; target varies	By virtue of the data it collects, the SOC may be able to report on IT statistics that other organizations cannot or are not sophisticated enough to gather and synthesize. There are many anecdotes of SOCs helping out IT operations with issues ranging from overloaded firewalls to departments printing too many pages per month. In any event, these should be used with care; the SOC should consider opening its access and letting other organizations self-service, assuming rights can be scoped appropriately.
Training completeness	Yearly cybersecurity training compliance; target 100%	In the event that the SOC is responsible for providing training services that are mandated, it should report on compliance results.
Phishing "as a service" results	Click rates for users clicking on phishing attacks (target: <5%); and participation rates for reporting said attack (target: >50%)	If the SOC is responsible for routine phishing training and testing campaigns, it is advisable for it to report on the results.
Use of other IT controls: Anti-Virus (AV), code signing	AV scanning and signature compliance (target: 100% up to date signatures and AV enablement); enforcement of executable and driver code signing (target: increasing year over year); other metrics as applicable	If the SOC operates a vulnerability management capability, or increasingly, gathers security hygiene information using other tools like EDR, it should capitalize on this data. For example, ordinary WEC will reveal how well system owners are using and enforcing AppLocker, CodeIntegrity and similar tools.
High risk services	Services with known Common Vulnerability Scoring System (CVSS) score 10 vulnerability and/or high-risk services facing the wider Internet (RDP, SSH, SQL, etc.); target: 0	If the SOC is performing routine port scanning, it is in a good position to spot and report on services and applications that are at high risk for exploitation, password spray, or account break-in.

Metric	Example Measure and Target	Remarks
High risk accounts	Breadth and number of permanent administrator access for given accounts (target: 0 persistent administrators)	Constituencies implementing zero persistent access/just in time access will likely wish to measure the extent to which this control is successful in limiting the damage that may be caused by takeover of a given account. If the SOC "scrapes" user directories such as active directory, it can furnish this data.

10.4 Data Trending

When defining targets for metrics and considering how to synthesize the data, do not forget to consider not only absolute thresholds (X minutes, Y alerts, Z percentage, etc.) but trends (increase month over month, decrease year over year, etc.). In some cases, trending will be less concerned with the specific number and more concerned with the movement of the trend line. Trend metrics are particularly helpful when changes are being implemented and the SOC needs to be able to show the impact of that change. For example, what are the effects of implementing a new internal cybersecurity training program? In this case, tracking incidents over time, both from before the training and afterwards will help leadership understand the immediate impact of the training. And continuing to track the trend in incidents after the training can help the leadership think about when refresher training may be appropriate.

Data trending over time can support both near term routine processes and longer-term situational awareness. With routine trending, changes can be monitored frequently to identify systemic problems more quickly. This can be particularly useful for something like monitoring data feed status or service level agreement compliance. In these cases, a single drop below a target number may not be cause for concern but a gradual decline over time might signal an issue. Situational awareness trend reports include things like developing year over year reports that show changes to a number of individual metrics to highlight key concerns or successes. The trend in data over time may highlight larger challenges or changes the SOC needs to address.

10.5 Not All Measures Result in Positive Outcomes

The closing thought on utilizing measures to improve SOC performance is that not all measure result in positive outcomes. There are certainly many positive reasons for a SOC to expend effort toward defining, gathering, synthesizing, and reporting on cybersecurity metrics. Executed thoughtfully and effectively, a SOC metrics program can drive both positive behaviors and positive business outcomes. That said, not all measures are good, and all too often, KPIs can drive undesired outcomes and behaviors if they are not thoughtfully utilized. Here are some motivations and outcomes that a metrics program should avoid:

- Driving negative behaviors among analysts, such as: competing to "win" in outcomes that do not align with strong service delivery, or cause team members to develop operational and risk blind spots; examples of negative behaviors and outcomes include:

- ◦ Analysts generating logs of superfluous detections created, without compensating controls for detection quality and relevancy
  - ◦ Fear of creating detection or analytics because they might not have a perfect true positive or false positive rate
  - ◦ Closing tickets out before investigation is concluded
  - ◦ Responding hastily, such as before root cause analysis has been performed, because an MTTR metric has been exceeded
  - ◦ Opening and closing tickets to boost productivity numbers
  - ◦ Not including patching and scanning numbers from services that are poorly managed
  - ◦ Onboarding exactly one well managed host from a given service, just to show compliance, while hiding 1000 other assets
- Take the fun out of cybersecurity analysis, detection writing, hunting and investigation by driving all analysts to the lowest common denominator of elementary service delivery and not leaving room for investigative creativity and innovation
- Pursuing a given cybersecurity metric that is either a) not strongly aligned to major areas of enterprise risk and/or b) to the exclusion of other metrics
- Distracting the SOC, executives and constituents away from core mission focus and toward the latest "metric of the day"

The SOC can avoid these outcomes through the following strategies:

- Seek team consensus when implementing new metrics, such that there is buy-in about both intent and implementation.
- Enact quality controls and checks to ensure specific metrics are not being manipulated through metric gaming, in particular with any metric that incorporates time (e.g., mean time to X and median time to X).
- Pursue a culture of transparency and openness about SOC maturity, capacity, and resourcing. Use poor performance against metrics as a method to highlight growth areas and resourcing requirements, de-emphasizing their use as means of penalizing team members
- Balance out metrics that measure basics of service delivery (such as ticket closure) with metrics that highlight growth and team contributions (SOP updates, analytic creation)
- Ensure metrics implement match visible and transparent investment areas, both down to the individual analyst and to cyber leadership, as appropriate
- Do not go overboard with a metrics program. It is possible to measure almost every aspect of what SOC team members do to an extreme degree. Consider keeping scope tight so the team doesn't feel overburdened with measures.

10.6 Summary – Strategy 10: Measure Performance to Improve Performance

10.1. SOCs of all sizes and levels of maturity may wish to implement a metrics program. Consider business outcomes, audience, and behaviors/morale in every metric.
   - Drive clarity, transparency, repeatability, professionalism, and focus on every metric chosen, particularly with SLAs and SLOs; simply measuring an aspect of operations will draw attention to that measure.

10.2. Leverage an outside organization if the SOC or its stakeholders feel a fresh or independent perspective is needed. Sometimes that is needed to confirm or bolster what the SOC already knew.

10.3. There are a range of example metrics the SOC can consider collecting and reporting on. Regardless of which they choose, metrics can be broken up into three groups: those that are meant for internal SOC consumption, those that describe the SOC's value and operating status to stakeholders, and then finally other things the SOC learns about the constituency's cybersecurity status that fall outside the SOC mission proper.

10.4. Be sure to review metrics routinely with the SOC and its stakeholders; they should demonstrate value and growth over time. Growth and progress on a trend are often as or more important than the absolute metric values themselves.

10.5. Not all measures result in positive outcomes. Choosing and monitoring the "wrong" measures can lead to wasted time or worse, a focus on harmful practices. Seek team consensus, compensating quality checks, and balance in metrics to emphasize not only basic service delivery, but growth in capabilities and a culture of transparency.

Strategy 11: Turn up the Volume by Expanding SOC Functionality

SOCs must continuously evolve in their quest to get ahead of the adversary. Having a solid incident response and detection function, along with basic CTI capabilities, is necessary but not sufficient for most SOCs given the adversaries' ease of hiding and shifting of techniques. Therefore, the SOC may find it necessary to incorporate additional functions which are designed to augment more routine detection and prevention techniques. The addition of these capabilities should be aligned to constituency needs and risk posture, not every SOC will need every one of these functions. This chapter describes expanded SOC functions that, when done effectively, can amp up a SOC's ability to detect and defend against more sophisticated attackers who often hide and quietly move in a constituency.

These additional functions include:

- Looking for the adversary in new ways through threat hunting
- Testing and enhancing the SOCs ability to detect the adversary through red teaming, purple teaming, and breach and attack simulation
- Concealing networks and assets, creating uncertainty and confusion, and/or influencing and misdirecting adversary perceptions and decisions through deception
- Advancing the SOCs knowledge of adversary actions, techniques, and tools through malware and digital forensic analysis
- Improving SOC operations through the use of tabletop exercises

11.1 Threat Hunting

When the constituency is ready for it, threat hunting is one of the first additional functions the SOC should consider putting in place. Many SOCs now consider this to be a foundational skill, but it does require solid IR processes and basic CTI capabilities to be in place first. Threat hunting, also called just hunting or hunt, is important because no set of detections will be complete when confronting adversaries operating in stealth mode. Hunting focuses on identifying new adversaries or previously undiscovered malicious actors already entrenched in the enterprise. Although definitions vary somewhat, it can generally be considered as:

Cyber threat hunting is a proactive security search through networks, endpoints, services, and data to discover malicious or suspicious activities that have evaded detection by existing, routine tools and monitoring [402].

Threat hunting differs from both threat detection and incident response. In "Strategy 7: Select and Collect the Right Data" and "Strategy 8: Leverage Tools to Support Analyst Workflow," threat detection was discussed, which is using monitoring and tools that alert when malicious activity raises an alarm. In "Strategy 5: Prioritize Incident Response," incident response was described as the way the SOC reacts to suspected and known malicious activity in the networks. In contrast, threat hunting is an active and proactive process that relies on skilled, intuitive experts to engage in detective and analytic activities not yet reduced to routine practice by the SOC [403]. To be effective, threat hunters create hypotheses based on adversary behavior, and search to validate by using intuition, logic and reasoning, and forensics [402], [404]. Perhaps most important, hunters take an assume breach mindset; they assume an adversary is already entrenched in the enterprise, so they are primarily interested in searching for evidence of identifying the adversary "right of hack" [405].

Threat hunters work between the seams of the SOC data being collected, routine detections, detected incidents and existing alerts, and CTI. Hunters need context, and more than just network or system data, they need visibility into mission areas, to see what adversaries might be doing across missions; they juxtapose this against what is important and normal for enterprise owners and users. For this reason, SOCs must attain a degree of maturity to best engage in threat hunting in a sustained manner. To assess if a SOC is ready for a threat-hunting activity and/or dedicated team, consider the following. Does the SOC:

- Effectively respond to incidents to address malicious activity?
- Address and close IR tickets accurately and in a timely fashion?
- Have data to see both network and host activity both broadly and with high fidelity?

Those that can effectively address the basics of incident response, and malicious adversary activity should consider threat hunting.

11.1.1 Why Hunt?

No matter how well instrumented an enterprise SOC is, or how expert the incident responders are, adversary activity will go undetected. Some of the business reasons of why SOCs find threat hunting of value include:

- Confirming and denying suspicions the adversary is on the network (and often, specific adversaries according to their ATT&CK techniques)
- Conventional means of detection are proving unsatisfactory
- Another organization provided a lead requiring a deeper look than what routine detections did not find
- CTI indicated an adversary may be targeting or interested in the constituency mission or data
- Vulnerabilities known to be in the constituency are being actively exploited

- The SOC wants to examine areas of investment for detections or more thorough instrumentation by applying advanced techniques, focused on specific adversaries, segments of the ATT&CK framework, lines of business, or some combination thereof
- Growing analytic techniques that can be feed back into routine operations

Most importantly, very sophisticated adversaries are extremely expensive to address; the longer they are in the network, the more expensive eviction will be as they become embedded, and traditional detections may not reveal their actions. The cost of a hunting team could be worth the cost avoidance of removing an adversary entrenched in a network, especially if rebuilding the user accounts and systems across the constituency becomes necessary. It is also wise to recognize what hunting is not. It is not:

- An opportunity for analysts to wander aimlessly through data for days on end. Generally, analysts should have specific objectives and/or hypotheses for a given hunt.
- An activity that goes on forever. As discussed below, a proper hunt program will organize hunts into distinct operations with clear planning, execution, and conclusion phases.
- Focused on finding vulnerabilities, misconfigurations, or missing internal security controls. Hunting may reveal poor hygiene and security practices along the way. However, analysts and the hunting program must be disciplined and structured to draw clear distinctions from hunt objectives and objectives of other teams such as pen testing or vulnerability scanning.

And, most importantly:

Hunting is not the same as routine incident investigation.

Hunting is a term often conflated with ordinary in-depth incident investigation. Indeed, both have huge overlap in tools, data, analytic technique, and staff. Hunting, in its purest sense, is an activity initiated outside routine incident investigation.

Every SOC performs hunting a little different, but at a high-level hunting consists of preparatory actions, hunt execution, and post hunt activities.

11.1.2 Preparing to Hunt

Prior to beginning hunting, the SOC should define the hunt program parameters including the generally expected outcomes, resourcing, rules of engagement, and authorization from SOC management to proceed with hunting activities. If not already established through a CTI program, the SOC may wish to establish guidelines that will help govern when it is acceptable to observe an adversary in situ, rather than moving to full response.

Once a program is established, various hunts can be developed and executed. The preparatory activities around hunt can be routinely conducted and are best done when the team is not under pressure to handle an imminent threat. Sources for inspiration of hunt activities include incidents within the enterprise, CTI, and adversary behavior experienced by others. To prepare, the following should be executed prior to each hunt:

- **Hypothesis generation:** SOC analysts will (hopefully) regularly come up with new ideas that may inspire future hunts. Just like with detection and tool planning, the SOC may wish to establish a backlog of work that can be pulled from.
- **Work planning and prioritization:** The SOC should pull hypotheses from its hunt work backlog and prioritize based on threat likelihood and mission. The hypotheses might come from quarterly or semi-annual planning, agile scrum, or both.
- **Resource identification:** The SOC needs to understand the personnel available, the systems, and tools that will be necessary. Some of the best hunts involve collaboration with security partners outside the SOC; for more, see Section 9.3 and 9.4.

11.1.3 Hunt Execution

To start hunting, most teams form an adversary scenario or hypothesis. This hypothesis could correlate to most any part of the kill chain or ATT&CK matrix: how an adversary might target the constituency, their lateral movement, what their actions on objectives might be, and so forth. A hypothesis might read like "I believe adversary X is present, as I think they are doing Y because they're ultimately trying to achieve Z." Based on the scenario, hunters develop TTPs and hypothetical values to use for forensic and other searching.

Each of the following activities will be executed for each "hunt." This level of structure will help the SOC, and the hunters establish clear scope, beginning, middle, and end of each hunt. Without this kind of structure, hunts can go on for months without a clear end in sight—this structure helps bound the resources and time involved.

1. **Plan the hunt:** The hunt team should clarify the hypotheses, goals, scope, timeline of the hunt, and nominate who will be involved in the hunt. If the hunt is focused on a specific set of assets, services, or business vertical, this may be an opportunity to draw in IT professionals from outside the SOC. At this state, the hunt team should also nominate the sources of data it has or needs to go get to satisfy the hunt. Finally, it is also helpful if the hunters gather any relevant cyber threat intelligence on the types of adversaries they are looking for, such as a named APT, FIN, or UNK group.
2. **Gather and gain access to necessary data:** The hunter should bring together the data nominated in Step 1. If the SOC is already gathering, curating, or has access to the data in question, this step is very quick.
3. **Perform iterative analysis:** This is the core execution of the hunt. During this state, hunters will write and execute various analytics against the data nominated from Step 1, which is usually but not always high-volume log data. During this stage, the SOC may also synthesize intermediate/summarized datasets just for the hunt, proactively pull in data from hosts and services, and discover new sorts of data it may need to leverage but did not realize it needed in Step 1. As hunters move through these activities, they should be recording the results of their analysis, query results, data gathered, etc. in a manner that is sharable across the team and can be recalled later.

This set of documentation should make the A-to-B-to-C "breadcrumb analysis" the hunter followed clear to the consumer of those documents, notes, and data. For more see, Section 11.1.4.

4. **Optional – plan and execute routine hunt analytics:** Some SOCs will set automated queries and other analytics to execute over the course of the hunt. If there is a new analytic developed, it may be helpful to run that every hour or every day rather than just once. For SOCs that have the flexibility, this can and should be implemented using its indigenous SIEM, SOAR, or big data detection platform.

5. **Respond or provide findings to the IR team:** If the hunters' hypothesis or hypotheses are found to be true, depending on the significance of the finding, the hunt may shift into incident response mode. This may cause all hunters to refocus and orient on the assets, services, TTPs, or adversaries associated with the true positive finding. If hunt is separate from the IR team, the findings might be handed to IR for remediation.

6. **Share results and synchronize operations:** The hunters should be routinely collaborating amongst themselves and checking in with their leads or SOC management on a cadence that works for the team. This ensures everyone is on the same page, and the hunt is going according to plan.

Successful hunts

Hunt team efforts vary in size and duration. They can be activated in response to leads or events or, the hunt team might generate hypotheses from enterprise available data. For example, asset-targeted hypothesis looks at the constituency's assets to identify what the business impact of specific threats and attacks may be first, rather than considering a specific adversary set of TTPs. This type of hypothesis creation is useful for vital services, industrial control systems (ICS), and other critical infrastructure missions, and begins with hunters looking at specialized system alerting, network activity, and system activity to find intrusions. No matter how hunt teams are activated or designed, here are some tips for success in carrying out hunts:

- **Focus on a specific hypothesis:** Develop a single or small related set of TTPs to start. Be clear on scope and avoid inadvertent scope creep.
- **Be curious:** Curious analysts are the most effective hunters. Technology can be taught, curiosity cannot. The subtlest of indicators can lead to big discoveries. (Read seminal 1989 book, *The Cuckoo's Egg* by Cliff Stoll, [406] if you are interested in an early example.)
- **Invest in people:** This includes hiring experts and giving them the time needed to perform this mission separate from other duties; threat hunting and interruptions are not compatible. This means that staff engaged in hunt will need several hours or days out of the week devoted to hunt activities.
- **Employ the beginners' mind:** Do not rely entirely on experience. The beginners' mind of "what is possible" can yield unexpected results of hiding actors.
- Build trust and learn what is normal in the enterprise: Trust and learning go together— if the hunters are not trusted, others will not share what is normal (or not).
- **Choose scenarios based on perceived mission interest to an adversary:** Sources of hypothesis creation include the following:

- ○ Organizational incident data and lessons learned
    - ○ Open source or purchased CTI
    - ○ ATT&CK TTPs
    - ○ SOC alerts, trends, and traffic
    - ○ Blogs and social media
    - ○ Asset, mission, or business data (demonstrated interest prioritized by importance the constituency)
- **Draw a clear distinction between confirmation of the hunt hypothesis, and other cybersecurity hygiene issues found:** It is very easy (and common!) for analysts to get excited about their findings, even if those findings have little to do with the original hunt hypothesis and scope. It is critical for hunters and their leads to draw a distinction between "hypothesis proven" and "random other bad things discovered."

Using ATT&CK

As described in "Strategy 6: Illuminate Adversaries with Cyber Threat Intelligence," ATT&CK provides a framework for categorizing adversary TTPs; this is a starting place to develop adversary scenarios and hypotheses. In creating the hypothesis, the team might choose a particular threat actor or group to start. For hunting, the team might pick one of the tactics and one of its associated techniques, based on the adversary, and the likelihood to see in the enterprise, or based on understanding SOC defenses and monitoring. The next step is to determine how an attacker would conduct the attack or understand the procedure. This involves identifying the tools, and the process or procedures for how to conduct the attack. In fact, hunters are creating the story or the chain of events that would lead to an adversary achieving their objectives. This is the same set of skills used in incident response to determine what happened and what an adversary was able to do successfully, only creating the story from the attacker's perspective. Once a set of TTPs is chained together, the hunt team can then derive the data necessary to detect the TTP activity. This will include sensors and their ability to observe different attributes of the TTPs. And, once the hunt team is executing, the set of detection techniques can be updated as gaps are identified, and current detections are validated.

Analysis

Based on the developed scenario(s), hunters then analyze networks, systems, services, and data to determine if adversaries are present. Threat hunting is often analyzed in three dimensions [407]:

- **Timing:** When an event occurs including the sequence and duration of events.
- **Behavior:** What events are occurring including relationships/correlation. Behavior can be malicious or legitimate user or application activity.
- **Terrain (cyber environment):** The systems, processes, applications, and networks in context of SOC monitoring and adversary movement/exploitation.

To expand the analysis, hunters explore the timing, terrain, and behavior built out in TTPs by developing queries. The queries can identify new associative attributes based on the hypotheses. These can include changeable characteristics such as domain names, network and host artifacts, hash values, malware, IP addresses, and other specifics that help identify an existing threat actor. These changeable attributes, when discovered, often lead to the development, expansion, and use of IOCs and TTPs once confirmed. Some of the network, EDR, and other data queries that might be developed to provide attributes, hints, and further leads to explore include the following [405]:

- Failed logins
- Hosts with new logins
- New users
- Uncommon processes (bottom 10 percent or so)
- Powershell downloads
- Windows: Recycle bin contents (malware, suspicious tools, files, etc.)
- Publicly facing Web site vulnerabilities and configuration weaknesses
- Compare DNS logs to CTI

The point of analysis is to conduct queries of data and logs that are not already being covered by existing detections. It means using creativity and deep forensics skills to find new adversarial behavior.

Tools

Several tools are useful, depending on the scenarios or hypotheses to be tested. In general, threat hunters need the ability to see network and host activity in detail. Being able to identify vulnerabilities, such as through vulnerability scanning results, may also help. As mentioned previously, the hunters can construct queries to examine the hypotheses and conduct forensics. The majority of a hunter's time is likely in designing and conducting various queries and correlations. For example, if examining lateral movement, the hunter might construct queries of logins (and time stamps) of several systems, where if a user is logged into multiple sessions simultaneously, this could be an alert (it can also be legitimate, so hunters are cautious). Table 27 lists the types of tools analysts might use as part of hunt. In addition, hunters can simulate malicious attacks to determine potential TTPs, and to detonate actual or simulated malicious code in sandbox environments (to limit any real damage to the enterprise).

Table 27. Cyber Threat Hunting Tools

Tool Type	Why Useful to Hunters
Log, data, and correlation engines (SIEM), big data platforms	The primary focus of hunting is creating queries and examining data in new ways and patterns. Any hunt team should start primarily with logs (system, proxy, DNS, web, etc.), and the tools to query them. For many SOCs, particularly those with an enterprise that makes extensive use of cloud technologies, log data may be exclusive forum for hunt activities.
Scripting and command line tools	These tools assist in many ways including analyzing logs and categorizing and describing classes and instances of malware (utilizing binary or text patterns). Examples include YARA, Threat Hunters forge tools: GitHub – "OTRF/ ThreatHunter-Playbook: A Threat hunter's playbook to aid the development of techniques and hypothesis for hunting campaigns [408]."

Tool Type	Why Useful to Hunters
Packet capture software	Enables hunters to view traffic, and aggregate, summarize, trend network traffic. It is indispensable in identifying insecure hosts and apps, dissect protocol traffic. Examples include Wireshark.
Customizable intrusion detection	Used to observe traffic across the network. Its usefulness is in the detailed logs. Example includes Zeek [252].
Vulnerability scanners	Vulnerability results assist hunters in identifying where the designated TTP or log queries/results might turn up with successful adversaries; it is useful to know what vulnerabilities are actively exploited. Example includes OpenVAS [409].
Persistence analysis and host information gathering	Specific to operating system environments used, and invaluable for analyzing processes host level, determining new and suspicious processes, persistence (such as process appears upon booting) and for malware analysis (including hunters' use of malware!). Examples include procmon, Sysmon, OS Query, etc.

11.1.4 Post-Hunt Activities

- **Share hunt results**: Regardless of whether the adversary was found or not, the hunters are likely to have findings that are of interest to various parties, depending on the results: other analysts in the SOC, SOC leadership, other cybersecurity leadership, and potentially stakeholders of the system and service owners that were the focus of the hunt.
- **Share and act on bad cybersecurity practices and improvements found**: Most hunts, even if unsuccessful at finding an adversary, are bound to uncover bad cyber hygiene. In fact, it would be surprising if a hunt did not "flip over a few rocks" regardless of the outcome. The hunt process should plug into the same routine vulnerability remediation, cyber hygiene, red team findings/remediation, and post incident response processes established by the SOC elsewhere.

11.1.5 Hunt Team Development

Hunt teams themselves are an iteratively evolving capability. Even experienced hunters have learning curves when joining new teams and new enterprises. The longer specific hunt team members work in an enterprise, the more effective they become. To become proficient in threat hunting for a specific environment, it is important each team member develops skills in the following areas:

- Train on malicious TTPs
- Develop and test hypotheses
- Learn how environments are instrumented, such as through sensoring and log collection
- Learn what normal activity is (usually means working with system administrators, detection analysts, and savvy system owners)

When a Hunt team starts out, the first learning curve is the discovery of the environment. Along with this, hunt teams often quickly find gaps in data collection, including alerts and sensor coverage. As the hunters advance as a team, members might specialize in areas, such as knowing how to evade detection and how to create a useful and specific hypothesis. As the teams evolve, the most effective hunters understand how to leverage each set of skills. Hunt teams that work together and collaborate closely are especially effective. The general evolution of a hunt team is described in Table 28.

Table 28. Evolution of a Cyber Threat Hunt Team

Phase Description	Output
Learn the enterprise environment	Augments existing enterprise understanding such as Internet-facing connections, important mission apps or platforms, cloud environments, isolated and not isolated hosts, new devices, IoT, Wireless Access Points, etc.
Learn about adversary TTPs	Understanding various threat actors, from basic to sophisticated, and which might target the enterprise. Understanding TTPs and basics for detection and alerts.
Identify gaps in data collection, including sensor detection, alerts	Recommendations for detection alerts, new sensors, tweaks on type of detection, such as anomaly or by TTPs.
Reuse others' adversary attack scenarios.	Identify incidents not previously detected by enterprise, but detected by other organizations, existing TTPs from others but applied to enterprise, detection configurations.
Develop original adversary attack scenarios	Identify new incidents, malicious actors, adversaries, TTPs, or detection strategies.
Constituency-specific adversary emulation and detection	Tailored detection alerts for adversaries. Effectively utilize advanced tools, including ML, and fast ability to detect and block potential adversaries before damage. Additional TTPs that might be injected into a BAS capability (See Section 11.4)

11.1.6 Maturing the Team

Hunt teams evolve when they spend the time to understand the "feedback" from their hunting. They look for how accurate and effective the hypotheses and create metrics around the data sources to determine high value to SOCs in detecting incidents. When hunters find unique malicious code, they can work with the SOC team and incident responders to understand the data sources that should have found the malware and adjust the defenses. For malicious activity found by hunters, the following should be discussed with SOC teams instrumenting SIEM, SOAR and other platforms:

- Which data sources would determine/confirm an incident or proactively stop the attack? Did they? Or did an analyst miss the event?
- Are there false positives that are obscuring SOC analysts view of real events? If so, lower the priority on these alerts.
- Are there false negatives that have missed the activity? Identify new or shifted TTPs to be used.

- How might the detected TTPs be translated into new analytics or detections?
- Was the activity already available in the data? Did the SOC lack the detections, time, expertise, or analytic rigor to find it until hunting started? How does this influence SOC investments moving forward?

The answers for each threat hunting discovery should be fed back into the SOC systems and monitoring and analysis playbooks.

11.2 Red Teaming

Another common proactive function in advanced SOCs is using an offensive approach to test defenses. Red teaming is the process of using offensive TTPs to emulate real-world threats to train and measure the effectiveness of the people, processes, and technology used to defend an environment [410]. In short, it is comprised of executing realistic adversary activity to test system or service vulnerabilities and strength of overall security mechanisms. In the context of the SOC, this often means exercising detection, analysis, and response capabilities.

Distinguishing red teaming from pen testing is not agreed upon in the industry. Some experts treat them interchangeably. Others would claim pen testing is simply exercising an aspect of the cyber defense technology, whereas red teaming is exercising the full cyber defense, including processes and people, simulating a real attack. Still others would claim red teaming is continuous. Most agree red teams are often more elaborate than pen testing, usually more resourced and comprehensive in approach. To simulate an external attack and realistic defense, the SOC defense teams are often unaware of the red team activities.

Red Teaming can be time and resource intensive, and yet, it is a valuable tool to the mature SOC and enterprise. In addition, the constituency can benefit from red teaming and measure the success in the following ways:

- **Increase the time and expense incurred by the adversary to successfully compromise:** As the Red team successfully infiltrates, the defense team learns to address the protection and monitoring accordingly.
- **Shorten the time for detection and increase the detection:** Red team assists in identifying false positives and false negatives, thus enabling defense teams to hone SOC monitoring.
- **Reduce ability for the adversary to pivot or move laterally:** Adversaries bank on increasing privileges and moving around undetected in the network; the red team can assist in identifying what this looks like, and the defense team can shore up protection and monitoring accordingly.
- **Increase visibility of SOC issues:** Red teaming can assist in highlighting issues the SOC might know about but is having difficulty in getting traction to resolve. For example, a system or service owner might believe their service is secure and devote minimal resourcing to proper security controls; a red team can dispel such a misconception.

11.2.1 Red Teaming as an Extension of the SOC

Red teaming can be scoped as an extension of the SOC or performed by an organization disparate from the SOC; both arrangements are common. If attached to the SOC, it might be conducted by experienced SOC generalists, or more preferably, executed by a dedicated red team attached to the SOC. When used as an extension of the SOC, the feedback loop between offensive and defensive teams can be more immediate and built in. In addition, when SOC personnel are used to staff the red team, the defense teaching and coordination is built in, and it can provide challenge and growth opportunity (most will agree acting as an adversary is fun). Of course, if advanced SOC personnel are used, they need to serve only as red team and need to be completely independent of the SOC, the enterprise, and organization and siloed off when on the red team.

There is benefit in outsourcing in this case, as the adversary would not necessarily have advanced knowledge of the enterprise defenses, and so a third party would start with no prior knowledge. Also, advanced SOC personnel are valuable to the defense, and may be needed. Using advanced SOC personnel, however, could simulate advanced adversary activity, as they know the defenses, and likely how to get around them. In short, there are advantages and disadvantages hosting the red team function in the SOC; some organizations may try both over time.

In any event, it is important to address the following:

- **Planning:** Red team activities should be resourced and scheduled in a way that reflects risk, criticality, red team capacity, production status, and SOC capacity.
- **Deconfliction:** Have a defined way to understand a real event versus red team.
- **Escalation:** Know the right management path and procedures for how and when the event needs to be brought to a higher level of management and of resourcing.
- **Resourcing:** Know which resources are red team, which are SOC, and how much of the SOC responds to a known red team activity. Clear lines should exist between them (internal or outsourced).

11.2.2 Getting Started in Red Teaming

Red teaming is a dynamic and large subject; there are a lot of different approaches. Some of the same resources for setting up exercises can be used for red teams for the logistics, scoping, engagement, and format portions of red team planning. True adversary emulation, by contrast, requires planning and understanding of the adversaries, including extensive knowledge of adversary TTPs, and exploits, and pulls on cyber threat intelligence ("Strategy 6: Illuminate Adversaries with Cyber Threat Intelligence") and adversary emulation, described next. Resources for red teaming include [411], [412], [413], [414].

Ideally, red teaming reveals security issues not discovered through other techniques like routine vulnerability scanning and proper secure system design, and before the adversary does. The longer a red team can stay undetected in an enterprise, the more lateral movement, the better for the team, just as for an adversary. Attributes of an effective red team include the following abilities [415] to:

- **Act as a true adversary:** Independent of the digital assets in scope, tools, and restrictions, meaning they do not have undue restrictions that distort the realism of their attack, or create artificial blind spots because they are not allowed to touch some systems they really should.
- **Teach the defense team:** The point of most red teams is to incorporate new insights, and improve defenses, both the personnel and the defenses.
- **Conduct up-to-date adversary emulation (optional):** The red team may wish to look like a real-world adversary known to the SOC, such as a named APT. If this option is being considered, see more in Section 11.2.3, below.
- **Measure the effectiveness:** The enterprise should be clear on what the red team will accomplish, and accordingly measure the success.
- **Bring attention to security deficiencies:** Often any kind of exercise is used as a "hammer" to bring resources and remediation to problems. Well-planned red team assessments will go for the most critical and weakest parts of the enterprise.

When undertaking red team activities, understanding the objectives and the rules of the engagement as well as having the infrastructure and tools are important. In addition, the red team will spend a lot of time planning and ensuring safeguards exist to not harm the production environment. The following tips will help the effectiveness:

- Employ a trusted agent or "white cell" or manager on the inside, such as the CIO or SOC Director who is aware of the red team activity and ensure response does not elevate beyond a certain level. This is an important role, as this person can stop the response to the red team activities before expending too many resources, taking attention away from real incidents, from being reported too high in an organization, or causing harm to production digital assets. Detection of red team activity often results in high emotions, and panic needs to be contained.
- Pick realistic scenarios that match up with the constituency's mission, previous cyber attacks, and the constituency's infrastructure.
- When starting out on red teaming, choose less complex scenarios to start, then evolve over time to more advanced scenarios and TTPs.
- Determine when the exercise stops ahead of time (or declared a success). Stick to the stopping point. Avoid scenario scope creep, as objectives can get muddy.
- Do not forget to clean up when a red team leaves by making sure all the durable changes such as persistence mechanisms and configuration changes are removed so they cannot be used by adversaries and do not threaten long term reliability of impacted systems and services.
- Spend time on the after-action or hot wash lessons learned to fix the gaps. The point of a red team is to improve defenses. Above all, red teams spend time understanding the aspects of adversary emulation they deploy for their targeted environment. Arguably, effective adversary emulation is a key to successful red teaming.

Once some understanding of the targeted enterprise environment is attained, and some vulnerabilities are identified as candidates to exploit, the team begins to create techniques that map to all phases of the kill chain.

There are many free and/or open-source tools to support red team efforts and also assist in the creation of exploits and attack sequences; some are used for threat hunting as well. Table 29 lists some of the popular red teaming tools.

Table 29. Red Teaming Tools

Tool Type	Description & Use
Detection tests and adversary techniques	Tools, scripts, and frameworks that assist in simulating APT and system breaches. Examples (these are not all interchangeable and have different features and emulate adversaries differently) include: Atomic: [417], Caldera [418], Endgame RTA [419], Uber Metta [420].
Malware Analysis Isolation Environments	Allow exploit creators to detonate malware and get details of what the malware did. Example includes open-source Cuckoo sandbox [421].
Vulnerability enumeration linked to exploits	Identifying vulnerabilities is key to successful exploits, and many tools link vulnerability enumeration with exploit execution. Examples of code and frameworks that link vulnerabilities and exploits include Metasploit [422], [423], sqlmap [424], Nikto [425] Cobalt Strike [426].
Reconnaissance	Tools that assist red teams in mapping out constituencies and related targeting. Tools include Spiderfoot (queries third party records for DNS, IP addresses, e-mails, names, etc.) [427]; nmap (networkmapper) [428]; OpenVAS (vulnerability scanner with over 50,000 vulnerability tests) [409]; Shodan [286]; MITRE ATT&CK Navigator: Web app that provides basic navigation and annotation of ATT&CK matrices [429]; Maltego open-source intelligence and graphical link analysis tool (available both in free and commercial editions) [430]
Custom Tools	Red teams are often advanced enough to create their own tools customized for a constituency's environment and optimized to not be caught.

11.2.3 Adversary Emulation

Advanced red teaming efforts sometimes include adversary emulation. Adversary emulation is an approach whereby a simulated attacker (such as a red team) mimics known threats and adversary-specific actions and behaviors. In contrast, more generalized red teaming utilizes any attack technique they are able to execute, unless specific rules of engagement limit their choices.

Is it red teaming or adversary emulation? If there is a need to distinguish between adversary emulation and other forms of red teaming, the following are distinct to adversary emulation:

- Is based on a real-world adversary known or likely to target your constituency
- Mimics the TTPs of this adversary
- Is not so generic that it could be multiple adversaries (in its entirely, many adversaries will have individual TTPs that overlap)

- The digital assets, services, lines of business, or users targeted are consistent with those targeted by the adversary in question.

The MITRE ATT&CK framework provides a means for understanding the phases of an attack as well as provides a curated set of adversary TTPs [418]. As a large knowledgebase, it provides the TTPs of real-world adversaries to teams, even if the constituency has not directly experienced an adversary's attack. A red team performing adversary emulation can leverage ATT&CK techniques or sub-techniques that match the campaign(s) perpetrated by the adversary they are emulating. For example, the team can match the same targeted CVEs, use the same RATs and bots, and demonstrate the ability to exfiltrate the same data as the adversary being emulated. As the red team gains experience, it might combine ATT&CK curated knowledge with additional cyber threat intelligence, and with incidents experienced by the SOC. In addition, can consider automation and reuse through breach and attack simulation tools, described in more detail in Section 11.4.

Finally, it is worth recognizing that some adversaries are known to perform destructive actions. However, there is no need for the adversary emulation team to take it that far. For example, emulating an adversary that "bricks" systems through scrambling BIOS/EFI widely is not constructive. Instead, disabling just one non-production system through a technique might be sufficient to show end-to-end success. Similarly, red teams should take care when compromising sensitive material or exfiltrating large amounts of data. Stealing one row out of a sensitive database is often enough to prove success and is preferable to the liability that can come from moving millions of rows of PII that must then be protected against theft from a true adversary.

11.2.4 Adversary and Defense Sophistication

As adversary emulators and red teams evolve their scenarios, becoming more sophisticated, defenses should likewise become more sophisticated, otherwise there is an effectiveness issue with the red team engagement. The Pyramid of Pain [431] provides a gauge for hunters, adversary emulators, and SOCs to use in measuring their detection. See Figure 40. This points out that if alerts are designed to detect on a hash value of malware, this is a low level of sophistication, an adversary (or red teamer, or emulator) can trivially alter a bit, re-compile, and is off and running with a new malware hash, now undetectable.

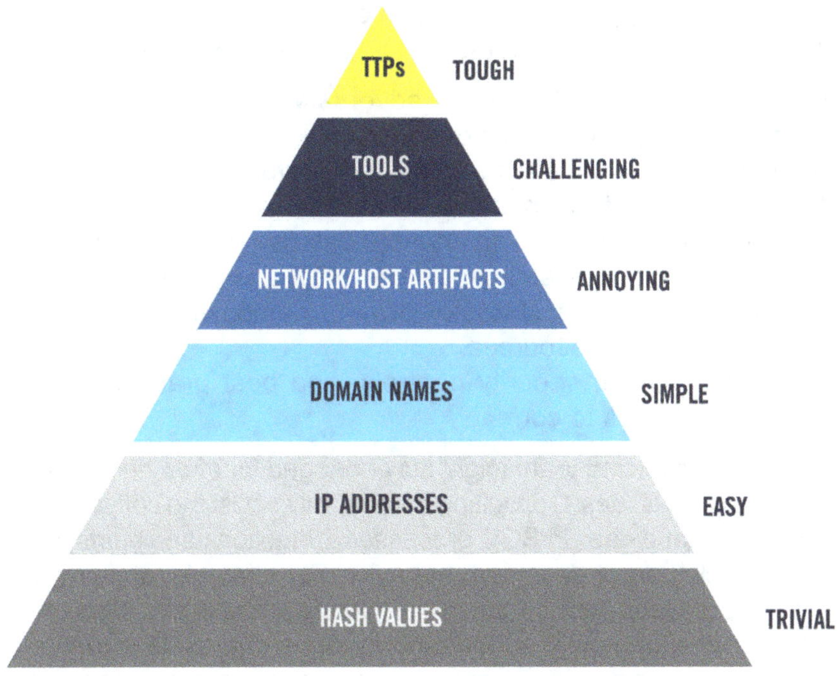

Figure 40. Pyramid of Pain [431]

Similarly, blocking IP addresses as a defense will simply encourage an adversary to use a different IP address. Moving up the pyramid, using TTPs to detect adversaries makes it tough (harder than other forms) for the adversaries to adjust and remain undetected. As red teaming and adversary emulation are discussed the pyramid of pain applies from the adversary's perspective. Applying this to red teaming:

- Red teams that use TTPs that minimize easily detectable tools, such as by using tooling, scripting and utilities that already exist on targeted hosts (otherwise known as "living off the land") to build out their scenarios are less likely to be detected.
- Likewise, defense teams consider set up alerts and correlated detections to address known TTPs.
- If the red team learns and proves quickly the SOC is not capable of advanced defense, re-scoping the exercise or increasing the "loudness" of their TTPs might be useful. For example, if the SOC is simply blocking IP addresses, and the red teams is running sophisticated and tailored stealthy attacks, the SOC is unlikely to adjust, or incorporate lessons learned.

11.2.5 Importance of Red Team Recommendations

Because of the significant resources expended in many red teaming efforts, the recommendations for mitigations and working with the SOC and defense teams to improve addressing adversaries is extremely important to add value to the constituency. The report is usually a basis for rapidly addressing critical deficiencies, a constituency spending more resources on security funding, and is used to inform senior leadership and others on the SOCs ability to detect and mitigate adversaries. Outcomes from red teams include:

- Highlight any method to identify an adversary. Focus on data feeds, sensoring techniques, and analytics as most bang for buck.
- Advance sophistication of detection (focus on detections and analytics that operate at the TTP level, not just "changeable" attributes like hash values and IP addresses.
- Change in participants' behavior to improve defense (i.e., a manager knows what not to do next time, or the defense learns a new TTP and employs it differently).
- Raising awareness to business/mission owners, demonstrating importance of security, and advocating for more resources.
- Highlight cyber hygiene, security controls, and best practices that the SOC has historically found difficult to correct.

In addition to a report, often the red team might stay engaged for after the actual exploitation and report to consult on fixing the issues. Consulting can not only assist the constituency in addressing the specific vulnerabilities and gaps in SOC processes, detections and alerts, and maintenance, but also in addressing classes of adversary behavior. For example, showing IT organizations and system owners the red team results can boost their awareness to cybersecurity. Sometimes system administrators change how they manage systems and system owners increase their savviness of suspicious cyber activity. Also, while training users not to open attachments from unknown sources will not sufficiently address sophisticated phishing attacks, focusing on e-mail and browser filters to block malicious attachments before they get to users can improve the constituency's stance against phishing. Another use is for red teams to re-test some of the conditions red teams exploited to measure improvement and execute some of the attacks to see if systems are less vulnerable, and if alerts are adjusted to improve detection.

11.3 Purple Teaming

The terms red team and blue team come from military exercises; blue teams are those defending the networks: the SOC as well as other members of the constituency's cybersecurity apparatus. There is not firm agreement on what constitutes a blue team, so for the purposes of introduction, a blue team is any group defending the network.

The term "purple teaming" refers to any sort of effort or operation where both blue and red team members come together to cooperatively execute both offensive and defensive activities in a way that helps both teams learn, iterate, and grow simultaneously.

The goal of the purple team is to improve a constituency's defenses against adversaries, including improving detection, response, and prevention through incident response plans and procedures, defense configurations, and playbooks through continuous and tested feedback loop. And often the purple team concept is not "one and done" but rather continual and routine engagement and collaboration over time.

One key defining difference between red team and purple team activities that a purple team is one where red and blue teams partner closely (usually every day during the assessment) and the activity is "open book" where blue team understands in detail what, where, and how the red team is doing what they do. Ordinary red team activities have neither of these qualities.

Purple teams can be assembled in several different configurations including:

- **Test/detect on a system or two:** Iterative testing detections by red team on a system or small test enclave setup and monitored by a blue team, for purposes of testing new blue team tools, sensors, detections, or analytics in a controlled setting
- **Blue team ride along:** Blue team rides along during a red team, meaning a blue teamer is temporarily on detail to the red team, watching them execute an operation for purposes of learning and training
- **Open knowledge vulnerability assessment, red team, or adversary emulation activity:** Pauses at each state stage of the operation so the blue team can "true up," or assess and improve their understanding of what the red team is doing, giving them full blow-by-blow visibility
- **Red/blue exercise on production systems:** Using production data where red emulates a specific named adversary
- **Cooperative deployment and execution of BAS tooling:** Red and blue teams work together (more on BAS in Section 11.6)

Purple teams can be the most effective when they focus on one or a few TTPs step by step. Choosing one set of TTPs can enable the blue team to evaluate how effective the analytic data and tweak it for higher fidelity of detection, resulting in higher quality of detection and response. Cooperative pacing ensures the blue team can "keep up" with each step. Also, the tests should be based on, and tailored to, the production environment being evaluated. For example, if the enterprise is not using PowerShell in the target environment (such as with macOS systems), the alerts on PowerShell may not be useful or good to analysts and might be a lower priority.

11.4 Breach and Attack Simulation

Red teaming, purple teaming, penetration testing, and vulnerability assessment activities tend to be both expensive to conduct, and thus have a very long revisitation rate for any given portion of the constituency or service. In addition, many of the TTPs conducted are bespoke and difficult to reproduce with full reliability. In between high-cost assessments and routine vulnerability scanning, there exists a set of products and tools that can reliably simulate some adversary activity through automation at scale: Breach as a Service (BaaS) or more commonly Breach and Attack Simulation (BAS) tools. BAS is a tool or system that enables the repeatable, measurable, and scalable testing of technical security controls by exercising those controls in an automated fashion. Automating repeatability is a major point of using BAS tools in contrast to analysts reproducing TTPs by hand, as in adversary emulation (See Section 11.2.3).

BAS tools typically take the form of an architecture that at first glance is not unlike an EDR or host-based vulnerability scanning tool. First, an agent is deployed on the end host(s) to be involved in the attack simulation. Second, there is a BAS server (or cloud SaaS offering) running software that controls the actions of all BAS agents. Third, a user or admin will connect and control the first two components by logging into the BAS interface with their web browser pointed at the BAS server. The user can schedule and automate the execution of attack activity on/against constituency hosts, networks, and services. High-quality BAS products generally feature a library and modular framework of various attack plugins and should cover adversary TTPs across the kill chain and ATT&CK framework.

For simplicity's sake, this section uses language with the assumption that the SOC is in the driver's seat for choosing, adopting, deploying, and operating a BAS solution. Often, an independent red team will either be the prime mover in this effort, or the SOC and red teams jointly operate a BAS capability. As will be discussed near the end, elements of the red and blue teams should partner in this venture, regardless of whether those that perform pen testing are located in the SOC organization or elsewhere in the constituency [432], [433].

11.4.1 Why Invest in BAS

Some of the rationale, use cases and business outcomes for BAS include:

- The constituency has no red team, its red team has limited capacity simply due to personnel and funding, its red team has a fixed scope perhaps due to outsourcing, or some combination thereof.
- The pen test team feels that a substantial portion of its work is a) relatively generic in TTPs/attack types and b) can be fully reduced to automation. As a result, implementing BAS can help increase focus to custom, bespoke activities.
- Cybersecurity stakeholders, including the SOC itself, wish to exercise SOC alert and incident handling tools, processes, and personnel, and/or to do so in a routine and repeatable fashion.
- The SOC and other stakeholders wish to evaluate and improve ATT&CK coverage for SOC detection and investigative tooling and data, such as their EDR.
- Cybersecurity stakeholders wish to measure and improve the effectiveness of cybersecurity controls outside the SOC and tighten the response time between test and improvement. This can include month over month, year over year evaluation of what attack techniques is detected and blocked, thereby objectively measuring cybersecurity investment.
- Testing detection and prevention products being considered or piloted, including both in-house and managed services.

11.4.2 BAS Requirements

As with any tool, the SOC should examine the following requirements and features when considering a BAS.

- **Coverage:**
  - One of the top ways most BAS tools are evaluated is the breadth of their attack technique library. Superficially, this can be done by raw attack module or script

count. However, as with early days of IDS systems, raw attack module or plugin count (as with signature count) does not necessarily address other factors, such as quality. Measuring ATT&CK coverage is a better approach, but care should be taken to evaluate the product claims.

- **How often the vendor releases module/plugin updates:**
  - Relevancy of those attack TTPs to the targeted constituency digital assets and mission.
  - Expect a best-of-breed BAS product will exercise host, firewall, email filtering controls thoroughly, but also look for products that have out of the box functionality for applications, web servers, databases, and cloud assets.
- **Pre-packaged and templated scenarios:**
  - A good BAS will not only have a wide ATT&CK coverage but be able to "string" or chain those modules together in a coherent manner, not only on one host or service, but across disparate IT/OT components. Properly implemented multi-stage attack simulation improves the fidelity of SIEM, SOAR, and ML detections in turn.
  - Consider BAS features that simulate a known adversary or group, particularly ones that are relevant to the consistency.
- **Transparency and extensibility:**
  - A strong BAS product will provide full transparency on their attack techniques. This can, for example, take the form of a well-documented plugin framework that uses a popular language like Python for scripting.
  - Modules should be customizable, either via code or point click in the BAS web UI, such that they can test thresholds around non-atomic behavior detections or ML models, or by copying and editing the module code.
  - Attack plugin code should be evaluated against SOC needs. Work with the vendor to understand the attack plugin code about the attacks themselves. Some plugins may simulate some behaviors of attacks, whereas others may execute them exactly as an adversary would. There are pros and cons to each, and the SOC should weigh this against the risk tolerance of their constituency.
  - Many BAS vendors feature robust integration with SIEM, SOAR, and EDR products. These integrations, in short, are the "easy" button for determining whether a given attack module was successfully detected. In addition, the BAS product should have an API and/or an open, queryable datastore that includes every action taken by every attack scenario run.
- **Risk reduction:**
  - Detailed logging around what the tool does every time, which is particularly helpful if the BAS tool is ever blamed for breaking a service or IT function.
  - The tool should be able to clean up any permanent changes it is made on the system, ranging from securely deleting dumped credentials, reverting registry changes, removal of user and group changes, etc.
  - The tool should support the equivalent of a "big red stop" button, meaning all attack activity can be stopped at once.
  - It may be helpful for the BAS vendor to demonstrate thorough due diligence in how it secures and tests the security of its own system—minimizing surface area for service exploitation, proper use of encryption around C2 protocols, tamper

resistance of the agents, and overall security of its own software supply chain such as upstream code reuse.
  - If the BAS uses a cloud-based backend, as with any other cloud-based SOC technology, the BAS vendor should demonstrate high confidence and best practices in how it secures their cloud service endpoint(s) and insulates BAS backend services that from breach of the vendor itself.
- **Common requirements:**
  - Support for the constituency's predominant operating systems and infrastructure
  - BAS server and agent manageability
  - Ability to track inventory/agent health
  - Existence and documentation around API/ability to export data
  - Reporting engine and trending over time (attack TTP and scenario success/fail)

11.4.3 Existing BAS Tools

There are a number of commercial and open-source BAS offerings available. Additionally, while some open-source tools are more limited in scope and intent, they may still be sufficient for what a SOC needs. These tools can assist with recreating TTPs, such as with adversary emulation. Here are some open-source and commercial examples to get started:

- MITRE's CALDERA for adversary emulation [418]
- Praetorian's Metasploit automation of MITRE ATT&CK TTPs [423]
- VECTR™ purple team platform with STIX/TAXII functionality [434]
- Uber Metta for adversary emulation [420]
- Endgame RTA [419]
- Common BAS offerings, including well-known commercial products [435], [436]

11.4.4 Succeeding with BAS

From product evaluation onward, the SOC should consider the following tips for making the most of a BAS capability:

The SOC and pen testers should partner closely
Generally, both SOC and the pen test team have a reasonable and recurring reason to utilize a BAS tool. Red team needs to perform testing on constituents, blue team needs to validate their own tools are working. If one team leverages a BAS solution it should consult and engage the other at a minimum.

Start small
The best way to get comfortable with BAS is to test it on a handful of hosts in a non-critical service or network segment. This will enable the SOC and other stakeholders to ensure it is using the tool responsibly before touching more critical digital assets.

Setup guard rails
If the SOC and red team do not already have notification, "cease fire" and deconfliction processes, designated POCs, and rules of engagement (ROE) sanctioned by executives, now is probably the time to codify these in writing.

Ensure the attacks being run are relevant to the target environment and the threats of concern
Not all attacks are equally relevant to all organizations. The SOC and red team should choose scenarios that are realistic and meaningful for their environment, and that stimulate movement around security controls and services of greatest concern.

Consider designating certain hosts for prototyping detection and prevention capabilities
This gives the SOC and red team the opportunity to closely collaborate around host instrumentation and attack techniques. A BAS tool can serve as a superb technological basis for some purple teaming activities, if used correctly and in partnership.

Be clear about how the results are (or are not) representative of the enterprise writ large
It may be easy to assume results from two hosts are representative of a much larger group, perhaps 1000 hosts. Sometimes this is true, and sometimes it is not. Results should be interpreted with care. With this said, blanketing every host in the enterprise can be expensive and usually unnecessary, especially in the context of other investment priorities.

You still need vulnerability scanners
Generally, BAS products are not focused on or intended to supply traditional vulnerability scanning. Vulnerability scanners focus on broad, comprehensive host and vulnerability coverage; they will test for the presence of thousands of different security controls and patches, whereas a BAS test may exercise dozens. Each have their own value and sweet spot.

Be clear about whether this can be used as evidence
The SOC and its executives should seek clarity in advance from legal counsel and/or regulatory auditors in advance as to whether the tool chosen is satisfactory for compliance, regulatory or legal requirements for routine pen testing of critical services.

BAS does not put the red team out of a job
There has and will always be a strong case for bespoke attack techniques, custom service/ application analysis, and the other things only a human can do in thoughtfully evaluating critical IT and OT systems. Just like a SIEM and SOAR enhance the SOC's ability to cut through billions of events, a BAS will free the red team from more routine parts of their job and enable them to do the things they would not have otherwise had time for.

11.5 Deception

Deception in cybersecurity is an effort to conceal networks and assets, create uncertainty and confusion, and/or influence and misdirect adversary perceptions and decisions [437]. Deception is also a means to perform advanced threat hunting, by enticing adversaries to interact with fake environments or data or to use fake information which can be tracked. For example, if a fake username password combination is planted somewhere and then is later used, it is likely to be unauthorized activity (sometimes called deception-based detection).

Deception can aid in early detection of adversaries and can provide valuable insight into adversary actions and behaviors which can be a form of developing internal threat intelligence. Often, deception techniques are used by advanced SOCs to augment signature and anomaly-based detection. If done well and thoughtfully, deploying deception technologies, and integrating the data they generate with other alerts and suspicious activity can assist mature SOCs on finding adversary activity that is otherwise subtle.

11.5.1 Deception Terminology

The cyber deception field continues to evolve, and terminology is still being normalized among practitioners and vendors. As a starting point, commercial deception technologies can be thought of as software which attempts to deceive, entice, detect, and analyze an attacker by distributing deception objects across the infrastructure. It has more advanced capabilities than traditional honeypots, such as deception lure, deception automation, attacker engagement, threat analysis, and threat hunting support. Below are some descriptions of terminology used in the deception space. Again, these are evolving concepts so as the cyber deception landscape matures the exact words and meaning are likely to evolve as well.

Deception objects are realistic-but-fake assets deployed throughout the constituency to lure attackers away from production assets. Different vendors use different terms; however, they can typically be placed in the following four categories:

- **Bait:** Something that the attacker finds worthwhile to steal (e.g., RDP credentials, Windows credentials, SSH Keys).
- **Breadcrumbs:** Planted to lead attacker towards decoys/traps (e.g., cookies, network shares, registry keys).
- **Lures:** Makes decoys more attractive to attacker (e.g., application with default credentials or vulnerable version of software).
- **Decoys (sometimes called Traps):** Systems, applications, or services that look like regular assets in a constituency's IT infrastructure. They can offer a safe sandbox environment to contain and analyze attackers.
  - **Decoy Types:**
    - **Emulated:** One or more VMs emulate the constituency's network, services, and operating systems based on that network's architecture. The VMs deceive the attacker into attacking the emulated decoys rather than the actual environment. Emulated services do not provide full engagement with an attacker, as they are not active by design, which is easier to identify since the attacker cannot complete the attack cycle. Emulation provides ease of deployment, a high-fidelity alarm system, scalability, and a mechanism for creating decoys that would otherwise be impossible to deploy, such as medical devices.

- **Real/Full OS:** Real operating systems, services, and applications that can provide full attacker engagement. This is the best option for authenticity and attacker analysis, but they are costly to scale and are resource intensive.
  - **Decoy Interaction Levels:**
    - **Low:** Gives an attacker limited access to the operating system. A low interaction decoy usually emulates a small amount of internet protocols and network services to deceive the attacker into thinking they are connecting to a real system. They are simple to deploy, do not consume significant resources, and are the easiest to maintain.
    - **Medium:** Emulated services where attacker communications are analyzed, and simulated responses designed to replicate a real service are returned to the attacker. If someone can logon to a decoy, then the decoy supports at least medium interaction.
    - **High:** Involves the use of real operating systems, services, and applications, where the attacker is given fully operational hosts or even networks to attack and engage with. Attacker behavior can be observed in a safe sandbox environment for threat intelligence and forensic data collection.

Deception deployments are a way of describing the "where" of a deployment. This includes:

- **Endpoint Deception:** Includes deceptive credentials, false shares, decoy documents, and other assets that appear on local systems.
- **Network Deception:** Includes decoy systems, servers, and services, that should appear indistinguishable from production assets.
- **Active Directory (AD) Deception:** Protects AD through techniques such as leading attackers to an AD decoy object via bait and breadcrumbs or populating the AD with fake, high privilege accounts linked to decoys that the attacker can obtain through querying AD.
- **Perimeter Deception:** Public-facing assets are placed at the network perimeter and mimic application responses to disrupt an attacker's reconnaissance attempts. Currently, most deception vendors focus on attacks inside the perimeter.

Additional Deception Terms:

- **Autonomous Deception:** Facilitates dynamic and automatic creation, deployment, maintenance, and adaptation of the deception environment through AI technology.
- **Moving Target Defense:** Creates uncertainty for attackers by moving or hiding targets to change the perception of the attack surface from the adversarial perspective.
- **Comprehensive Deception Suite:** Describes products that combine Endpoint, Network, and Active Directory deception into an active defense solution.

11.5.2 Managing Deception Risks

While highly valuable when executed well, there are also risks with employing deception capabilities, ranging from not tricking adversaries and having expended valuable time and resources without benefit, to adversaries hijacking environments and using them as launching points into real attacks in enterprises. Setting up environments to engage adversaries in a more active way is one of the riskiest initiatives an advanced SOC might employ. Therefore, deception should be used carefully, and only by highly mature SOCs with advanced expertise. Pre-requisites for starting to incorporate deception into the environment include:

- **Maturity of cyber defenses:** The SOC should be good at incident response, detection, and threat hunting before considering deception techniques.
- **Adequate resources:** Advanced analysts with time to specifically dedicate to deploying, maintaining, and monitoring deception capabilities, sufficient funding such that deploying a deception program does not take away from other SOC activities.
- **Rules of engagement:** The SOC may need to make decisions about when and how to allow or terminate adversary access to either fake or real resources. Plans of action for expected scenarios should be agreed to prior to the deployment of any deception technology.
- **Approval by lawyers:** Deploying deception capabilities will likely require consideration for privacy laws, user consent, etc. and risks discussed above; all should be discussed in context with legal counsel.
- **Approval by upper management:** This includes SOC leadership as well as others who are responsible for cyber risk across the constituency.

11.5.3 Succeeding with Deception

Some tips for SOCs considering deception include:

It is important to develop a set of objectives before deploying deception technologies
Deception can assist with several strategic outcomes. Developing or purchasing the right deception technology will be dependent on the type of outcome the constituency wants to achieve. MITRE Engage is a framework for discussing and planning adversary engagement, deception, and denial activities and it provides three high level strategic goals to consider as well as providing additional planning information The three goals outlined by MITRE Engage are: to expose adversaries on the network, to affect adversaries on the network, or to elicit new information about adversaries [438].

Consider using commercial products if the SOC is just getting started
Commercial deception products can be used to set up fake accounts, services, and files within a real network environment in a structured manner. For example, fake CEO calendars can be created, fake e-mail, and planting fake files for attackers to click and download. Since these are not real, any interactive activity may be considered malicious. The commercial products' main value is around automating the creation, deployment, and monitoring of deception lures, a process that would otherwise be very time consuming.

Deception deployments must be dynamically changing and mirror the characteristics of the real environment
Look for products with changing attributes; adversaries are not fooled by a static honey pot, where the characteristics do not change over time. Because of the lack of change, they are readily identifiable to many adversaries. A proper deception environment needs to look

realistic in several ways: it must appear to be complex; "lived in," meaning it appears like it is undergoing real use; and it must not show obvious signs of being built on top of a deception framework.

Deception requires care and feeding

Besides the need to mirror reality of changing data on the network, deception needs to be carefully monitored to ensure adversaries are constrained within expected environments [439].

11.6 Malware Analysis Capability

SOCs routinely encounter suspect files whose purpose, contents, provenance, and pedigree need to be evaluated. One of the most common methods of course are to use local anti-virus and cloud-based malware analysis tools. As a SOC matures and its needs progress, it requires malware analysis capability with increasing frequency and importance. Malware analysis is the process of seeking understanding of the behavior and purpose of a suspicious file to aid in detection and threat reduction [440]. For SOCs that are well-resourced, forming an advanced malware analysis capability can be not only rewarding but necessary. Malware analysis may be used to achieve one or more of the following objectives for the SOC:

- Breaking down a piece of malware and formulating a working understanding of its attack vector and likely purpose, such as qualities that may link it to known malware families or actors.
- Evaluating malware behavior, such that its persistence, stealth, and C2 capabilities may be understood (and thus more easily detected), such as with specific IOCs.
- Feeding both back into not only the incident currently being worked, but long term understanding and assessment of threats, campaigns, and actors, and in support of proactive hunt.

The difference between a basic malware analysis capability, such as that recommended for incident response and CTI, and the more advanced capability is the deep dive into really understanding a particular piece of malware, and/or adversary's movements through networks and host environments. This requires both advanced ability and significant time and personnel resources which may not be appropriate for every SOC. Figure 41 shows the progression of malware analysis expertise from least to greatest required.

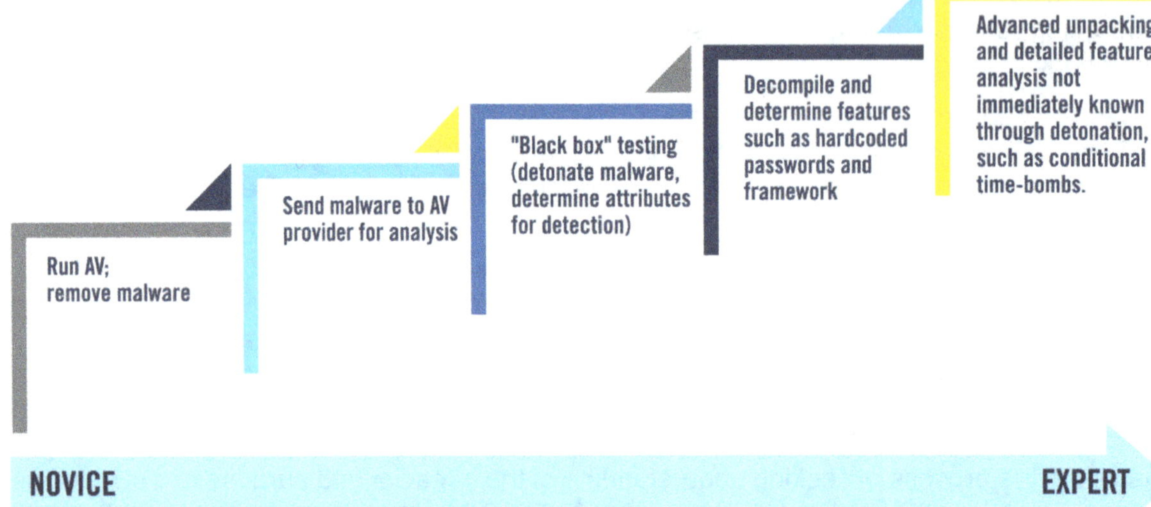

Figure 41. Progression of Malware Analysis Expertise

As experts perform even more advanced malware analysis, the payoff can be big in the greater SOC and security community. SOCs with the most advanced malware analysis are invariably assisting others. Those that produce and share useful malware analysis, such as TTPs and adversary association, with others become community touchstones in tracking down adversaries. Producing and sharing results get the team invited to the big, highly visible incidents to assist. When maintaining this capability in-house, the SOC has a much higher likelihood of detecting and countering adversaries in the constituency, because the IOCs, the TTPs, and any other resulting findings are specific to the constituency.

> *When maintaining malware analysis capability in-house, the SOC has a higher likelihood of detecting and countering adversaries, because results are specific to the constituency.*

Malware writing and usage is like handwriting analysis, there are clues to be uncovered about potential individual or group threats. Thoroughly breaking down malware could answer the following questions:

- Is the malware targeted to the constituency, or is it more general, sent to many enterprises?
- Does the malware "phone home" and how, (i.e., what port is used, does it communicate back to another server, and is C2 established)?
- Are passwords, certificates, or keys hardcoded and reused in the malware, (i.e., does it require credentials, to what, and are these used elsewhere)?
- Is the code a derivative of other code, (i.e., where did the malware come from, what group, has it been seen before in slightly different form)?

- What does the code do (i.e., what happens, are tools downloaded, does it log keystrokes, etc.)?
- What does the sophistication, stealth, and novelty of the code indicate about the resourcing and motivation of the actors?

SOCs considering an in-house malware analysis investment past ordinary AV and automated malware detonation platforms like Remnux and Cukoo should take the following into consideration:

- Frequency of malware analysis services needed. How many suspect files is the SOC encountering every day, week, or month? And of those, how many needed through analysis beyond simple detonation?
- If expert analysis is not frequently needed, will outsourcing work? Could the SOC bridge the gap between outsourced analysis and causal support from one or two SOC analysis who have passing interest or aspirations in this field?
- For the suspect files being encountered, is automated tooling able to determine their provenance and pedigree reliably? Or may the SOC be in a position where it is hit with targeted attacks and custom malware on an ongoing basis?

Malware analysis is a topic that can span several books by itself and is comprehensively documented in existing literature [441], [442], [443], [444]. As a starting point, this section addresses some of the essential elements of how to initiate a malware analysis capability in the SOC.

11.6.1 Getting Started

Establishing and maintaining an advanced malware analysis capability requires a commitment on the part of the SOC, especially since the skills needed for this capability are difficult to attract and to maintain and many SOCs will not need someone working in this space full time. SOCs wishing to grow their malware analysis capabilities have the following choices:

- Verify they have made the most with existing automated malware detonation and other tooling; maybe there is unrealized value
- Grow existing staff who have a fledgling interest in the field; although there are risks associated with applying less experiences talent to a problem
- Hire staff with veteran experience in the area who are willing to perform other duties when not inspecting malware
- Leverage other SOCs proficient in malware who might be willing to help, such as in federated or hierarchical SOC arrangements, or SOCs who have strong intel sharing arrangements
- Contract and procure managed services for malware analysis
- Some hybrid of the above

SOCs protecting particularly attractive, high-value constituencies often have access to malware-rich network and host traffic. This can in turn attract talent who are more likely to stay "entertained" protecting that constituency. The environments for analyzing malware are best isolated from other enterprise network activities, but often need Internet access for runtime analysis. Because malware analysis depends highly on the skills of the analysts, tools vary extensively and are personal choices. Most analysts do not agree about every single tool, and each analyst will purposefully maintain toolkits with a lot of overlap because different scenarios call for different but very similar tools. Table 30 offers just a few of the classes of tools used.

Table 30. Advanced Malware Analysis Tools

Tool Type	Use Description
Malware analysis appliances and software environments	Environments and appliances enable analysts to focus on the analysis. Examples: Cuckoo [421], REMnux [445].
General purpose virtual environments, and customizable and modular analysis environment tools	Enables malware to be detonated and analyzed in contained or otherwise isolated environments. Examples: VMWare [147], VirtualBox [446], AWS [447], Azure [448].
Decompilers and analyzers	Used to decompose executables and provide text output code to be analyzed. The output is structured and readable source code digestible by developers/ analysts. Most of these are not free (except for educational use). Example: IDA Pro [449].
Executable header and binary analyzers, such as for ELF and PE executable formats	ELF and PE analyzers assist in understanding characteristics of execution or linking, including static and dynamic binaries, libraries, and files for Linux and Windows, respectively. Examples include Radare2, eflfutils, elfkickers (Some of these may come with Linux, for example); Portable Executable (PE) header analysis is used in Windows systems, and is useful for analyzing DLLs, object code, core dumps, and executables.
Unpackers, decompression and decoding utilities	There are frequently tools needed to unpack, decompress, and decode specific formats and malware families. They are too numerous to list here, but like other tools, change frequently, and new versions and variants are needed on the fly.
Debuggers	These help the analyst execute a program under controlled conditions, enabling them to monitor things like memory contents, program variables and data structures, API calls, and CPU registers. Examples include: Windbg [450] and Ollydbg [451].
Windows utilities for diagnosing systems, processes, and applications	Microsoft provided utilities that will assist analysts in diagnosing Windows systems and applications. Example: Sysinternals [452].
Hex Editors	Various editors exist allowing analysts to view and edit binaries, and to view hard drive physical spaces (even when pointers are not available).
Cloud-based malware analysis services and environments	Leverages the scalability, virtualization, and integration possibilities of the cloud to support malware analysis in a controlled environment.

11.6.2 Succeeding with Malware Analysis

SOCs pursuing an investment in malware analysis should consider the following tips for success in their journey, many of which mirror other SOC investments and best practices:

Fully assess the SOCs need for a malware capability
Be clear that the conditions for standing up a malware analysis capability are met, and there is work to keep staff engaged. By contrast, SOCs that do not see a routine cadence of advanced malware or targeted attacks may not be good candidates.

Identify the return on investment
Another challenge to maintaining advanced malware capability is sometimes the return on investment (ROI) to budget decision-makers is not obvious and can be difficult to convince those outside the cybersecurity field. Accordingly, the SOC should be able to tell the story akin to "before we were blind to what our adversaries were doing to us, but now we understand" and how that influences everything from incident response to remedial cyber hygiene.

Ensure malware analysts feel plugged into three groups (See Sections 9.2, 9.3, and 9.4.1)
- Incident investigators and responders who need their help and feedback
- Hunters and threat intel analysts who should integrate knowledge of malware pedigree, behaviors, etc. into their routine operations, and with whom the malware analysis will likely collaborate on both authoring hunt hypothesis and intel reporting
- Other malware analysts, such as those in neighboring, partner, and federated SOCs

Support publishing and sharing
Malware analysts are much more likely to stay when they are able to publish fruits of their work, recognizing the SOC must also respect publishing requirements of their constituency, and the SOC may need to navigate this. The goodwill the SOC gains from sharing information will reinforce its ability to both attract and retain talent.

Give malware analysts control over their own platform
Analysts will frequently need random tools, utilities, updates, and dependencies; leveraging a static toolset and platform they must seek external approval to update is a non-starter. While there can and should be deliberate core common tooling, they will each also need to download tools and utilities to a) suit their own tastes and preferences and b) because they need it for a given incident *right now*. Do not fight this as it will alienate staff and encumber timely analysis.

Isolate the malware analysis environment from all other constituency resources, including the SOC's analytic platforms
It is important that the risk of "leaks" from the malware analysis lab be minimized. This can include techniques such as network segmentation, trusted file transfer mechanisms, easy "pave"/refresh of the analysis environment, and various procedural controls, the details of which are beyond the scope of this book.

Get the SOCs rules, authorities, and procedures related to malware handling down on paper
For example, common advanced malware analysis practices such as watching adversaries in action, trading indicators of compromise, and detonating malware may be unsettling to

some stakeholders. So, if planning to set up this capability, be clear about the benefits to the constituency, and codify their acceptance in writing.

11.7 Digital Forensics

All SOCs will perform some level of forensic review of suspicious or potentially compromised systems and data. However, there is a difference between conducting forensics simply to understand what an adversary did during so the SOC can perform IR, and law enforcement-caliber forensics that may be used to support criminal proceedings or legal action. Digital forensics is formally described as the "application of science to the identification, collection, examination, and analysis, of data while preserving the integrity of the information and maintaining a strict chain of custody for the data [453]."

The SOC needs to be clear with itself about what are its goals when investing in any digital forensic capability. As mentioned in "Strategy 5: Prioritize Incident Response," some basic media and memory forensics are necessary to perform effective IR. Having a basic forensic capability internal to the SOC means being able to understand for example what files an adversary exfiltrated from a system or what RAT was loaded into memory. Often times a SOC will follow legally sound procedures for forensics only in terms of evidence prevention and chain of custody such that that a) they can achieve their own IR goals while b) supporting future legal proceedings in the unlikely event they are necessary.

For advanced SOCs, directly arming legal counsel or law enforcement may indeed be their goal. Important aspects of forensics involve following proper search authority, chain of custody, validation with mathematics, use of validated tools, repeatability, reporting, and possibly expert testimony [127]. When the SOC determines an incident is serious enough to consider legal repercussions, the team should seek legal counsel to determine what tools and procedures are needed to perform legally sound handling of digital evidence to support potential future use in criminal, civil, or administrative proceedings. For many teams, this includes means for maintaining evidence.

There is a significant amount of published information in this field, this section provides some general information for consideration. For more in-depth treatment of this topics see: [453], [454], [455], [456], [457], [458], [459], [460], [461], [462].

11.7.1 Getting Started

Historically, digital forensics focused on analyzing media or hard-drive images in-depth. However, there are now several different types and sub-fields, including memory, operating systems (log & file system focused), media, mobile, and network analysis. Because digital forensics is highly detailed work, techniques, and structures vary depending on the platform, which include traditional enterprise and cloud environments, and can include analyzing entire filesystems or devices, such as Windows, Mac, Linux (and Unix). Table 31 describes some of these types of digital forensics and descriptions.

Table 31. Cyber Forensic Types

Digital Forensic Type	Description
Memory	Enables investigators to find and understand suspicious activity that is detectable primarily in volatile memory (such as RAM), as it is running through temporary memory caches or processes that are no longer present when the system is shut down. More sophisticated attacks are not easily discoverable through other SOC detection but are identifiable in memory. Volatile memory forensic interests include currently running: • System and application processes • Temporary memory caches • CPU registers • Routing tables, ARP cache • Kernel (or other OS) statistics • Port communications, executable temporary files
Media, files, and operating system	Investigators examine the hard drive of a system, along with system logs, processes, and filesystems. Filesystems are useful in recovering suspicious files and traces of attackers; also, investigators often look at activity that has been "deleted" by attackers (where pointers are removed, but the file of deleted content is still resident on the hard drive). Common Filesystems include: • Windows: FAT, NTFS • macOS: HFS+, APFS • Linux: EXT
Network	Network forensics consists of analyzing network packets and traffic to determine possible attack vectors, exploitation of network protocols, and the scope and magnitude of an attack. For example, it assists with answering "where else?" It also includes examining data in motion for evidence [463]. This consists of examining IP traffic, communications between systems and over networks including: • NetFlow • DNS logs • Firewall logs
Mobile	Mobile forensics has become a subfield of expertise. Many aspects of operating system, memory, and network forensics are conducted on mobile technologies as well. The difference is in the system internals and hardware coupled to filesystems. SD Cards, data and cache configurations vary in mobile forensics. Data of interest, such as geolocation, cellular communications, social media, and operating system variations differ for mobile. Common operating systems and filesystems for mobile include: • Apple: iOS, APFS • Android: Linux, Android File System • Google: Chrome OS (or AndroidOS; Linux-based), Chrome Filesystem

Digital Forensic Type	Description
Cloud storage	Cloud forensics vary and is more complex from other forensics types because of the distributed nature of virtualizing environments and filesystems. Cloud providers often have their own structures and even different types of filesystems, requiring investigators to learn about the specific systems. Filesystems include file-based, object (data structure, focused on data as files), and block (where data is stored in blocks) filesystem (or filesystem-like) environments. Common cloud filesystems include: • Amazon Elastic File System (EFS) [464], FSx [465] • Oracle CloudFS (ASM Cluster File System (ACFS), ASM Dynamic Volume Manger (ADVM) [466] • Microsoft Azure File Share (FSLogix) [467] • Google Persistent Disk, Colossus (replacing Google File System) [468]
Data	Although the other types of forensics include aspects of data, increasingly data is considered separately. As this trend continues, data forensics could become a substantial subfield in forensics, and may include applications and data-specific filesystems. Filesystems and applications with, and specific to, data include: • Databases (including SQL and noSQL technologies) • E-mail (including calendars, deleted e-mails and contacts) • Web and browser data

11.7.2 Forensic Tools

Forensic software varies; a full accounting of all the software types used in each category of forensics is outside the scope of this of this book. However, Table 32 provides a brief description of some well-known types of tools and examples of software. Notably forensic software often accomplishes more than one of the functions in Table 32. The general process for conducting forensics on media or memory is:

- Establish the goal of the forensics first and understand what the analyst is looking to discover. Choose the software needed.
- Create a copy without disturbing the files or memory. Volatile memory is especially transient.
- Analyze the contents of the drive, memory, and/or filesystems of interest. Drive reconstruction enables the analyst to more thoroughly review the items in question.
- Extract items of interest. Drive reconstruction software also enables the analyst to methodically capture the important pieces of information (or evidence), enumerate them, and make notes.
- Create reports of the items of interest. Many of the software options either plug into report-creating software or have the built-in ability for analysts to create reports. If analysts need to customize reports, consider this as software is being chosen, as some forensics software does not enable report customization natively.

Some SOCs will purchase purpose-built forensics workstations that have write blockers and mass storage built into them. In addition to tools listed in other sections (Wireshark is great for packet capture in network forensics, for example), Table 32 lists some of the functionality of tools used by investigators in conducting forensics. Many digital forensics tools (open-source and commercially available) combine functions to provide analysts easily accessed forensics results.

Table 32. Common Digital Forensics Tools

Tool functionality	Purpose
Disk, media, and data capture tools	Software that scans or captures disk images, make copies of media. Examples include Autopsy [469] and Sleuth Kit [470].
Read-only hard drive imagers or write blockers	In hardware drive and media forensics, important capability that blocks investigators' activities from inadvertently changing the files, filesystems, or otherwise changing the drive or system being investigated. Even if a SOC does not do in-depth forensics or malware analysis, the most cursory inspection of media involved in an incident requires making a copy of the original. This hardware device (along with media image analysis tools) allows copying a hard drive without performing any write operations against it [471].
Extractors	Often combined with write blocker functionality, extractors enable analysts to analyze the device or data in question and pull out potentially important forensic evidence. Examples include Bulk Extractor, Redline, and the Linux command dd.
Drive reconstruction	Once a hard drive or filesystem is acquired, the analyst can then start examining its contents. Tools like FTK [472], EnCase [473], Sleuthkit/Autopsy [470], and SANS Sift Workstation [474] are essential for performing this task, from examining filesystem tables, to looking for deleted files, to performing timeline analysis.

11.7.3 Succeeding with Digital Forensics

Tips for successful forensics activities include:

Know the SOC's goals for forensics
Is the goal of the forensics to be conducted more for IR or for legal purposes? The goal of forensics should be clear because how the forensics are conducted will be scrutinized in some situations. The goal drives the tools that are chosen and how the data and analysis are handled. For example, in some cases analysts may be asked to testify and describe the tools they used and demonstrate the efficacy of the process and software in a court of law. If the analyst used experimental software (which analysts do for memory and more transient data), they must explain how the method was sound.

Ensure the SOC has the proper authorities
Before beginning forensics, review and ensure the authorities needed are obtained; also review and ensure processes are in place to perform the type of forensics they are focused on and have repeatable technical process in place to minimize data changes. These might include:

- Collect and conduct forensics working from most volatile to least volatile [454].
- Keep systems running (avoid shutdown) until evidence collection is complete [454].
- Work from copies whenever possible.
- Protect state as much as possible.

- Do not run access altering programs, such changing the date/time a file was accessed [454].

Minimize exposure to sensitive information

Recognize that performing forensics may mean an analyst is exposed to sensitive or disturbing content. Some data found may be offensive, illegal, or otherwise traumatic to analysts. Have processes in place to minimize exposure or support analysts who perform this role.

Acknowledge that forensics is a niche skill set

Unless the SOC is very large, it is very easy for one or a few members of the SOC to become single points of failure, especially in specialty skill areas such as mobile. It can be helpful to cross-train, there is a lot a novice can accomplish if properly trained on just a few things and with process guardrails in place. Review "Strategy 4: Hire AND Grow Quality Staff" for ideas on retaining staff as well as planning for staff turn-over.

Plan for an increased workload during a major incident

Forensic analysis is incredibly important during a major incident, but it can be challenging to balance the desire to move quickly to respond with the time it takes to perform adequate analysis. Ensure there is someone in charge of prioritizing and tracking forensic work, so chain of custody is not broken, and work is correctly triaged and prioritized. This will enable the forensics experts to focus on their work and less on playing traffic cop with all the requests that may come in. Additionally, be very judicious about when to say stop. Analyzing a single hard drive or mobile device can be a several week journey, however a quick inspection might be all that is needed. Also, consider having an external entity on retainer, such as a managed service, or other SOC, that can pick up some of the burden if needed.

11.8 Tabletop Exercises

Discussion-based, or tabletop exercises are commonly used when time and resources are limited and typically are used to go over cyber incident scenarios. In cybersecurity, a TTX is a structured discussion- and scenario-based exercise where personnel with relevant roles and responsibilities meet to practice incident response preparedness including plans, policies, and procedures. They do this by being presented with certain response scenarios and artificial stimuli; they assume their roles and play out their responses [475], [476].

TTX are different from other forms of exercises in that they do not fully exercise the technical aspects of responding to an incident; systems are not usually examined for forensics and full analysis is not conducted. TTXs focus instead on the processes, the roles, judgement, and ensuring participants are familiar with what to do, who to call, and what facilities and tools are available to them for analysis. Familiarity saves time in the event of real incidents, as participants are not as likely to scramble to find somewhere to work and locate contact information.

An important objective is that the people who are not familiar with incident response in the constituency become familiar, as well as develop an organizational understanding of what decisions will be made and with what resources. They are used to:

- Ensure all groups that should be involved in incident response are correctly accounted for, present, and know what to expect.
- Enhance and enforce participant's knowledge of their roles and responsibilities in an incident.
- Check that processes are in place and identify role, resource, and process gaps for the SOC and the cybersecurity apparatus.
- Assist constituents in ensuring they too have appropriate resourcing, processes, and the right participants to support effective incident response.
- Help the SOC and other cybersecurity stakeholders bring attention and resources to incident scenarios of concern.
- Help anticipate how participants might behave in a real incident, thus supporting proactive corrections and improvements.

No matter what the TTX will accomplish, the overarching goal is to prepare involved parties for a real incident, and find gaps, not to ensure everyone perfectly responds.

A successful tabletop exercise identifies gaps in roles tools, tooling and processes, and other issues to be addressed, not that a team can execute perfectly.

11.8.1 Effective TTXs

Because they are not fully testing the capabilities and skills of participants, it is very important to provide clear, well-thought-out prompts and communication at each stage or "inject" of the TTX. Qualities that ensure a TTX will be effective include:

- Make it relevant to the constituency and the participants
- Make it as realistic as possible within the bounds of the resources allocated to the TTX
- Separate participation in the TTX from time spent doing other SOC functions
- Include and engage all relevant participants, organizations, and processes
- Respect the participants' time by providing information prior to the event, having a plan for the event, and setting expectations for follow up after the event
- Make it engaging and entertaining; draw the interest of the participants through choice of scenario, ways in which participants interact, and use of language
- Provide realistic TTX prompts and injects that will help participants confront the challenges and confusion that may have to be confronted during the middle of a real incident response

Realistic scenarios are the secret to ensuring a TTX meets the objectives for a constituency. Planning and running a TTX for a full-blown cyber apocalypse, where the main SOC is rendered a smoldering crater, by exercising a fully staffed hot site with all SOC functionality in a different location 200 miles away might not be realistic for most constituencies. Instead, start with a business objective. For example, a Board of Trustees wants assurance the company can handle highly visible and possibly embarrassing cyber incidents. For this, several scenarios might be considered for a TTX:

- Adversaries steal the CEO's or other high-profile executives' unencrypted mobile devices
- An insider has provided intellectual property or data to an authorized third party
- Adversaries have gained access to payroll systems, and are collecting pay (an interesting twist is the same scenario for payroll that is outsourced to an external company, and how that incident is handled jointly)
- Ransomware is used to encrypt systems with important intellectual property, and demands cryptocurrency for payment
- Intellectual property stored in cloud is leaked to business competitors

What makes these kinds of TTX prompts worthwhile is that they engage audience that often is not technical and helps them understand their role in cybersecurity. TTXs tend to shift attention away from the bits and bytes of security operations, and toward risk and consequence management.

Other common business objectives are to train participants for an actual event, to exercises changes in an incident response plan after a reorganization or other major company shift, or to raise awareness to executives of their role in a cyber event. A TTX can be instrumental in helping executives and non-cyber IT stakeholders understand and prepare for just how critical they are to cyber and incident response specifically.

For chosen scenarios, consider the number and type of constituents affected, the number and types of systems affected, and the impact such as a scenario would have on a business or mission. Also consider which roles will be exercised as central to the response. For example, incident responders will have different roles when third parties, such as cloud providers or business services are involved or outsourced; and the incident response plan might want to consider Service Level Agreements and other understandings between business and mission partners. Often the TTX is where these gaps and challenges are identified. A great benefit of the TTX is therefore to address the findings for a relatively low cost and while not in the fog of actual incidents.

11.8.2 Succeeding with TTX

As with all exercises, setting up a relevant and successful TTX entails planning and preparation. Most of the work is done before the TTX occurs. These considerations are best addressed before commencing the TTX:

Ensure ownership
Someone needs to be assigned responsibility for the overall success of the TTX. That person will be responsible for exercise logistics, ensuring the right scenario(s) are picked, picking the right physical locations, picking which organizations to involve, and running TTX planning meetings. They will need to weigh issues like exercise dynamics and determining if technical

or non-technical resources are involved. Note: this may not be the person running the tabletop exercise itself.

Find the right person to run it

Exercises need one person in charge; this person needs to have an experienced background in the scenarios determined, and ability to facilitate and coordinate across varied populations, including executives and technical teams. As stated just below, the person running the TTX will need to show good ability to control the room and keep the group on track and on time.

Present injects (if part of the TTX)

Determine if and when injects are used, and how they appear; consider the technical participants' roles. Injects can include descriptions of developments that occur during the TTX.

Determine the audience and maintain control

Define who the audience is, as well as who the audience is not, and ensure the exercise is scoped and stays in boundaries. For example, if the scenario is for the internal IR team and immediate management, ensure guard rails are in place so that law enforcement is not notified by mistake. This is more an issue with real exercises than TTX, but ensure participants understand the audience.

Capture findings and develop after action reports

Determine who is responsible for this, and what the report will look like upon completion; determine what is important to capture, and ensure the right people are included to capture the findings. It is most helpful if someone with a non-speaking role can be responsible for capturing minutes, findings, and remediation items as the TTX progresses.

Understand the timeframe and maintain pacing

Does the exercise take place over 3 hours or 2 weeks, for example? When do injects occur (if used)? Is the exercise close to real time of true events? The exercise lead is normally responsible to keep the exercise moving along and using a timeline and understanding the timeframe facilitates success.

Be flexible

Understand if the timeframe and/or timeline is adjustable ahead of time. The exercise may unearth delays and parts of the TTX may take longer than others, which may not be anticipated ahead of time. This is good and may be a lesson learned; the exercise may need to be shifted one way or another to respect the schedule of the participants.

A limitation of TTX is that participants are less likely to identify shortfalls in data collection and fidelity, alerts, and analysis and forensics tool shortcomings. This is more likely to be identified by other types of exercises and activities, including threat hunting and pen testing. More reading and resources on TTX include the following: [477], [478], [479], [480].

11.9 Summary – Strategy 11: Turn Up the Volume by Expanding SOC Functionality

11.1. **Threat Hunting** is one of the best ways to find adversaries that otherwise would elude detection. When implemented effectively, a hunting program will bring structure, discipline, and repeatability to the SOC's efforts outside routine incident handling necessary to find adversaries across the kill chain.

11.2. **Red teaming** is the process of using offensive Tactics, Techniques, and Procedures (TTPs) to emulate real-world threats to train and measure the effectiveness of the people, processes, and technology used to defend an environment [410].

11.3. **Purple teaming** is the process of closely partnering between the SOC and red teamers to iteratively learn and improve SOC and other defensive capabilities.

11.4. **Breach and Attack Simulation (BAS)** is a tool or system that enables the repeatable, measurable, and scalable testing of technical security controls by exercising those controls in an automated fashion.

11.5. **Deception** is an effort to conceal networks and assets, create uncertainty and confusion, and/or influence and misdirect adversary perceptions and decisions. Deception can also be seen as a means to enhance and perform types of threat hunting, by enticing adversaries with fake environments, data, or use fake information that can then be tracked.

11.6. **Malware analysis** is the process of seeking understanding of the behavior and purpose of a suspicious file to aid in detection, response, and threat reduction [440].

11.7. **Digital forensics** includes the application of investigative procedures involving the rigorous, repeatable examination of digital evidence. This includes analyzing media or hard-drive images in-depth. There are several different types and sub-fields, including memory, operating systems (log & file system focused), media, mobile, and network analysis.

11.8. **Tabletop exercise (TTX)** is a discussion-based exercise where personnel with relevant roles and responsibilities meet to practice incident response preparedness including plans, policies, and procedures by inhabiting roles in response to one or more incident scenarios. A successful tabletop exercise identifies gaps primarily in roles and processes.

References

[1]   Federal Bureau of Investigation (FBI), "The Morris Worm: 30 Years Since First Major Attack on the Internet," 2 November 2018. [Online]. Available: https://www.fbi.gov/news/stories/morris-worm-30-years-since-first-major-attack-on-internet-110218.

[2]   Cisco, "What Is an Advanced Persistent Threat (APT)?," 29 October 2021. [Online]. Available: https://www.cisco.com/c/en/us/products/security/advanced-persistent-threat.html.

[3]   Committee on National Security Systems (CNSS), "CNSS Instruction No. 4009 Glossary," 6 April 2015. [Online]. Available: https://www.cnss.gov/CNSS/openDoc.cfm?YV5eCARxrtHkRwkrYRARpQ==.

[4]   N. Brownlee and E. Guttman, "Request for Comments: 2350 Expectations for Computer Security Incident Response," June 1998. [Online]. Available: http://www.ietf.org/rfc/rfc2350.txt.

[5]   G. Killcrece, K.-P. Kossakowski, M. Zajicek and R. Ruefle, "Organizational Models for Computer Security Incident Response Teams (CSIRTs)," December 2003. [Online]. Available: https://kilthub.cmu.edu/articles/report/Organizational_Models_for_Computer_Security_Incident_Response_Teams_CSIRTs_/6575921/1.

[6]   M. West-Brown, D. Stikvoort, K. Kossakowski, G. Killcrece, R. Ruefle and M. Zajicekm, "Handbook for Computer Security Incident Response Teams (CSIRTs)," April 2003. [Online]. Available: http://resources.sei.cmu.edu/library/asset-view.cfm?assetid=6305.

[7]   R. Shirey, "RFC4949, Internet Security Glossary, Version 2," August 2007. [Online]. Available: http://tools.ietf.org/html/rfc4949.

[8]   Cisco, "Network Management System: Best Practices," January 2015. [Online]. Available: https://www.cisco.com/c/en/us/support/docs/availability/high-availability/15114-NMS-bestpractice.pdf.

[9]   National Institute of Standards and Technology (NIST), "Information Security Continuous Monitoring (ISCM) for Federal Information Systems and Organizations, NIST SP 800-137," September 2011. [Online]. Available: https://csrc.nist.gov/publications/detail/sp/800-137/final.

[10]   Cybersecurity and Infrastructure Security Agency (CISA), "CISA Insider Threat Mitigation," November 2021. [Online]. Available: https://www.cisa.gov/insider-threat-mitigation.

[11]   National Institute of Standards and Technology (NIST), "Security and Privacy Controls for Information Systems and Organizations," December 2020. [Online]. Available: https://csrc.nist.gov/publications/detail/sp/800-53/rev-5/final.

[12]   Forum of Incident Response and Security Teams (FIRST), "Computer Security Incident Response Team (CSIRT) Services Framework, V2.1," FIRST, November 2019. [Online]. Available: https://www.first.org/standards/frameworks/csirts/FIRST_CSIRT_Services_Framework_v2.1.0.pdf.

[13]   National Park Service, "Opana Radar Site," August 2019. [Online]. Available: https://www.nps.gov/articles/opana-radar-site.htm.

[14]   National Institute of Standards and Technology (NIST), "NIST Computer Security Resource Center Glossary," 6 April 2015. [Online]. Available: https://csrc.nist.gov/glossary/term/cyber_incident.

[15]   M. R. Endsley, "Toward a Theory of Situation Awareness in Dynamic Systems," *Human Factors,* 1995.

[16]   J. Boyd, "The Essence of Winning and Losing," 1995. [Online]. Available: http://pogoarchives.org/m/dni/john_boyd_compendium/essence_of_winning_losing.pdf.

[17]   J. Boyd, "Destruction and Creation," 1976. [Online]. Available: http://www.goalsys.com/books/documents/DESTRUCTION_AND_CREATION.pdf.

[18]   Center for Disease Control and Prevention (CDC), "Health Insurance Portability and Accountability Act of 1996," CDC, 14 September 2018. [Online]. Available: https://www.cdc.gov/phlp/publications/topic/hipaa.html.

[19]   Federal Trade Commission, "Gramm-Leach-Bliley Act," Federal Trade Commission (FTC), 2 July 2002. [Online]. Available: https://www.ftc.gov/tips-advice/business-center/privacy-and-security/gramm-leach-bliley-act.

[20]   Congress, 107th, "H.R.5005 - Homeland Security Act of 2002," [Online]. Available: https://www.congress.gov/bill/107th-congress/house-bill/5005/text.

[21]   European Union Parliment, "Directive (EU) 2016/1148 of the European Parliament and of the Council concerning measures for a high common level of security of network and information systems across the Union," 6 July 2016. [Online]. Available: https://eur-lex.europa.eu/legal-content/EN/TXT/?uri=uriserv:OJ.L_.2016.194.01.0001.01.ENG.

[22]   European Union - Horizon 2020 Programme Framework, "General Data Protection Regulation (GDPR) Compliance Guidelines," November 2021. [Online]. Available: https://gdpr.eu/.

[23]   PCI Security Standards Council, "PCI Security Standards Overview," November 2019. [Online]. Available: https://www.pcisecuritystandards.org/pci_security/standards_overview.

[24] International Organization for Standards (ISO), "ISO/IEC 27001 Information Security Management," 2013. [Online]. Available: https://www.iso.org/isoiec-27001-information-security.html.

[25] Ponemon Institute, "The State of SOC Effectiveness: Signs of Progress but More Work Needs to Be Done," 1 June 2020. [Online]. Available: https://www.ponemon.org/research/ponemon-library/security/the-state-of-soc-effectiveness-signs-of-progress-but-more-work-needs-to-be-done.html.

[26] Gartner, "Enterprise Asset Management (EAM) Software Reviews and Ratings," November 2021. [Online]. Available: https://www.gartner.com/reviews/market/enterprise-asset-management-software.

[27] M. Stone, C. Irrechukwu, H. Perper, D. Wynne and L. E.-i.-C. Kauffman, "NIST Special Publication 1800-5, IT Asset Management," National Institute of Standards and Technology (NIST, September 2018. [Online]. Available: https://nvlpubs.nist.gov/nistpubs/SpecialPublications/NIST.SP.1800-5.pdf.

[28] B. Hollandsworth, "10 Tips on How to Win the IT asset Management Challenge," NetworkWorld, 12 December 2012. [Online]. Available: https://www.networkworld.com/article/2162121/10-tips-on-how-to-win-the-it-asset-management-challenge.html.

[29] S. K. White, "IT Asset Management (ITAM): A Centralized Approach to Managing IT Systems and Assets," CIO, 11 September 2019. [Online]. Available: https://www.cio.com/article/3437476/it-asset-management-itam-a-centralized-approach-to-managing-it-systems-and-assets.html.

[30] N. Dvir, "IT Asset Naming Conventions," The ITAM Review, 7 January 2011. [Online]. Available: https://www.itassetmanagement.net/2011/01/17/asset-naming-conventions/.

[31] J. Goodchild, "Asset Management Mess? How to Get Organized," 6 October 2020. [Online]. Available: https://www.darkreading.com/edge/theedge/asset-management-mess-how-to-get-organized/b/d-id/1338043.

[32] SANS, "Security Policy Templates," [Online]. Available: https://www.sans.org/information-security-policy/.

[33] C. Crowley, "Common and Best Practices for Security Operations Centers: Results of the 2019 SOC Survey," 2019. [Online]. Available: https://www.sans.org/media/analyst-program/common-practices-security-operations-centers-results-2019-soc-survey-39060.pdf.

[34] J. Oltsik, "The Cybersecurity Skills Shortage is Getting Worse," CSO Online, 21 August 2020. [Online]. Available: https://www.csoonline.com/article/3571734/the-cybersecurity-skills-shortage-is-getting-worse.html.

[35] M. Loukides, "What is DevOps?," O'Reilly Radar, 7 June 2012. [Online]. Available: http://radar.oreilly.com/2012/06/what-is-devops.html.

[36]   Agile Alliance, "What is Agile?," November 2021. [Online]. Available: https://www.
         agilealliance.org/agile101/.

[37]   AWS, "What is Continuous Integration?," 12 November 2021. [Online]. Available:
         https://aws.amazon.com/devops/continuous-integration/.

[38]   Google Cloud, Deloitte, "Future of the SOC: SOC People: Skills not Tiers," 2020.
         [Online]. Available: https://www2.deloitte.com/content/dam/Deloitte/us/Documents/
         about-deloitte/Deloitte_and_Chronicle_Future_of_the_SOC-Skills_Before_Tiers.pdf.

[39]   K. Higgins, "Death of the Tier 1 SOC Analyst," DarkReading, 16 November 2017.
         [Online]. Available: https://www.darkreading.com/analytics/death-of-the-tier-1-soc-
         analyst.

[40]   CriticalStart, "The Impact of Security Alert Overload," CRITICALSTART, April 2019.
         [Online]. Available: https://www.criticalstart.com/wp-content/uploads/CS_MDR_
         Survey_Report.pdf.

[41]   R. Henderson, "Reinventing the Role of the Tier 1 SOC Analyst," VMWare Lastline,
         19 May 2020. [Online]. Available: https://www.lastline.com/blog/reinventing-the-role-
         of-the-tier-1-soc-analyst/.

[42]   OASIS Cyber Threat Intelligence (CTI) Technical Committee, "Introduction to
         TAXII," OASIS, 15 October 2021. [Online]. Available: https://oasis-open.github.io/cti-
         documentation/taxii/intro.html.

[43]   OASIS Cyber Threat Intelligence (CTI) Technical Committee, "Introduction to
         STIX," OASIS, 5 October 2021. [Online]. Available: https://oasis-open.github.io/cti-
         documentation/stix/intro.

[44]   Software Engineering Institute, "National Computer Security Incident Response
         Teams (CSIRTs)," Carnegie Mellon Universiry, November 2021. [Online]. Available:
         https://www.sei.cmu.edu/our-work/cybersecurity-center-development/national-csirts/.

[45]   R. Van Os, "Measuring Capability Maturity in Security Operations Centers," SOC
         CMM, November 2021. [Online]. Available: https://soc-cmm.com.

[46]   Department of Defense (DOD) Acquisition and Sustainment, "Securing the Defense
         Industrial Base CMMC 2.0," Department of Defense (DOD), November 2021. [Online].
         Available: https://www.acq.osd.mil/cmmc/index.html.

[47]   C. Crowley, "2020 SOC Survey," 2020. [Online]. Available: https://soc-survey.com/.

[48]   Ponemon Institute, "Improving the Effectiveness of the Security Operations Center,"
         2019. [Online]. Available: https://www.devo.com/wp-content/uploads/2019/07/2019-
         Devo-Ponemon-Study-Final.pdf.

[49]   Datashield, "Pros and Cons of Outsourced SOC," February 2018. [Online]. Available:
         https://www.datashieldprotect.com/blog/pros-and-cons-of-an-outsourced-soc.

[50]   J. Lopez, "What is outsourcing? What does it mean for companies?," Medium. com, November 2019. [Online]. Available: https://medium.com/coderslink/what-is-outsourcing-what-does-it-mean-for-companies-eff73fe60372.

[51]   Reportlinker, "The Global Managed Security Services (MSS) Market Size is Expected to Grow," PRNeweWire.com, 18 April 2018. [Online]. Available: https://www.prnewswire.com/news-releases/the-global-mss-market-size-is-expected-to-grow-from-usd-24-05-billion-in-2018-to-usd-47-65-billion-by-2023--at-a-compound-annual-growth-rate-cagr-of-14-7-300632466.html.

[52]   S. Deshpande, "Market Share: Managed Security Services, Worldwide, 2018," Gartner, 6 May 2019. [Online]. Available: https://www.gartner.com/en/documents/3913258/market-share-managed-security-services-worldwide-2018.

[53]   A. DeNisco Rayome, "How to Reverse the Cybersecurity Staffing Shortage: 5 Tips," 19 July 2019. [Online]. Available: https://www.techrepublic.com/article/how-to-reverse-the-cybersecurity-staffing-shortage-5-tips/.

[54]   ISC2, "Cybersecurity Professionals Focus on Developing New Skills as Workforce Gap Widens: (ISC)[2] CYBERSECURITY WORKFORCE STUDY, 2018," 2018. [Online]. Available: https://www.isc2.org/-/media/ISC2/Research/2018-ISC2-Cybersecurity-Workforce-Study.ashx?la=en&hash=4E09681D0FB51698D9BA6BF13EEABFA48BD17DB0%5Ch.

[55]   J. A. Lewis, "The Cybersecurity Workforce Gap," 29 January 2019. [Online]. Available: https://www.csis.org/analysis/cybersecurity-workforce-gap.

[56]   Wall Street Journal, "Beat the Resource Crunch by Outsourcing Cybersecurity," [Online]. Available: https://deloitte.wsj.com/cio/2019/05/30/beat-the-resource-crunch-by-outsourcing-cybersecurity/.

[57]   T. Meek, "Outsourcing Cybersecurity: When And How To Bring In Contractors," 27 March 2017. [Online]. Available: https://www.forbes.com/sites/eycybersecurity/2017/03/27/outsourcing-cybersecurity-when-and-how-to-bring-in-contractors/#43d4b33b6ca1.

[58]   M. Baker, "Five Best Practices for Outsourcing Cybersecurity," 19 July 2016. [Online]. Available: https://www.datacenterknowledge.com/archives/2016/07/19/five-best-practices-for-outsourcing-cybersecurity.

[59]   M. Russinovich and T. Garnier, "Sysmon v13.21," Microsoft, 1 June 2021. [Online]. Available: https://docs.microsoft.com/en-us/sysinternals/downloads/sysmon.

[60]   Osquery, "OS Query Performant Endpoint Visibility," 2019. [Online]. Available: https://osquery.io/.

[61]   Elastic, "Elasticsearch," GitHub, 2021. [Online]. Available: https://github.com/elastic/elasticsearch.

[62]   D. Barney, "22 Critical Metrics and KPIs for MSPs," 23 February 2017. [Online]. Available: https://www.kaseya.com/blog/2017/02/23/22-critical-metrics-and-kpis-for-msps/.

[63]   J. Russell, "10 Tips for selecting a Managed Security Services Provider (MSSP)," 10 January 2021. [Online]. Available: https://www.harmony-tech.com/10-tips-for-selecting-a-managed-security-services-provider-mssp/.

[64]   N. Lord, "How to Hire & Evaluate Managed Security Service Providers (MSSPS)," 27 July 2017. [Online]. Available: https://digitalguardian.com/blog/how-hire-evaluate-managed-security-service-providers-mssps.

[65]   M. Swanson, P. Bowen, A. Phillips, D. Gallup and D. Lynes, "NIST Special Publication 800-34 Rev. 1: Contingency Planning Guide for Federal Information Systems," May 2010. [Online]. Available: https://csrc.nist.gov/publications/detail/sp/800-34/rev-1/final.

[66]   S. Mondal, "Diversity And Inclusion: A Complete Guide For HR Professionals," 21 May 2021. [Online]. Available: https://ideal.com/diversity-and-inclusion/.

[67]   E. Washington and C. Patrick, "3 Requirements for a Diverse and Inclusive Culture," 17 September 2018. [Online]. Available: https://www.gallup.com/workplace/242138/requirements-diverse-inclusive-culture.aspx.

[68]   (ISC)2, "Strategies for Building and Growing Strong Cybersecurity Teams: (ISC)2 CYBERSECURITY WORKFORCE STUDY, 2019," 2019. [Online]. Available: https://www.isc2.org/-/media/ISC2/Research/2019-Cybersecurity-Workforce-Study/ISC2-Cybersecurity-Workforce-Study-2019.ashx?la=en&hash=D087F6468B4991E0BEFFC017BC1ADF59CD5A2EF7.

[69]   J. Oltsik, "The Life and Times of Cybersecurity Professionals 2018," April 2019. [Online]. Available: https://www.esg-global.com/hubfs/pdf/ESG-ISSA-Research-Report-Life-of-Cybersecurity-Professionals-Apr-2019.pdf.

[70]   M. Aiello, "Four (Self-Inflicted) Roadblocks To Finding Quality Cyber Professionals," 12 November 2019. [Online]. Available: https://www.forbes.com/sites/forbestechcouncil/2019/11/12/four-self-inflicted-roadblocks-to-finding-quality-cyber-professionals/?sh=6ec0cb631fba.

[71]   S. S. David Guest, "Introduction to a T-Shape for Product Designers," 1 March 2016. [Online]. Available: https://medium.com/the-edge-of-a-void/introduction-to-t-shaped-individuals-interdisciplinary-work-1db3a09c2aac#:~:text=The%20T-shaped%20person%20is%20a%20metaphor%20used%20to,force%20behind%20it%20and%20was%20actively%20promoting%20it.

[72]   L. Zhang, "30 Behavioral Interview Questions You Should Be Ready to Answer," [Online]. Available: https://www.themuse.com/advice/30-behavioral-interview-questions-you-should-be-ready-to-answer.

[73] A. Sachdeva, "41 Behavioural Interview Questions You Must Know (Best Answers Included)," 21 July 2018. [Online]. Available: https://www.themartec.com/insidelook/behavioral-interview-questions.

[74] S. Caltagirone, A. Pendergast and C. Betz, "Diamond Model of Intrusion Analysis," Center for Cyber Threat Intelligence and Threat Research, Hanover, MD, 5 July 2013.

[75] Lockheed Martin, "Cyber Kill Chain - Lockheed Martin," 2020. [Online]. Available: https://www.lockheedmartin.com/en-us/capabilities/cyber/cyber-kill-chain.html.

[76] NIST, "National Initiative for Cybersecurity Education (NICE) Cybersecurity Workforce Framework," August 2017. [Online]. Available: https://nvlpubs.nist.gov/nistpubs/SpecialPublications/NIST.SP.800-181r1.pdf.

[77] Software Engineering Institute - Carnegie Mellon University, "What Skills Are Needed When Staffing Your CSIRT?," 18 March 2016. [Online]. Available: http://resources.sei.cmu.edu/asset_files/whitepaper/2017_019_001_485684.pdf.

[78] OST2, "Open Security Training 2," OpenSecurityTraining2, 2021. [Online]. Available: https://opensecuritytraining.info/.

[79] GIAC, "Global Information Assurance Certification: Find Certifications by Focus Area," 2021. [Online]. Available: https://www.giac.org/.

[80] Offensive Security Ltd., "Kali Linux Penetration Testing and Ethical Hacking Linux Distribution," 2020. [Online]. Available: http://www.kali.org/.

[81] National Initiative for Cybersecurity Careers and Studies (NICCS), "National Initiative for Cybersecurity Careers and Studies Education and Training Catalog," 2021. [Online]. Available: https://niccs.cisa.gov/training/search.

[82] Informatech, "Blackhat: The World's Leading Information Security Event Series," Informatech, 2021. [Online]. Available: https://www.informatech.com/brands/black-hat.

[83] Defcon.org, "Defcon," 2020. [Online]. Available: https://defcon.org/index.html.

[84] RSAC, "RSA Conference: Where the world talks security," 2021. [Online]. Available: https://www.rsaconference.com/.

[85] Shmoocon, "Shmoocon," 2021. [Online]. Available: https://shmoocon.org/.

[86] B-Sides, "B-Sides Front Page," 2021. [Online]. Available: http://www.securitybsides.com/w/page/12194156/FrontPage.

[87] Layer One, "Layer One," 2021. [Online]. Available: https://www.layerone.org/.

[88] Software Engineering Institute, "FloCon," 2021. [Online]. Available: https://www.sei.cmu.edu/news-events/news/article.cfm?assetId=735822.

[89] Phreaknic, "Phreaknic," 2021. [Online]. Available: https://phreaknic.info/.

[90] "Welcome to Hackers on Planet Earth!," 2021. [Online]. Available: https://hope.net/.

[91] Hacker Halted, "Hacker Halted," 2021. [Online]. Available: https://www.hackerhalted.com/.

[92] Thotcon, "Thotcon," 2021. [Online]. Available: https://www.thotcon.org/.

[93] SANS Institute, "SANS Cyber Security Summit," 2021. [Online]. Available: https://www.sans.org/cyber-security-summit/.

[94] IEEE Computer Society, "IEEE Computer Society's Technical Committee on Security and Privacy," 2021. [Online]. Available: https://www.ieee-security.org/index.html.

[95] Forum of Incident Response and Security Teams (FIRST), "Annual Conferences: Annual FIRST Conference on Computer Security Incident Handling," FIRST, November 2021. [Online]. Available: https://www.first.org/conference/.

[96] USENIX, "Upcoming Usenix Conferences," 2021. [Online]. Available: https://www.usenix.org/conferences.

[97] W. Johnson, "How to Lose Your Best Employees," 20 April 2018. [Online]. Available: https://hbr.org/2018/04/how-to-lose-your-best-employees.

[98] SANS Institute, "SANS Institute: Cyber Security Skills Roadmap," 2021. [Online]. Available: https://www.sans.org/cyber-security-skills-roadmap/.

[99] "Open Security Training Learning Paths," October 2021. [Online]. Available: https://opensecuritytraining.info/Learning%20Paths.html.

[100] S. E. Page, The Diversity Bonus: How Great Teams Pay Off in the Knowledge Economy, Princeton University Press, 2017.

[101] K. Brown, "To Retain Employees, Focus on Inclusion – Not Just Diversity," 5 December 2018. [Online]. Available: https://hbr.org/2018/12/to-retain-employees-focus-on-inlusion-not-just-diversity.

[102] H. Devlin, "Unconscious Bias: What is it and Can it be Eliminated?," December 2018. [Online]. Available: https://www.theguardian.com/uk-news/2018/dec/02/unconscious-bias-what-is-it-and-can-it-be-eliminated.

[103] S. Florentine, "Employee Retention: 8 Strategies for Retaining Top Talent," CIO, 27 February 2019. [Online]. Available: https://www.cio.com/article/2868419/how-to-improve-employee-retention.html.

[104] L. Zelster, "The Big Picture of the Security Incident Cycle," SANS, 27 September 2010. [Online]. Available: https://www.sans.org/blog/the-big-picture-of-the-security-incident-cycle/.

[105] P. Stamp, *Building a World-Class Security Operations Function*, Cambridge, MA: Forrester Research, Inc., 2008.

[106] A. Rzasa, "SOC Events Per Analyst Hour (EPAH)," 14 May 2018. [Online]. Available: https://respond-software.com/soc-events-per-analyst-hour-aka-epah/.

[107] E. Schultz, "DOE's Computer Incident Advisory Capability (CIAC)," 26 October 1990. [Online]. Available: https://www.osti.gov/biblio/6054719-doe-computer-incident-advisory-capability-ciac.

[108] P. Cichonski, K. Masone, T. Grance and K. Scarfone, "NIST Special Publication 800-61 Rev 2: Computer Security Incident Handling Guide," August 2012. [Online]. Available: https://nvlpubs.nist.gov/nistpubs/SpecialPublications/NIST.SP.800-61r2.pdf.

[109] M. Bartock, J. Cichonski, M. Souppaya, M. Smith, G. Witte and K. Scarfone, "Guide for Cybersecurity Event Recovery," December 2016. [Online]. Available: https://nvlpubs. nist.gov/nistpubs/SpecialPublications/NIST.SP.800-184.pdf.

[110] P. Toth, "NIST Recovering from a Cyber Security Incident," 1 December 2017. [Online]. Available: https://www.nist.gov/system/files/documents/2017/12/01/recovery-webinar.pdf.

[111] VERIS, "VERIS: The Vocabulary for Event Recording and Incident Sharing," 2021. [Online]. Available: http://veriscommunity.net/index.html.

[112] ENISA, "Reference Incident Classification Taxonomy," 26 January 2018. [Online]. Available: https://www.enisa.europa.eu/publications/reference-incident-classification-taxonomy/.

[113] MISP Project, "MISP Taxonomies and Classification as Machine Tags," 22 October 2021. [Online]. Available: https://www.misp-project.org/taxonomies.html.

[114] T. Sager, "The Cyber OODA Loop: How Your Attacker Should Help You Design Your Defense," October 2020. [Online]. Available: https://csrc.nist.gov/CSRC/media/Presentations/The-Cyber-OODA-Loop-How-Your-Attacker-Should-Help/images-media/day3_security-automation_930-1020.pdf.

[115] Cisco, "What Is an Incident Response Plan for IT?," 18 July 2021. [Online]. Available: https://www.cisco.com/c/en/us/products/security/incident-response-plan.html.

[116] AT&T Cybersecurity, "Security Incidents: Types of Attacks and Triage Options," 2021. [Online]. Available: https://cybersecurity.att.com/resource-center/ebook/insider-guide-to-incident-response/types-of-security-incidents.

[117] The MITRE Corporation, "ATT&CK Enterprise Matrix," 29 April 2021. [Online]. Available: https://attack.mitre.org/.

[118] MS-ISAC, "Ransomeware Guide," September 2020. [Online]. Available: https://www.cisa.gov/sites/default/files/publications/CISA_MS-ISAC_Ransomware%20Guide_S508C.pdf.

[119] Incident Response Consortium, "Incident Response Consortium Playbooks," 2017. [Online]. Available: https://www.incidentresponse.com/playbooks/.

[120] J. Creasey, "CREST Cyber Security Incident Response Guide," 2013. [Online]. Available: https://www.crest-approved.org/wp-content/uploads/2014/11/CSIR-Procurement-Guide.pdf.

[121] U.S. Environmental Protection Agency (EPA), "EPA Incident Action Checklist - Cybersecurity," February 2021. [Online]. Available: https://www.epa.gov/sites/default/files/2017-11/documents/171013-incidentactionchecklist-cybersecurity_form_508c.pdf.

[122] A. Chuvakin and L. Zeltser, "SANS Critical Log Review Checklist for Security Incidents," [Online]. Available: https://www.sans.org/brochure/course/log-management-in-depth/6?msc=Cheat+Sheet+Blog.

[123] SANS Instiute, "SCORE: Checklists & Step-by-Step Guides," November 2021. [Online]. Available: https://www.sans.org/media/score/checklists/APT-IncidentHandling-Checklist.pdf.

[124] Microsoft, "Microsoft Cloud Incident Response Playbooks," 9 November 2021. [Online]. Available: https://docs.microsoft.com/en-us/security/compass/incident-response-playbooks.

[125] Creative Commons Attribution-ShareAlike, "AWS Security Incident Response Playbook Templates," September 2021. [Online]. Available: https://github.com/aws-samples/aws-incident-response-playbooks.

[126] R. Pherson and R. Heuer, "Chapter 3, Criteria for Selecting Structured Techniques," in *Structured Analytic Techniques*, CQ Press; Third edition, pp. 30-33.

[127] National Institute of Standards and Technology, "NIST Computer Security Resource Center Glossary," 6 April 2015. [Online]. Available: https://csrc.nist.gov/glossary/term/digital_forensics.

[128] Applied Incident Response, "Applied Incident Response References," 2020. [Online]. Available: https://www.appliedincidentresponse.com/resources/.

[129] SANS Institute, "SANS Cyber Security Blog," October 2021. [Online]. Available: https://www.sans.org/blog/?focus-area=digital-forensics.

[130] SANS Institute, "SANS Cyber Security Tools - Digital Forensis," October 2021. [Online]. Available: https://www.sans.org/tools/?focus-area=digital-forensics.

[131] Microsoft, "Microsoft Security Incident Response Blog," 2021. [Online]. Available: https://www.microsoft.com/security/blog/incident-response/.

[132] Red Hat, "10.4.2. Gathering Post-Breach Information," 2021. [Online]. Available: https://access.redhat.com/documentation/en-us/red_hat_enterprise_linux/4/html/security_guide/s2-response-invest-tool.

[133] S. Axelsson, "The Base-Rate Fallacy and its Implications for the Difficulty of Intrusion Detection," *Recent Advances in Intrusion Detection,* 1999.

[134] National Oceanic Atmospheric Administration (NOAA), "NOAA Gulf Spill Restoration," 2020. [Online]. Available: https://www.gulfspillrestoration.noaa.gov/.

[135] D. S. Hilzenrath, "Technician: Deepwater Horizon Warning System Disabled," *Washington Post,* 23 July 2010.

[136] D. Miessler, "The Definition of a Green Team," 2019. [Online]. Available: https://danielmiessler.com/blog/the-definition-green-team-how-different-red-team/.

[137] IBM, "Incident Management Architecture," 2021. [Online]. Available: https://www.ibm.com/cloud/architecture/architectures/incidentManagementDomain.

[138] Cloud Security Alliance, "Cloud Incident Response Framework – A Quick Guide," 21 April 2020. [Online]. Available: https://cloudsecurityalliance.org/artifacts/cloud-incident-response-framework-a-quick-guide.

[139] Amazon Web Services (AWS), "AWS Security Incident Response Guide," 23 November 2020. [Online]. Available: https://docs.aws.amazon.com/whitepapers/latest/aws-security-incident-response-guide/welcome.html.

[140] Amazaon Web Services (AWS), "Building a cloud-specific incident response plan," 18 August 2017. [Online]. Available: https://aws.amazon.com/blogs/publicsector/building-a-cloud-specific-incident-response-plan/.

[141] Google Cloud, "Data Incident Response Process," 2021. [Online]. Available: https://cloud.google.com/security/incident-response/.

[142] IBM, "IBM Resilient Incident Response Platform On Cloud," February 2018. [Online]. Available: http://www-03.ibm.com/software/sla/sladb.nsf/pdf/7534-03/$file/i126-7534-03_02-2018_en_US.pdf.

[143] Microsoft, "Security Control: Incident Response," 31 March 2021. [Online]. Available: https://docs.microsoft.com/en-us/security/benchmark/azure/security-control-incident-response#101-create-an-incident-response-guide.

[144] The MITRE Corporation, "Cloud Matrix," 29 April 2021. [Online]. Available: https://attack.mitre.org/matrices/enterprise/cloud/.

[145] S. Basu, "Guidance for Setting Up a Cloud Security Operations Center (cSOC)," 27 September 2018. [Online]. Available: https://blogs.oracle.com/cloud-infrastructure/guidance-for-setting-up-a-cloud-security-operations-center-csoc.

[146] Oracle, "Oracle Cloud Infrastructure and the GDPR," 3 February 2021. [Online]. Available: https://docs.oracle.com/en-us/iaas/Content/Resources/Assets/whitepapers/oci-gdpr.pdf#:~:text=Oracle%20Cloud%20Infrastructure%20has%20incident%20response%20mechanisms%20and,in%20the%20Data%20Processing%20Agreement%20for%20Oracle%20Services.

[147] VMWare, "Operations Management," 13 December 2019. [Online]. Available: https://docs.vmware.com/en/VMware-Cloud-on-Dell-EMC/services/vmc.dell.emc.security/GUID-64D4E823-2BBA-4F32-A0C0-38990A808184.html.

[148] B. Eichorst, "How to automate incident response in the AWS Cloud for EC2 instances," 29 October 2020. [Online]. Available: https://aws.amazon.com/blogs/security/how-to-automate-incident-response-in-aws-cloud-for-ec2-instances/.

[149] J. Jennis and C. Fernandes, "Incident Response in the Cloud," 13 June 2017. [Online]. Available: https://www.youtube.com/watch?v=ZyeTSI900zw&list=PLhr1KZpdzukePsKIUofhgp50b63-5yr1V.

[150] Microsoft, "Automate Incident Handling in Azure Sentinel with Automation Rules," 13 March 2021. [Online]. Available: https://docs.microsoft.com/en-us/azure/sentinel/automate-incident-handling-with-automation-rules.

[151] McAfee, "A Step-By-Step Guide to Cloud Security Best Practices," 2021. [Online]. Available: https://www.mcafee.com/enterprise/en-us/security-awareness/cloud/cloud-security-best-practices.html.

[152] The MITRE Corporation, "Mobile Matrices," 9 February 2021. [Online]. Available: https://attack.mitre.org/matrices/mobile/.

[153] Norton, "What is Smishing?," 18 January 2018. [Online]. Available: https://us.norton.com/internetsecurity-emerging-threats-what-is-smishing.html.

[154] M. Souppaya and K. Scarfone, "Guidelines for Securing Wireless Local Area Networks (WLANs)," February 2012. [Online]. Available: https://nvlpubs.nist.gov/nistpubs/Legacy/SP/nistspecialpublication800-153.pdf.

[155] Santoku, "Santoku - Packaged and Delivered," 2021. [Online]. Available: https://santoku-linux.com/.

[156] Micosoft, "Mobile device management," 25 March 2021. [Online]. Available: https://docs.microsoft.com/en-us/windows/client-management/mdm/.

[157] Microsoft, "Mobile Threat Defense Integration with Intune," 18 December 2020. [Online]. Available: https://docs.microsoft.com/en-us/mem/intune/protect/mobile-threat-defense.

[158] Public Safety Canada, "Developing an Operational Technology and Information Technology Incident Response Plan," 27 November 2020. [Online]. Available: https://www.publicsafety.gc.ca/cnt/rsrcs/pblctns/dvlpng-ndnt-rspns-pln/index-en.aspx.

[159] American Public Power Association, "Public Power Cyber Incident Response Playbook," August 2019. [Online]. Available: https://www.publicpower.org/system/files/documents/Public-Power-Cyber-Incident-Response-Playbook.pdf.

[160] The MITRE Corporation, "ATT&CK® for Industrial Control Systems," 29 April 2021. [Online]. Available: https://collaborate.mitre.org/attackics/index.php/Main_Page.

[161] J. Baines, "Findings From Examining More Than a Decade of Public ICS/OT Exploits," 5 August 2021. [Online]. Available: https://www.dragos.com/resource/findings-from-examining-public-ics-ot-exploits/.

[162] Intel.gov, "How the IC Works," 2021. [Online]. Available: https://www.intelligence.gov/how-the-ic-works.

[163] E. Hutchins, M. Cloppert and R. Amin, "Intelligence-Driven Computer Network Defense Informed by Analysis of Adversary Campaigns and Intrusion Kill Chains," 2010. [Online]. Available: https://www.lockheedmartin.com/content/dam/lockheed-martin/rms/documents/cyber/LM-White-Paper-Intel-Driven-Defense.pdf.

[164] D. Chismon and M. Ruks, "Threat Intelligence: Collecting, Analysing, Evaluating," 2015. [Online]. Available: https://www.foo.be/docs/informations-sharing/Threat-Intelligence-Whitepaper.pdf.

[165] Office of the Director of National Intelligence, "A White Paper on the Key Challenges in Cyber Threat Intelligence: Explaining the "See it, Sense it, Share it, Use it" approach to thinking about Cyber Intelligence," 30 October 2018. [Online]. Available: https://www.dni.gov/files/CTIIC/documents/White_paper_on_Cyber_Threat_Intelligence_ODNI_banner_10_30_2018.pdf.

[166] J. Ettinger, A. Galyardt, R. Gupta, D. DeCapria, E. Kanal, D. J. Klinedinst, D. Shick, S. J. Perl, G. B. Dobson, G. T. Sanders, D. L. Costa and L. Rogers, "Cyber Intelligence Tradecraft Report: The State of Cyber Intelligence Practices in the United States (Study Report and Implementation Guides)," 2019. [Online]. Available: https://resources.sei.cmu.edu/library/asset-view.cfm?assetID=546578.

[167] ISACA Now, "11 Cyber Threat Intelligence Tips," [Online]. Available: https://www.isaca.org/resources/news-and-trends/isaca-now-blog/2016/11-cyber-threat-intelligence-tips.

[168] CREST, "What is Cyber Threat Intelligence and How is it Used?," 2019. [Online]. Available: https://www.crest-approved.org/wp-content/uploads/CREST-Cyber-Threat-Intelligence.pdf.

[169] R. J. Heuer, "Psychology of Intelligence Analysis," 1999. [Online]. Available: https://www.cia.gov/static/9a5f1162fd0932c29bfed1c030edf4ae/Pyschology-of-Intelligence-Analysis.pdf.

[170] K. P. R. Pherson, *Critical thinking for Strategic Intelligence*, Thousand Oaks, California: CQ Press, 2020.

[171] E. Glaser, *An Experiment in the Development of Critical Thinking*, New York: AMS Press, 1972.

[172] E. J. Glantz, "A Tradecraft Primer: Structured Analytic Techniques for Improving Intelligence Analysis," 2013. [Online]. Available: http://www.personal.psu.edu/ejg8/class/sra111/StructuredAnalytics_v1.pdf.

[173] R. . J. J. Heuer, "Taxonomy of Structured Analytic Techniques," 2008. [Online]. Available: http://www.pherson.org/wp-content/uploads/2013/06/03.-Taxonomy-of-Structured-Analytic-Techniques_FINAL.pdf.

[174] GCHQ National Cyber Security Centre, "Advisory: APT29 targets COVID-19 Vaccine Developments," 16 July 2020. [Online]. Available: https://media.defense.gov/2020/Jul/16/2002457639/-1/-1/0/NCSC_APT29_ADVISORY-QUAD-OFFICIAL-20200709-1810.PDF.

[175] Mandiant, "APT1: Exposing One of China's Cyber Espionage Units," 2013. [Online]. Available: https://www.mandiant.com/resources/apt1-exposing-one-of-chinas-cyber-espionage-units.

[176] Mandiant, "Mandiant Threat Intelligence," 2021. [Online]. Available: https://www.mandiant.com/advantage/threat-intelligence/subscribe.

[177] Crowdstrike, "Hat-tribution to PLA Unit 61486," 9 June 2014. [Online]. Available: https://www.crowdstrike.com/blog/hat-tribution-pla-unit-61486/.

[178] A. Kozy, "Two Birds, One STONE PANDA," 30 August 2018. [Online]. Available: https://www.crowdstrike.com/blog/two-birds-one-stone-panda/.

[179] National Cybersecurity and Communications Integration Center (NCCIC) - FBI, "GRIZZLY STEPPE – Russian Malicious Cyber Activity," 29 December 2016. [Online]. Available: https://us-cert.cisa.gov/sites/default/files/publications/JAR_16-20296A_GRIZZLY%20STEPPE-2016-1229.pdf.

[180] Office of Director of National Intelligence, "National Intelligence Strategy," 2019. [Online]. Available: https://www.dni.gov/files/ODNI/documents/National_Intelligence_Strategy_2019.pdf.

[181] Financial Services Information Sharing and Analysis Center (FS-ISAC), "Sharing Timely, Relevant and Actionable Intelligence Since 1999," 2021. [Online]. Available: https://www.fsisac.com/who-we-are.

[182] M. Kadoma, "Intelligence Goals Library: All Your Use Cases in One Place," 24 April 2019. [Online]. Available: https://www.recordedfuture.com/intelligence-goals-library-overview/.

[183] B. Gourley, "Security Intelligence at the Strategic, Operational and Tactical Levels," 19 March 2018. [Online]. Available: https://securityintelligence.com/security-intelligence-at-the-strategic-operational-and-tactical-levels/.

[184] Joint Chiefs of Staff, "Joint Intelligence," 22 October 2013. [Online]. Available: https://www.jcs.mil/Portals/36/Documents/Doctrine/pubs/jp2_0.pdf.

[185] Anomali, "Anomali Preferred Partner Store," 2021. [Online]. Available: https://www.anomali.com/marketplace/threat-intelligence-feeds.

[186] Alienvault, "Alienvault.com," 2021. [Online]. Available: https://otx.alienvault.com/adversary/.

[187] Various, "awesome-threat-intelligence," 20 December 2020. [Online]. Available: https://github.com/hslatman/awesome-threat-intelligence.

[188] Dragos, "Threat Activity Groups," 2021. [Online]. Available: https://www.dragos.com/threat-activity-groups/.

[189] MISP, "MISP - Open Source Threat Intelligence Platform & Open Standards For Threat Information Sharing," 2021. [Online]. Available: https://www.misp-project.org/.

[190] GitHub, "mitre/cti: Cyber Threat Intelligence Repository expressed in STIX 2.0," 2021. [Online]. Available: https://github.com/mitre/cti.

[191] Proofpoint, Inc., "Proofpoint Emerging Threats Rules," 2021. [Online]. Available: https://rules.emergingthreats.net/.

[192] SANS Institute, "SANS Internet Storm Center," 2021. [Online]. Available: https://isc.sans.edu/.

[193] ThreatConnect, "Blog: Risk-Threat-Response," 2021. [Online]. Available: https://threatconnect.com/blog/.

[194] IBM, "IBM X-Force Exchange," 2021. [Online]. Available: bmcloud.com.

[195] Crowdstrike, "Crowdstrike Adversary Universe," 2021. [Online]. Available: https://adversary.crowdstrike.com/en-US/.

[196] Digital Shadows, "Digital Shadows blog: Threat Intelligence," 2021. [Online]. Available: digitalshadows.com.

[197] FireEye, "Advanced Persistent Threat Groups (APT Groups)," 2021. [Online]. Available: https://www.fireeye.com/current-threats/apt-groups.html.

[198] FireEye threat Research Blog, "Threat Research," 2021. [Online]. Available: https://www.fireeye.com/blog/threat-research.html.

[199] Proofpoint, Inc., "Blog: Threat Insight Blog (proofpoint.com)," 2021. [Online]. Available: https://www.proofpoint.com/us/blog/threat-insight.

[200] Recorded Future, "Adversary Infrastructure Report 2020: A Defender's View," 7 January 2021. [Online]. Available: https://www.recordedfuture.com/2020-adversary-infrastructure-report/.

[201] Digital Shadows, "Threat Intelligence: A Deep Dive," 12 December 2019. [Online]. Available: https://www.digitalshadows.com/blog-and-research/threat-intelligence-a-deep-dive/.

[202] ScienceDirect, "Building an Intelligence-Led Security Program," 2014. [Online]. Available: https://www.sciencedirect.com/book/9780128021453/building-an-intelligence-led-security-program.

[203] T. Rid and B. Buchanan, "Attributing Cyber Attacks," *Journal of Strategic Studies,* 23 December 2014.

[204] K. Vanderlee, "DebUNCing Attribution: How Mandiant Tracks Uncategorized Threat Actors," 17 December 2020. [Online]. Available: https://www.fireeye.com/blog/products-and-services/2020/12/how-mandiant-tracks-uncategorized-threat-actors.html.

[205] R. Downs, "Adversaries and Their Motivations (Part 1)," 23 October 2014. [Online]. Available: https://unit42.paloaltonetworks.com/adversaries-and-their-motivations-part-1/.

[206] E. Hutchins, M. Cloppert and R. Amin, "Intelligence-Driven Computer Network Defense Informed by Analysis of Adversary Campaigns and Intrusion Kill Chains," 2010. [Online]. Available: https://www.lockheedmartin.com/content/dam/lockheed-martin/rms/documents/cyber/LM-White-Paper-Intel-Driven-Defense.pdf.

[207] Defense Science Board, "Resilient Military Systems and the Advanced Cyber Threat," January 2013. [Online]. Available: https://nsarchive2.gwu.edu/NSAEBB/NSAEBB424/docs/Cyber-081.pdf.

[208] The MITRE Corporation, "Getting Started with ATT&CK," October 2019. [Online]. Available: https://www.mitre.org/sites/default/files/publications/mitre-getting-started-with-attack-october-2019.pdf.

[209] The MITRE Corporation, "MITRE ATT&CK Design and Philosophy," March 2020. [Online]. Available: https://attack.mitre.org/docs/ATTACK_Design_and_Philosophy_March_2020.pdf.

[210] K. Nickels, "Putting MITRE ATT&CK™ into Action with What You Have, Where You Are," MITRE and Sp4rkCon by Walmart, 7 June 2019. [Online]. Available: https://www.youtube.com/watch?v=bkfwMADar0M.

[211] The MITRE Corporation, "Accessing ATT&CK Data," 2021. [Online]. Available: https://attack.mitre.org/resources/working-with-attack/.

[212] J. DeGroot , "50 Threat Intelligence Tools for Valuable Threat Insights," Digital Guardian, 31 December 2020. [Online]. Available: https://digitalguardian.com/blog/50-threat-intelligence-tools-valuable-threat-insights.

[213] Gartner Peer Insights, "Security Threat Intelligence Products and Services Reviews and Ratings," 2021. [Online]. Available: https://www.gartner.com/reviews/market/security-threat-intelligence-services.

[214] Forum of Incident Response and Security Teams (FIRST), "Traffic Light Protocol: FIRST Standards Definitions and Usage Guidance - Version 1.0," June 2016. [Online]. Available: https://www.first.org/tlp/.

[215] Homeland Security Digital Library, "A Compendium of Analytic Tradecraft Notes," February 1997. [Online]. Available: https://www.hsdl.org/?abstract&did=442801.

[216] A. Linoski and T. Walczyk, "Federated Search 101," *Net Connect,* vol. 133, 15 July 2008.

[217] R. Schultz, "What is Federated Search," 16 June 2020. [Online]. Available: https://blog.coveo.com/what-is-federated-search/.

[218] P. J. Sadalage and M. Fowler, *NoSQL Distilled: A Brief Guide to the Emerging World of Polyglot Persistence*, Addison-Wesley Professional, 2013.

[219] R. Bace, *Intrusion Detection*, Indianapolis: Macmillan Technical Publishing, 2000.

[220] S. Halevi, H. Krawczyk, D. Boneh and W. Shao, "Stanford.edu: Randomized Hashing for Digital Certificates: Halevi-Krawczyk Hash," 12 February 2014. [Online]. Available: https://crypto.stanford.edu/firefox-rhash/.

[221] The MITRE Corporation, "Pre-OS Boot: Bootkit," 2021. [Online]. Available: https://attack.mitre.org/techniques/T1542/003.

[222] N. L. Petroni, T. Fraser, J. Molina and W. A. Arbaugh, "Copilot - a Coprocessor-based Kernel Runtime Integrity Monitor," *USENIX Security,* 2004.

[223] Trusted Computing Group, "Trusted Platform Module," 2019. [Online]. Available: https://trustedcomputinggroup.org/work-groups/trusted-platform-module/.

[224] Trusted Computing Group, "Trusted Boot," 2021. [Online]. Available: https://trustedcomputinggroup.org/resource/trusted-boot/.

[225] G. Hoglund and J. Butler, *Rootkits: Subverting the Windows Kernel*, Boston: Addison-Wesley.

[226] A. Chuvakin, "Named: Endpoint Threat Detection & Response," 26 July 2013. [Online]. Available: https://blogs.gartner.com/anton-chuvakin/2013/07/26/named-endpoint-threat-detection-response/.

[227] CrowdStrike, "Endpoint Detection and Response (EDR)," 6 February 2020. [Online]. Available: https://www.crowdstrike.com/cybersecurity-101/endpoint-security/endpoint-detection-and-response-edr/.

[228] VMware: Endpoint Security, "What is Endpoint Security?," [Online]. Available: https://www.vmware.com/topics/glossary/content/endpoint-security.

[229] The MITRE Corporation and MITRE Engenuity, "Open and fair evaluations based on ATT&CK®," 2021. [Online]. Available: https://attackevals.mitre-engenuity.org/index.html.

[230] VirusTotal, "Welcome to YARA's Documentation," 2021. [Online]. Available: https://yara.readthedocs.io/en/stable/index.html.

[231] Microsoft, "Use Windows Event Forwarding to Help with Intrusion Detection," 2019. [Online]. Available: https://docs.microsoft.com/en-us/windows/security/threat-protection/use-windows-event-forwarding-to-assist-in-intrusion-detection.

[232] Microsoft, "Windows Defender Application Control and AppLocker Overview," September 2020. [Online]. Available: https://docs.microsoft.com/en-us/windows/security/threat-protection/windows-defender-application-control/wdac-and-applocker-overview.

[233] Microsoft, "Microsoft Windows CodeIntegrity," 30 April 2018. [Online]. Available: https://docs.microsoft.com/en-us/windows-hardware/customize/desktop/unattend/microsoft-windows-codeintegrity.

[234] Apple, "Safely Open Apps on Your Mac," 30 April 2021. [Online]. Available: https://support.apple.com/en-us/HT202491.

[235] Tripwire, Inc., "Open Source Tripwire," 16 March 2019. [Online]. Available: https://github.com/Tripwire/tripwire-open-source.

[236] Canonical Ltd, "Security – Firewall Ubuntu," 2019. [Online]. Available: https://ubuntu.com/server/docs/security-firewall.

[237] Crowdstrike, "How to Manage a Host Firewall with CrowdStrike," March 2020. [Online]. Available: https://www.crowdstrike.com/blog/tech-center/manage-host-firewall/.

[238] SentinelOne, "Firewall Control – Feature Spotlight," December 2018. [Online]. Available: https://www.sentinelone.com/blog/feature-spotlight-firewall-control/.

[239] D. Goodin, "Anti-virus Protection Gets Worse," 21 December 2007. [Online]. Available: http://www.channelregister.co.uk/2007/12/21/dwindling_antivirus_protection/.

[240] R. McMillan, "Is Antivirus Software a Waste of Money?," 2 March 2012. [Online]. Available: http://www.wired.com/wiredenterprise/2012/03/antivirus/.

[241] J. Fruhlinger, "What is DLP? How Data Loss Prevention Software Works and Why you Need it," July 2020. [Online]. Available: https://www.csoonline.com/article/3564589/what-is-dlp-how-data-loss-prevention-software-works-and-why-you-need-it.html.

[242] J. I. Breeden, "4 Top Deception Tools and How They Ensnare Attackers," November 2020. [Online]. Available: https://www.csoonline.com/article/3596076/4-top-deception-tools-and-how-they-ensnare-attackers.html.

[243] Gartner Research, "Hype Cycle for Information Security," 10 July 2006. [Online]. Available: https://www.gartner.com/en/documents/493842-hype-cycle-for-information-security-2006.

[244] Vectra, "Vectra AI," 2021. [Online]. Available: https://www.vectra.ai/.

[245] Darktrace, "Darktrace Technology," 2021. [Online]. Available: https://www.darktrace.com/en/enterprise-immune-system/.

[246] Corelight, "Corelight: Faster Investigations, More Effective Threat Hunts," 2021. [Online]. Available: https://corelight.com/.

[247] M. Heckathorn, "Network Monitoring for Web-Based Threats," February 2011. [Online]. Available: http://www.sei.cmu.edu/reports/11tr005.pdf.

[248] Cisco, "Cisco IOS NetFlow," 2021. [Online]. Available: http://www.cisco.com/en/US/products/ps6601/products_ios_protocol_group_home.html.

[249] B. Claise, "Request for Comments: 3954 Cisco Systems NetFlow Services Export Version 9," October 2004. [Online]. Available: http://www.ietf.org/rfc/rfc3954.txt.

[250] 1. U. C. §. 3127, "18 U.S. Code § 3127 - Definitions for Chapter, (3) "Pen Register"," 2001. [Online]. Available: https://www.law.cornell.edu/uscode/text/18/3127.

[251] Carnegie Mellon University Software Engineering Institute, "CERT NetSA Security Suite: SiLK," 2020. [Online]. Available: https://tools.netsa.cert.org/silk.

[252] Zeek Project, "The Zeek Network Security Monitor," The International Computer Science Institute, 2021. [Online]. Available: http://www.zeek.org/.

[253] Zeek Project, "Zeek Package Browser," 2021. [Online]. Available: https://packages.zeek.org/.

[254] The Tcpdump Group, "Tcpdump/Libpcap public repository," 2021. [Online]. Available: http://www.tcpdump.org/.

[255] D. Miessler, "A tcpdump Tutorial and Primer," June 2021. [Online]. Available: http://danielmiessler.com/study/tcpdump/.

[256] The Wireshark Foundation, "Wireshark Go Deep," August 2021. [Online]. Available: https://www.wireshark.org/.

[257] NetWitness, an RSA Business, "NetWitness® Network," 2021. [Online]. Available: https://www.netwitness.com/en-us/products/network-security-network-monitoring.

[258] Arkime, "Arkime: Full Packet Capture," 2021. [Online]. Available: https://arkime.com/.

[259] FireEye, "Advanced Malware Analysis Tool," 2021. [Online]. Available: https://www.fireeye.com/products/malware-analysis.html.

[260] Broadcom, "Content Analysis and Sandboxing," 2021. [Online]. Available: https://www.broadcom.com/products/cyber-security/network/gateway/atp-content-malware-analysis.

[261] Palo Alto Networks, "Wildfire Malware Analysis Engine," 2021. [Online]. Available: https://www.paloaltonetworks.com/products/secure-the-network/wildfire.

[262] Cisco, "Cisco Adaptive Wireless Intrusion Prevention System Configuration Guide, Release 7.3.101.0," 18 February 2018. [Online]. Available: https://www.cisco.com/c/en/us/td/docs/wireless/mse/3350/7-.

[263] Cisco, "Cisco Adaptive WIPS Deployment Guide," 16 March 2017. [Online]. Available: https://www.cisco.com/c/en/us/td/docs/wireless/technology/wips/deployment/guide/WiPS_deployment_guide.html.

[264] E. Valente, "Capturing 10G versus 1G Traffic Using Correct Settings!," 2021. [Online]. Available: https://www.sans.org/reading-room/whitepapers/detection/capturing-10g-1g-traffic-correct-settings-33043.

[265] It's FOSS, "What is Z File System (ZFS)," March 2021. [Online]. Available: https://itsfoss.com/what-is-zfs/.

[266] Linux Raid Wiki, "A Guide to Mdadm," May 2018. [Online]. Available: https://raid.wiki.kernel.org/index.php/A_guide_to_mdadm.

[267] Open Information Security Foundation (OISF), "Suricata," 2021. [Online]. Available: https://suricata.io/.

[268] Vmware, "What is ESXi?," 2021. [Online]. Available: https://www.vmware.com/products/esxi-and-esx.html.

[269] Xen Project, "The Xen Project," 2021. [Online]. Available: http://www.xenproject.org/.

[270] Network Critical, "Passive Fiber TAPs," 2021. [Online]. Available: https://www.networkcritical.com/fiber-taps.

[271] Netscout, "NETSCOUT TAPs for Enterprise," 2021. [Online]. Available: https://www. netscout.com/product/enterprise/netscout-taps.

[272] Gigamon, "Passive Fiber Optic Network Tap," 2021. [Online]. Available: https://www. gigamon.com/products/access-traffic/network-taps/g-tap-m-series.html.

[273] Microsoft, "Resource Providers for Azure Services," 24 June 2021. [Online]. Available: https://docs.microsoft.com/en-us/azure/azure-resource-manager/management/ azure-services-resource-providers.

[274] Amazon Web Services (AWS), "AWS Resource and Property Types Reference," 21 June 2021. [Online]. Available: https://docs.aws.amazon.com/AWSCloudFormation/ latest/UserGuide/aws-template-resource-type-ref.html.

[275] Microsoft, "Azure Resource Provider Operations," 25 June 2021. [Online]. Available: https://docs.microsoft.com/en-us/azure/role-based-access-control/resource-provider- operations.

[276] Tripwire, Inc., "AWS vs. Azure vs. Google – What's the Difference from a Cloud Security Standpoint?," 29 December 2019. [Online]. Available: https://www.tripwire. com/state-of-security/security-data-protection/cloud/aws-azure-google-difference- cloud-security-standpoint/.

[277] Amazon Web Services (AWS), "AWS Security Hub," 2021. [Online]. Available: https://aws.amazon.com/security-hub/?aws-security-hub-blogs.sort-by=item. additionalFields.createdDate&aws-security-hub-blogs.sort-order=desc.

[278] Amazon Web Services (AWS), "Amazon GuardDuty," 2021. [Online]. Available: https:// aws.amazon.com/guardduty/2021.

[279] Microsoft Azure, "Azure Security Center," 2021. [Online]. Available: https://azure. microsoft.com/en-us/services/security-center/.

[280] Google Cloud,"Trust and Security," 2021. [Online]. Available: https://cloud.google. com/security.

[281] J. Kindervag, "'Zero Trust': The Way Forward in Cybersecurity," 1 October 2017. [Online]. Available: https://cloud.google.com/security.

[282] National Institue of Standards and Technology (NIST), "SP 800-207 Zero Trust Architecture," August 2020. [Online]. Available: https://csrc.nist.gov/publications/ detail/sp/800-207/final.

[283] E. Gilman and D. Barth, Zero Trust Networks, OReilly Media, July 2017.

[284] National Security Agency (NSA), "NSA and CISA Recommend Immediate Actions to Reduce Exposure Across all Operational Technologies and Control Systems," July 2020. [Online]. Available: https://media.defense.gov/2020/Jul/23/2002462846/-1/-1/1/ OT_ADVISORY-DUAL-OFFICIAL-20200722.PDF.

[285] Source Forge, "NetworkMiner Packet Analyzer," 7 August 2015. [Online]. Available: https://sourceforge.net/projects/networkminer/.

[286] Shodan, "Search Engine for the Internet of Things," 2021. [Online]. Available: https://www.shodan.io/.

[287] National Security Agency (NSA), "GitHub – nsacyber/GRASSMARLIN," 2021. [Online]. Available: https://github.com/nsacyber/GRASSMARLIN.

[288] Dragos, "Tools for the ICS Cybersecurity Community," 2021. [Online]. Available: https://www.dragos.com/community-tools/.

[289] Microsoft, "Azure Defender for IOT," 2021. [Online]. Available: https://azure.microsoft.com/en-us/services/azure-defender-for-iot.

[290] ICS-CERT, "Industrial Control Systems, CISA," 2021. [Online]. Available: https://us-cert.cisa.gov/ics.

[291] North American Electric Reliability Corporation, "E-ISAC Home," 2021. [Online]. Available: https://www.eisac.com/.

[292] D. Haye, "Harness the Power of SIEM," 15 April 2009. [Online]. Available: http://www.sans.org/reading_room/whitepapers/detection/harness-power-siem_33204.

[293] A. A. Chuvakin, "Anton Chuvakin Homepage," 2021. [Online]. Available: http://www.chuvakin.org/.

[294] K. Kent and M. Souppaya, "NIST Special Publication 800-92: Guide to Computer Security Log Management," September 2006. [Online]. Available: https://csrc.nist.gov/publications/detail/sp/800-92/final.

[295] A. A. Chuvakin and K. J. Schmidt, *Logging and Log Management: The Authoritative Guide to Dealing with Syslog, Audit Logs, Events, Alerts and other IT 'Noise'*, Boston, MA: Syngress, 2012.

[296] M. Nicolett and K. M. Kavanagh, "Critical Capabilities for Security Information and Event Management," 21 May 2012. [Online]. Available: https://www.gartner.com/doc/2022315.

[297] ISACA, "Security Information and Event Management: Business Benefits and Security, Governance and Assurance Perspective," 28 December 2010. [Online]. Available: http://www.isaca.org/Knowledge-Center/Research/ResearchDeliverables/Pages/Security-Information-and-Event-Management-Business-Benefits-and-Security-Governance-and-Assurance-Perspective.aspx.

[298] Gartner, "Security Information and Event Management Reviews and Ratings," 2021. [Online]. Available: https://www.gartner.com/reviews/market/security-information-event-management.

[299] B. Barlow, "The S Curve of Business," 2021. [Online]. Available: https://www.rocketsource.co/blog/s-curve-of-business/.

[300] K. Jackson Higgins, "Death of the Tier 1 SOC Analyst," 16 November 2017. [Online]. Available: https://www.darkreading.com/analytics/death-of-the-tier-1-soc-analyst/d/d-id/1330446.

[301] Siemplify, "Should You Stop Hiring Tier 1 SOC Analysts?," 17 April 2018. [Online]. Available: https://www.siemplify.co/blog/changing-role-tier-1-soc-analyst/.

[302] hadoop, "MapReduce Tutorial," 2 May 2021. [Online]. Available: https://hadoop.apache.org/docs/r1.2.1/mapred_tutorial.html.

[303] M. Drake, "A Comparison of NoSQL Database Management Systems and Models," 9 August 2019. [Online]. Available: https://web.archive.org/web/20190813163612/https://www.digitalocean.com/community/tutorials/a-comparison-of-nosql-database-management-systems-and-models.

[304] MongoDB, "Key-Value Databases," 2021. [Online]. Available: https://www.mongodb.com/databases/key-value-database.

[305] S. tavros Harizopoulos, D. Abadi and P. Boncz, "VLDB 2009 Tutorial: Column-Oriented Database Systems," 2009. [Online]. Available: http://www.cs.umd.edu/~abadi/talks/Column_Store_Tutorial_VLDB09.pdf.

[306] Amazon Web Services (AWS), "Querying Data with Federated Queries in Amazon Redshift," 2021. [Online]. Available: https://docs.aws.amazon.com/redshift/latest/dg/federated-overview.html.

[307] Splunk, "Data Fabric Search," 12 September 2019. [Online]. Available: https://docs.splunk.com/Documentation/DFS/1.1.0/DFS/Federatedsearch.

[308] Microsoft, "Cross-Database and Cross-Cluster Queries," 13 February 2020. [Online]. Available: https://docs.microsoft.com/en-us/azure/kusto/query/cross-cluster-or-database-queries?pivots=azuredataexplorer.

[309] MongoDB, "Segmenting Data by Location," 2021. [Online]. Available: https://docs.mongodb.com/manual/tutorial/sharding-segmenting-data-by-location/.

[310] A. Chuvakin, "Security Correlation Then and Now: A Sad Truth About SIEM," 19 December 2019. [Online]. Available: https://medium.com/anton-on-security/security-correlation-then-and-now-a-sad-truth-about-siem-fc5a1afb1001.

[311] X. Waibel, "How to Pick the Right Notebook for Data Science," February 2020. [Online]. Available: https://towardsdatascience.com/how-to-pick-the-right-notebook-for-data-science-7dc418c4da57.

[312] J. Gray, A. Bosworth, A. Layman and H. Pirahesh, "OLAP Data Cube: A Relational Aggregation Operator Generalizing Group-By, Cross-Tab, and Sub-Total," in *Proceedings of the International Conference on Data Engineering (ICDE). pp. 152–159*, doi:10.1109/ICDE.1996.492099, 1996.

[313] Decision Making Confidence, "Kepner Tregoe Decision Making The Steps, The Pros and The Cons," [Online]. Available: https://www.decision-making-confidence.com/kepner-tregoe-decision-making.html.

[314] G. Kenny, "What Is the Role of SIEM in the Fusion Center Era?," 12 April 2019. [Online]. Available: https://securityintelligence.com/what-is-the-role-of-siem-in-the-fusion-center-era/.

[315] Kralanx Cyber Security, "The SIEM Is Dead. MDR is The New SIEM!," 2021. [Online]. Available: https://www.kralanx.com/siem-is-dead/?doing_wp_cron=1583711923.479 13599014282222656250.

[316] C. Saunderson, "SIEM is Dead?," 10 November 2016. [Online]. Available: https://www.linkedin.com/pulse/siem-dead-craig-saunderson/.

[317] J. Oltsik, "Goodbye SIEM, Hello SOAPA," 29 November 2018. [Online]. Available: https://www.csoonline.com/article/3145408/goodbye-siem-hello-soapa.html.

[318] FireEye, "Cyber Defense Summit 2019 Breakout Sessions Executive Track," 9 October 2019. [Online]. Available: https://summit.fireeye.com/learn/tracks.html#executive-soc-metrics.

[319] K. Schwaber and J. Sutherland, "The 2020 Scrum Guide: The Definitive Guide to Scrum: The Rules of the Game," November 2020. [Online]. Available: https://scrumguides.org/scrum-guide.html.

[320] Information Sciences Institute, "Common Intrusion Detection Framework," 10 September 1999. [Online]. Available: http://gost.isi.edu/cidf/.

[321] R. Danyliw, J. Meijer and Y. Demchenko, "The Incident Object Description Exchange Format," December 2007. [Online]. Available: http://www.ietf.org/rfc/rfc5070.txt.

[322] IBM Cloud Education, "IT Service Management (ITSM)," 21 December 2020. [Online]. Available: http://www-01.ibm.com/software/tivoli/features/cei/.

[323] Splunk, "Common Information Model Add-on Manual," 23 June 2021. [Online]. Available: https://docs.splunk.com/Documentation/CIM/4.15.0/User/Overview.

[324] Hewlett Packard Enterprise, "HPE Security ArcSight Common Event Format," May 2016. [Online]. Available: https://www.secef.net/wp-content/uploads/sites/10/2017/04/CommonEventFormatv23.pdf.

[325] The MITRE Corporation, "CEE: Commonn Event Expression," 28 November 2014. [Online]. Available: http://cee.mitre.org/.

[326] Microfocus, "ArcSight Connector Supported Products," May 2020. [Online]. Available: https://www.microfocus.com/media/flyer/arcsight_connector_supported_products_flyer.pdf.

[327]   Microsoft, "Appendix L: Events to Monitor," 26 July 2018. [Online]. Available: https://docs.microsoft.com/en-us/windows-server/identity/ad-ds/plan/appendix-l--events-to-monitor.

[328]   GitHub, "Bro Cheatsheets," 27 October 2018. [Online]. Available: https://github.com/corelight/bro-cheatsheets.

[329]   Cornell Law School, "Federal Rules of Evidence," 1 December 2020. [Online]. Available: https://www.law.cornell.edu/rules/fre.

[330]   GitHub, "Google/GRR Rapid Response: Remote Live Forensics for Incident Response," 2021. [Online]. Available: https://github.com/google/grr.

[331]   Multiple, "MicrosoftDocs/windows-itpro-docs," March 2021. [Online]. Available: https://github.com/MicrosoftDocs/windows-itpro-docs/blob/public/windows/security/threat-protection/use-windows-event-forwarding-to-assist-in-intrusion-detection.md.

[332]   Man7, "Auditd(8) — Linux Manual Page," September 2013. [Online]. Available: http://man7.org/linux/man-pages/man8/auditd.8.html.

[333]   Elastic, "LogStash Centralize, Transform & Stash Your Data," 2021. [Online]. Available: https://www.elastic.co/logstash.

[334]   Fluentd, "Fluentd is an open source data collector for unified logging layer," 2021. [Online]. Available: https://www.fluentd.org/.

[335]   Elastic, "Beats Lightweight data Shippers," elastice, 2021. [Online]. Available: https://www.elastic.co/beats/.

[336]   The Apache Software Foundation, "Apache Kafka," 2017. [Online]. Available: https://kafka.apache.org/.

[337]   The Apache Software Foundation, "Apache Nifi," 2018. [Online]. Available: https://nifi.apache.org/.

[338]   Graylog, Inc, "Graylog Log Management," 2021. [Online]. Available: https://www.graylog.org/.

[339]   Microsoft Azure, "Azure Data Explorer," 2021. [Online]. Available: https://azure.microsoft.com/en-us/services/data-explorer/#overview.

[340]   The Apache Software Foundation, "Open Source NoSQL Database," Cassandra, 2021. [Online]. Available: http://cassandra.apache.org/.

[341]   Project Juptyer, "Jupyter," 25 October 2021. [Online]. Available: https://jupyter.org/.

[342]   Elastic, "Kbana," Elastic, 2021. [Online]. Available: https://www.elastic.co/kibana.

[343]   The Apache Software Foundation, "Apache Flink® — Stateful Computations over Data Streams," 2021. [Online]. Available: https://flink.apache.org/.

[344]   Samza, "Apache Samza: A Distributed Stream Processing Framework," 2021. [Online]. Available: http://samza.apache.org/.

[345] The Apache Software Foundation, "Apache Spark: Unified engine for large-scale," 2018. [Online]. Available: https://spark.apache.org/.

[346] Apache Software Foundation, "Apache Storm," 14 October 2021. [Online]. Available: https://storm.apache.org/.

[347] Hadoop, "HDFS Architecture Guide," 10 October 2020. [Online]. Available: https://hadoop.apache.org/docs/r1.2.1/hdfs_design.html.

[348] Databricks Inc., "Databricks," 2021. [Online]. Available: https://databricks.com/try-databricks?itm_data=NavBar-TryDatabricks-Trial.

[349] The R Foundation, "The R Project for Statistical Computing," R Project, October 2021. [Online]. Available: https://www.r-project.org/.

[350] G. Sadowsk, A. Litan, T. Bussa and T. Phillips, "Market Guide for User and Entity Behavior," 23 April 2018. [Online]. Available: https://www.cbronline.com/wp-content/uploads/dlm_uploads/2018/07/gartner-market-guide-for-ueba-2018-analyst-report.pdf.

[351] Imperva, "What is User and Entity Behavior Analytics (UEBA)," 2021. [Online]. Available: https://www.imperva.com/learn/data-security/ueba-user-and-entity-behavior-analytics/.

[352] Aruba, "CISO's Guide to Machine Learning & User and Entity Behavioral Analytics," 2 June 2017. [Online]. Available: https://www.arubanetworks.com/assets/CisoGuide.pdf.

[353] A. Chuvakin, "Ok, So Who Really MUST Get a UEBA?," 24 January 2017. [Online]. Available: https://blogs.gartner.com/anton-chuvakin/2017/01/24/ok-so-who-really-must-get-a-ueba/.

[354] BMC, "BMC Helix ITSM is the Next Generation of Remedy," BMC, 2021. [Online]. Available: https://www.bmc.com/it-solutions/remedy-itsm.html.

[355] ServiceNow, "Security Operations," Servicenow, 2021. [Online]. Available: https://www.servicenow.com/products/security-operations.html.

[356] TheHive Project, "TheHive Project: Security Incident Response for the Masses," 2020. [Online]. Available: http://thehive-project.org/.

[357] Sandia National Laboratories, "Get SCOT," 2021. [Online]. Available: https://getscot.sandia.gov/.

[358] Best Practical Solutions, LLC , "RT for Incident Response," 2021. [Online]. Available: https://bestpractical.com/rtir.

[359] IBM, "Case Management," 2021. [Online]. Available: https://www.ibm.com/docs/en/qsip/7.4?topic=administration-case-management.

[360] LogRhythm, "Case Management," LogRhythm, 2021. [Online]. Available: https://logrhythm.com/products/features/case-management/.

[361] The VERIS Community, "VERIS: The Vocabulary for Event Recording and Incident Sharing," 2021. [Online]. Available: http://veriscommunity.net/.

[362] Siemplify, "ThreatFuse, E-book: Blue Print for Modern SOC Operations," Anomali, 2021. [Online]. Available: https://www.siemplify.co/.

[363] Palo Alto Networks, "Cortex: Security Operations Workflow Automation," 2021. [Online]. Available: https://www.paloaltonetworks.com/cortex/soar.

[364] DFLabs, "Sumo Logic Cloud SOAR," DFLabs, 2021. [Online]. Available: https://www.sumologic.com/solutions/cloud-soar/.

[365] Johns Hopkins Applied Physics Laboratry (APL), "IACD: Integrated Adaptive Cyber Defense," 2019. [Online]. Available: https://www.iacdautomate.org/integrate-soar.

[366] IF This Then That (IFTTT), "IFTTT Pro and Pro+: Seriously Powerful Tools for Creators," 2021. [Online]. Available: https://ifttt.com/.

[367] Zapier, Inc., "Zapier," 2021. [Online]. Available: https://zapier.com/how-it-works.

[368] Microsoft, "Microsoft Power Automate," [Online]. Available: https://flow.microsoft.com/.

[369] S. Rouiller, "Virtual LAN Security: Weaknesses and Countermeasures (VLAN Hopping)," 19 June 2003. [Online]. Available: https://www.sans.org/white-papers/1090/.

[370] Microsoft, "AppLocker," 16 October 2017. [Online]. Available: https://docs.microsoft.com/en-us/windows/security/threat-protection/windows-defender-application-control/applocker/applocker-overview.

[371] SELinux Project, "Security-Enhanced Linux (SELinux)," 2021. [Online]. Available: https://github.com/SELinuxProject.

[372] Microsoft, "Privileged Access Management for Active Directory Domain Services," 5 January 2021. [Online]. Available: https://docs.microsoft.com/en-us/microsoft-identity-manager/pam/privileged-identity-management-for-active-directory-domain-services.

[373] National Council of ISACs, "National Council of ISACs," [Online]. Available: https://www.nationalisacs.org/.

[374] InfraGuard, "InfraGuard, Partnership for Protection," 2021. [Online]. Available: https://www.infragard.org/Application/Account/Login.

[375] WiCySC, "Women in Cybersecurity, WiCySC," 2021. [Online]. Available: https://www.wicys.org/.

[376] Cloud Security Alliance, "Cloud Security Alliance," 2021. [Online]. Available: https://cloudsecurityalliance.org/.

[377] Cybersecurity and Infrastructure Security Agency (CISA), "Cybersecurity & Infrastructure Security Agency Home Page," 2021. [Online]. Available: https://www.cisa.gov/.

[378] Federal Bureau of Investigation (FBI), "FBI Cybercrime," 2021. [Online]. Available: https://www.fbi.gov/investigate/cyber.

[379] European Union Agency for Cybersecurity, "European Union Agency for Cybersecurity," 2021. [Online]. Available: https://www.enisa.europa.eu/.

[380] Canadian Center for Cyber Security, "Canadian Center for Cyber Security," 2021. [Online]. Available: https://cyber.gc.ca/en/.

[381] CSA Singapore, "CSA Singapore," 2021. [Online]. Available: https://www.csa.gov.sg/who-we-are/our-organisation.

[382] Australian Cyber Security Centre, "Australian Cyber Security Centre," 2021. [Online]. Available: https://www.cyber.gov.au/.

[383] C. Snyder, "SANS 2018 Survey Results are In: Top Three Takeaways and How To Improve Your SOC," 14 August 2018. [Online]. Available: https://www.extrahop.com/company/blog/2018/sans-2018-soc-survey-results-optimize-your-soc/.

[384] A. Hidalgo, *Implementing Service Level Objectives: A Practical Guide to SLIs, SLOs, and Error Budgets*, O'Reilly Media, 2020.

[385] National Institute of Standards and Technology (NIST), "Cybersecurity Framework," 2021. [Online]. Available: https://www.nist.gov/cyberframework.

[386] Defense Security Information Agency (DISA), "Defense Security Information Agency, Security Technical Implementation Guides, 2021," 2021. [Online]. Available: https://public.cyber.mil/stigs/.

[387] Amazon Web Services (AWS), "AWS Quicksight," [Online]. Available: https://aws.amazon.com/quicksight/.

[388] Microsoft, "Microsoft Power BI," 2021. [Online]. Available: https://powerbi.microsoft.com/en-us/.

[389] Tableau, "Tableau, Business Intelligence and Analytics Software," 2021. [Online]. Available: https://www.tableau.com/.

[390] Grafana Labs, "Grafana: The Open Observability Platform," 2021. [Online]. Available: https://grafana.com/.

[391] Elastic, "Elastic Kibana: Explore, Visualize, Discover Data," 2021. [Online]. Available: https://www.elastic.co/kibana.

[392] Verizon, "2021 Data Breach Investigations Report," [Online]. Available: https://www.verizon.com/business/resources/reports/dbir/.

[393] FireEye, Inc., "M-Trends Cyber Security Trends," 2021. [Online]. Available: https://www.fireeye.com/current-threats/annual-threat-report/mtrends.html.

[394] P. Fasulo, "Top 20 Cybersecurity KPIs to Track in 2021," 8 July 2019. [Online]. Available: https://securityscorecard.com/blog/9-cybersecurity-metrics-kpis-to-track.

[395] E. Zhang, "Security and Analytics Experts Share the Most Important Cybersecurity Metrics and KPIs," 8 December 2017. [Online]. Available: https://digitalguardian.com/blog/what-are-the-most-important-cybersecurity-metrics-kpis.

[396] Security Magazine, "10 Most Popular Cybersecurity Metrics," 10 June 2021. [Online]. Available: https://www.securitymagazine.com/articles/95391-most-popular-cybersecurity-metrics.

[397] P. Paul E. Black, K. Scarfone and M. Souppaya, "Cyber Security Metrics and Measures," National Institute of Standards and Technology (NIST), 2 March 2009. [Online]. Available: https://www.nist.gov/publications/cyber-security-metrics-and-measures.

[398] D. Bodeau, R. Graubart, R. McQuad and J. Woodill, "Cyber Resiliency Metrics, Measures of Effectiveness, and Scoring," The MITRE Corporation, 2018. [Online]. Available: https://www.mitre.org/sites/default/files/publications/pr-18-2579-cyber-resiliency-metrics-measures-of-effectiveness-and-scoring.pdf.

[399] C. Zimmerman and C. Crowley, "Practical SOC Metrics," 17 December 2019. [Online]. Available: https://www.youtube.com/watch?v=brRCMgXfTHA.

[400] Q. Al Harfi Albluwi, "Framework for Performance Evaluation of Computer Security Incident Response Capabilities," 2017. [Online]. Available: https://digitalcommons.uri.edu/cgi/viewcontent.cgi?article=1658&context=oa_diss.

[401] Educause, "Effective Security Metrics," March 2017. [Online]. Available: https://www.educause.edu/focus-areas-and-initiatives/policy-and-security/cybersecurity-program/resources/information-security-guide/toolkits/effective-security-metrics.

[402] McAfee, "What Is Cyber Threat Hunting?," 2021. [Online]. Available: https://www.mcafee.com/enterprise/en-us/security-awareness/operations/what-is-cyber-threat-hunting.html.

[403] Cisco, "Hunting for Hidden Threats," 2019. [Online]. Available: https://www.cisco.com/c/dam/en/us/products/se/2019/8/Collateral/cybersecurity-series-2019-threat-hunting.pdf.

[404] Cisco, "What Is Threat Hunting?," [Online]. Available: https://www.cisco.com/c/en/us/products/security/endpoint-security/what-is-threat-hunting.html#~related-topics.

[405] J. Trull , "Threat hunting: Part 1—Why your SOC Needs a Proactive Hunting Team," March 2020. [Online]. Available: https://www.microsoft.com/security/blog/2020/03/10/threat-hunting-part-1-why-your-soc-needs-a-proactive-hunting-team/.

[406] C. Stoll, *The Cuckoo's Egg*, Simon & Schuster, 1989.

[407] R. Daszczyszak, D. Ellis, S. Luke and S. Whitley, "TTP-Based Hunting," The MITRE Corporation, Annapolis Junction, MD, 2019.

[408] R. Rodriguez, "OTRF/ThreatHunter-Playbook: A Threat Hunter's Playbook to Aid the Development of Techniques and Hypothesis for Hunting Campaigns," 2021. [Online]. Available: https://github.com/OTRF/ThreatHunter-Playbook.

[409] Greenbone Networks, "OpenVAS – Open Vulnerability Assessment Scanner," 2021. [Online]. Available: https://openvas.org/.

[410] RedTeam Guide, "Red Team Development and Operations: Definitions," 12 April 2021. [Online]. Available: https://redteam.guide/docs/definitions.

[411] RedTeam Guide, "Red Team Development and Operations: Planning," 5 January 2017. [Online]. Available: https://redteam.guide/docs/planning/.

[412] Threatexpress, "Red Team and Security Professional Blog: Threatexpress," 2021. [Online]. Available: https://threatexpress.com/.

[413] Council on Foreign Relations, "How to Use Red Teams and Be a Red Teamer: A Conversation with Mark Mateski," 27 February 2017. [Online]. Available: https://www.cfr.org/podcasts/how-use-red-teams-and-be-red-teamer-conversation-mark-mateski.

[414] B. Clark, "Red Team Field Manual," 11 February 2014. [Online]. Available: https://ebooks-it.org/1494295504-ebook.htm.

[415] D. Miessler, "Five Attributes of an Effective Corporate Red Team," 17 December 2019. [Online]. Available: https://danielmiessler.com/blog/five-attributes-effective-corporate-red-team/.

[416] B. Strom, "Getting Started with ATT&CK: Adversary Emulation and Red Teaming," 17 July 2019. [Online]. Available: https://medium.com/mitre-attack/getting-started-with-attack-red-29f074ccf7e3.

[417] Red Canary, "Atomic Red Team™ Library," 2021. [Online]. Available: https://atomicredteam.io/.

[418] The MITRE Corporation, "Scalable Automated Adversary Emulation Platform," GitHub, 6 July 2021. [Online]. Available: https://github.com/mitre/caldera.

[419] D. Kerr, "Introducing Endgame Red Team Automation," 19 March 2019. [Online]. Available: https://www.elastic.co/blog/introducing-endgame-red-team-automation.

[420] C. Gates, "Introducing Metta: Uber's Open Source Tool for Adversarial Simulation," 13 May 2018. [Online]. Available: https://medium.com/uber-security-privacy/uber-security-metta-open-source-a8a49613b4a.

[421] Stichting Cuckoo Foundation, "Cuckoo Automated Malware Analysis," 2019. [Online]. Available: https://cuckoosandbox.org/.

[422] Rapid7, "Get Metasploit," 2021. [Online]. Available: https://metasploit.com/.

[423] V. Motos, "Vmotos Purple Team Attack Automation," GitHub, 16 April 2010. [Online]. Available: https://github.com/vmotos/purple-team-attack-automation.

[424] B. Damele A. G. and M. Stampar, "sqlmap Automatic SQL injection and database takeover tool," 11 October 2021. [Online]. Available: https://sqlmap.org/.

[425] C. Sullo, "Nikto," 2021. [Online]. Available: https://github.com/sullo/nikto.

[426] Cobaltstrike, "Software for Adversary Simulations and Red Team Operations," 2021. [Online]. Available: https://www.cobaltstrike.com/.

[427] V. Wills, "OSINT for Cyber Threat Intelligence," 8 September 2020. [Online]. Available: https://www.spiderfoot.net/osint-for-cyber-threat-intelligence#:~:text=OSINT%20 for%20Cyber%20Threat%20Intelligence%201%20Cyber%20Threat,With%20 OSINT.%20...%204%20Value%20of%20Automation.%20.

[428] G. F. Lyon, "Nmap," 3 October 2020. [Online]. Available: https://nmap.org/.

[429] The MITRE Corporation, "MITRE ATT&CK® Navigator," 2021. [Online]. Available: https://mitre-attack.github.io/attack-navigator/.

[430] Maltego, "Maltego," 2021. [Online]. Available: https://www.maltego.com/.

[431] D. Bianco, "The Pyramid of Pain," 17 January 2014. [Online]. Available: http://detect-respond.blogspot.com/2013/03/the-pyramid-of-pain.html.

[432] C. Harvey, "Breach and Attack Simulation: Find Vulnerabilities Before the Bad Guys Do," 8 October 2018. [Online]. Available: https://www.esecurityplanet.com/threats/ breach-and-attack-simulation-find-vulnerabilities-before-the-bad-guys-do/.

[433] A. Arbuckle, "Fact vs Fiction: The Truth About Breach and Attack Simulation Tools," 25 July 2019. [Online]. Available: https://www.securityweek.com/fact-vs-fiction-truth-about-breach-and-attack-simulation-tools.

[434] VECTR, "VECTR™ Purple Team Platform with STIX/TAXII Functionality," VECTR, 2021. [Online]. Available: https://vectr.io/.

[435] PenTestIT, "List of Adversary Emulation Tools," 6 August 2020. [Online]. Available: https://pentestit.com/adversary-emulation-tools-list/.

[436] S. Ingalls, "1 Top Breach and Attack Simulation (BAS) Vendors," 6 May 2021. [Online]. Available: https://www.esecurityplanet.com/products/breach-and-attack-simulation-bas-vendors/.

[437] D. Climek, A. Macera and W. Tirenin, "Cyber Deception," Cyber Security Information Systems and Information Analysis Center, 8 March 2016. [Online]. Available: https:// www.csiac.org/journal-article/cyber-deception/.

[438] The MITRE Corporation, "Welcome to MITRE Engage™," 2021. [Online]. Available: https://engage.mitre.org/.

[439] L. Zhang and L. L. Vrizlynn, "Three decades of deception techniques in active cyber defense - Retrospect and outlook," *Computers & Security,* p. 23, 3 June 2020.

[440] K. Baker, "Malware Analysis," 26 March 2020. [Online]. Available: https://www. crowdstrike.com/cybersecurity-101/malware/malware-analysis/.

[441] Forum of Incident Response and Security Teams (FIRST), "Malware Analysis Resources," 2021. [Online]. Available: https://www.first.org/global/sigs/malware/ resources/.

[442] Digitial Forensics-Incident Response (DIFR), "Malware Analysis Tutorials: a Reverse Engineering Approach," 2021. [Online]. Available: https://www.dfir.training/jreviews/search-results?keywords=reverse+engineering.

[443] Ethical Hacking Guru, "The Complete Malware Analysis Tutorial," 18 July 2019. [Online]. Available: https://ethicalhackingguru.com/the-complete-malware-analysis-t

[444] M. Sikorski and A. Honig, "Practical Malware Analysis The Hands-On Guide to Dissecting Malicious Software," 2020. [Online]. Available: https://lib-ebooks.com/practical-malware-analysis-the-hands-on-guide-to-dissecting-malicious-software/.

[445] L. Zeltser, "REMnux: A Linux Toolkit for Malware Analysis," 2021. [Online]. Available: https://docs.remnux.org/.

[446] Oracle, "VirtualBox," 22 November 2021. [Online]. Available: https://www.virtualbox.org/.

[447] Amazon Web Services (AWS), "AWS Free Tier," 2021. [Online]. Available: https://aws.amazon.com/free/?all-free-tier.sort-by=item.additionalFields.SortRank&all-free-tier.sort-order=asc&awsf.Free%20Tier%20Types=*all&awsf.Free%20Tier%20Categories=categories%23compute&trk=ps_a134p000006pkmaAAA&trkCampaign=acq_paid_search_brand&sc_.

[448] Microsoft, "Microsoft Azure Virtual Machines," 2021. [Online]. Available: https://azure.microsoft.com/en-us/free/virtual-machines/search/?OCID=AID2200277_SEM_9ebbc8301c751aad0894a994194e0361:G:s&ef_id=9ebbc8301c751aad0894a994194e0361:G:s&msclkid=9ebbc8301c751aad0894a994194e0361.

[449] Hex-rays, "IDA as a debugger," Hex Rays, 2021. [Online]. Available: https://hex-rays.com/ida-pro/ida-debugger/.

[450] Microsoft, "Download Debugging Tools for Windows," 16 June 2021. [Online]. Available: https://docs.microsoft.com/en-us/windows-hardware/drivers/debugger/debugger-download-tools.

[451] O. Yuschuk, "Ollydbg," 26 November 2021. [Online]. Available: https://www.kali.org/tools/ollydbg/.

[452] M. Russinovich, "Sysinternals," 26 October 2021. [Online]. Available: https://docs.microsoft.com/en-us/sysinternals/.

[453] K. Kent, S. Chevalier, T. Grance and H. Dang, "Special Publication 800-86, Guide to Integrating Forensic Techniques into Incident Response," August 2006. [Online]. Available: https://nvlpubs.nist.gov/nistpubs/legacy/sp/nistspecialpublication800-86.pdf.

[454] D. Brezinski and T. Killalea, "Guidelines for Evidence Collection and Archiving," February 2002. [Online]. Available: https://datatracker.ietf.org/doc/html/rfc3227.

[455] GitHub, "Volatility Foundation/Volatility," 7 April 2020. [Online]. Available: https://github.com/volatilityfoundation/volatility/wiki.

[456]   Forensic Focus, "Forensic Focus for Digital Forensics & E-Discovery Professionals," Forensic Focus, 2021. [Online]. Available: https://www.forensicfocus.com/articles/.

[457]   DFRWS, "Digital Forensic Research Workshop Conferences," DFRWS, 2021. [Online]. Available: https://dfrws.org/.

[458]   D. Farmer and W. Venema, *Forensic Discovery 1st Edition*, Boston: Addison Wesley Professional Computing.

[459]   E. Casey, *Handbook of Digital Forensics and Investigation*, E. Casey, Ed., Academic Press, 2010.

[460]   B. Middleton, *Cyber Crime Investigator's Field Guide* (2nd, 05) Hardcover, Auerbach s, Hardcover, 2004.

[461]   M. H. Ligh, A. Case, J. Levy and A. Walters, *The Art of Memory Forensics: Detecting Malware and Threats in Windows, Linux and Mac Memory*, Indianopolis: Wiley, 2014.

[462]   M. Sikorski and A. Honig, *Practical Malware Analysis*, San Francisco: No Starch Press, 2012.

[463]   E. Conrad and J. Feldman, "Network Forensics," 2016. [Online]. Available: https://www.sciencedirect.com/topics/computer-science/network-forensics.

[464]   Amazon Web Services (AWS), "Amazon Elastic File System (EFS)," 2021. [Online]. Available: https://aws.amazon.com/efs/.

[465]   Amazon Web Services (AWS), "Amazon FSx," 2021. [Online]. Available: https://aws.amazon.com/fsx/.

[466]   Oracle, "Oracle ACFS 12c Release 2: Introduction and Technical Overview," November 2017. [Online]. Available: https://www.oracle.com/technetwork/database/database-technologies/cloud-storage/cloudfs-12c-overview-wp-1965426.pdf.

[467]   Multiple, "What is FSLogix?," 15 July 2021. [Online]. Available: https://docs.microsoft.com/en-us/fslogix/overview.

[468]   D. Hildebrand and D. Serenyi, "Colossus Under the Hood: A Peek into Google's Scalable Storage System," 19 April 2021. [Online]. Available: https://cloud.google.com/blog/products/storage-data-transfer/a-peek-behind-colossus-googles-file-system.

[469]   B. Carrier, "Autopsy Digital Forensics," 2021. [Online]. Available: https://www.autopsy.com/.

[470]   Basis Technology, "Open Source Digital Forensics," 2020. [Online]. Available: http://sleuthkit.org/.

[471]   Opentext Security, "Tableau Forensic Imagers & Duplicators," 2021. [Online]. Available: https://security.opentext.com/tableau/hardware/forensic-imagers-duplicators.

[472]   AccessData Group, LLC, "Forensic Toolkit (FTK) | AccessData," 2020. [Online]. Available: http://www.accessdata.com/products-services/forensic-toolkit-ftk.

[473] Guidance Software, Inc., "EnCase Forensic Software – Top Digital Forensics & Investigations Solution," 2020. [Online]. Available: http://www.guidancesoftware.com/encase-forensic.

[474] R. Lee, "SIFT Workstation," 2021. [Online]. Available: https://www.sans.org/tools/sift-workstation/.

[475] T. Grance, T. Nolan, K. Burke, R. Dudley, G. White and T. Good, "NIST Special Publication 800-84 Guide to Test, Training, and Exercise Programs for IT Plans and Capabilities," September 2006. [Online]. Available: https://nvlpubs.nist.gov/nistpubs/Legacy/SP/nistspecialpublication800-84.pdf.

[476] RedLegg, "Tabletop Exercise: Pretty Much Everything You Need to Know," 2021. [Online]. Available: https://www.redlegg.com/solutions/advisory-services/tabletop-exercise-pretty-much-everything-you-need-to-know.

[477] E. Ouzounis, P. Trimintzios and P. Saragiotis, "National Exercise - Good Practice Guide," 17 December 2009. [Online]. Available: https://www.enisa.europa.eu/publications/national-exercise-good-practice-guide.

[478] J. Kick , "Cyber Exercise Playbook," MITRE, Wiesbaden, Germany, November 2014.

[479] Cybersecurity and Infrastructure Security Agency (CISA), "CISA Tabletop Exercise Package: Exercise Planner Handbook," 2020. [Online]. Available: https://www.cisa.gov/sites/default/files/publications/2%20-%20CTEP%20Exercise%20Planner%20Handbook%20%282020%29%20FINAL_508.pdf.

[480] Center for Internet Security, "Six Tabletop Exercises (cisecurity.org)," October 2018. [Online]. Available: https://www.cisecurity.org/wp-content/uploads/2018/10/Six-tabletop-exercises-FINAL.pdf.

[481] R. Bejtlich, *The Tao of Network Security Monitoring: Beyond Intrusion Detection*, Boston: Pearson Education, 2005.

[482] R. Bejtlich, *Practice of Network Security Monitoring*, San Francisco: No Starch Press, 2013.

[483] J. Muniz, *The Modern Security Operations Center*, Addison-Wesley Professional, 2021.

[484] D. Murdoch, *Blue Team Handbook: Incident Response Edition*, CreateSpace, 2014.

[485] D. Murdoch, *Blue Team Handbook: SOC, SIEM, and Threat Hunting*, CreateSpace, 2019.

[486] S. Northcutt, *Network Intrusion Detection (3rd Edition)*, Indianapolis: New Riders Publishing, 2002.

[487] D. Nathans, *Designing and Building Security Operations Center*, Syngress, 2014.

[488] C. Prosise, K. Mandia and M. Pepe, *Incident Response and Computer Forensics, 2nd ed.*, Osborne: McGraw-Hill, 2003.

[489] E. E. Schultz and R. Shumway, *Incident Response: A Strategic Guide to Handling System and Network Security Breaches*, Sams, 2001.

[490] S. Roberts and R. Brown, *Intelligence-Driven Incident Response: Outwitting the Adversary*, O'Reilly Media, 2017.

[491] K. R. Van Wyk and R. Forno, *Incident Response*, Sebastopol, CA: O'Reilly Media, Inc., 2001.

[492] A. White and B. Clark, *Blue Team Field Manual*, CreateSpace Independent Publishing Platform, 2017.

[493] Software Engineering Institute, "Create a CSIRT," January 2017. [Online]. Available: https://resources.sei.cmu.edu/library/asset-view.cfm?assetid=485693.

[494] E. Zhang, "How to Build a Security Operations Center (SOC): Peoples, Processes, and Technologies," 1 December 2020. [Online]. Available: https://digitalguardian.com/blog/how-build-security-operations-center-soc-peoples-processes-and-technologies.

[495] S. Salinas, "Security Operations Center: Ultimate SOC Quick Start Guide," Exabeam, 24 January 2019. [Online]. Available: https://www.exabeam.com/security-operations-center/security-operations-center-a-quick-start-guide/.

[496] M. Jacka and P. Keller, *Business Process Mapping: Improving Customer Satisfaction*, John Wiley and Sons, 2009.

[497] J. Doerr, *Measure What Matters*, Penguin, 2018.

[498] S. Young, "Improving the Federal Government's Investigative and Remediation Capabilities Related to Cybersecurity Incidents," 27 August 2021. [Online]. Available: https://www.whitehouse.gov/wp-content/uploads/2021/08/M-21-31-Improving-the-Federal-Governments-Investigative-and-Remediation-Capabilities-Related-to-Cybersecurity-Incidents.pdf.

[499] N. Brownlee and E. Guttman, "Request for Comments 2350 Expectations for Computer Security Incident Response," 1998. [Online]. Available: http://www.ietf.org/rfc/rfc2350.txt.

[500] Wikipedia, "Red Queen hypothesis," January 2021. [Online]. Available: https://en.wikipedia.org/wiki/Red_Queen_hypothesis.

[501] W. R. Stevens and K. R. Fall, *TCP/IP Illustrated Vol 1: The Protocols, 2nd ed.*, Boston: Addison-Wesley Professional, 2011.

[502] G. R. Wright and W. R. Stevens, *TCP/IP Illustrated, Vol 2: The Implementation, 2nd ed.*, Boston: Addison-Wesley Professional, 1995.

[503] H. Carvey, *Windows Forensic Analysis Toolkit, 3rd Edition: Advanced Analysis*, Waltham: Syngress, 2012.

[504] K. J. Jones, R. Bejtlich and C. W. Rose, *Real Digital Forensics: Computer Security and Incident Response,* Boston: Addison-Wesley Professional, 2005.

[505] B. Carrier, *File System Forensic Analysis*, Boston: Addison-Wesley Professional, 2005.

[506] D. Farmer and W. Venema, *Forensic Discovery*, Boston: Addison-Wesley Professional, 2005.

[507] C. L. Brown, *Computer Evidence: Collection and Preservation*, Boston: Course Technology, 2009.

[508] P. Engebretson, *The Basics of Hacking and Penetration Testing: Ethical Hacking and Penetration Testing Made Easy*, Syngress, 2011.

[509] D. Stuttard and M. Pinto, T*he Web Application Hacker's Handbook: Finding and Exploiting Security Flaws*, Hoboken: Wiley, 2011.

[510] J. Scrambray, V. Liu and C. Sima, *Hacking Exposed Web Applications*, New York: McGraw-Hill Osborne Media, 2010.

[511] C. Anley, J. Heasman, F. Lindner and G. Richarte, *The Shellcoder's Handbook: Discovering and Exploiting Security Holes*, Hoboken: Wiley, 2007.

[512] C. Eagle, *The IDA Pro Book: The Unofficial Guide to the World's Most Popular Disassembler*, San Francisco: No Starch Press, 2011.

[513] E. Eilam, *Reversing: Secrets of Reverse Engineering*, Hoboken: Wiley, 2005.

[514] L. Spitzner, *Honeypots: Tracking Hackers*, Boston: Addison-Wesley Professional, 2002.

[515] R. Bejtlich, *Extrusion Detection: Security Monitoring for Internal Intrusions*, Boston: Addison-Wesley Professional, 2005.

[516] D. Climek, A. Macera and W. Tirenin, "Cyber Deception," Cyber Security Information Systems and Information Analysis Center, 8 March 2016. [Online]. Available: https://www.csiac.org/journal-article/cyber-deception/.

[517] P. Cichonski, T. Millar, T. Grance and K. Scarfone, "Computer Security Incident Handling Guide," [Online]. Available: https://nvlpubs.nist.gov/nistpubs/SpecialPublications/NIST.SP.800-61r2.pdf.

[518] SANS Institute, "SANS Institute: Reading Room | Incident Handling," [Online]. Available: https://www.sans.org/reading-room/whitepapers/incident/paper/33901.

[519] Cybersecurity & Infrastructure Security Agency (CISA), "US-CERT Federal Incident Notification Guidelines," 1 April 2017. [Online]. Available: https://us-cert.cisa.gov/incident-notification-guidelines.

[520] M. Bimfort, "A Definition of Intelligence," 18 September 1995. [Online]. Available: https://www.cia.gov/static/554d7d05a62d7d6de84b5b84ae6702ae/A-Definition-Of-Intelligence.pdf.

[521] S. Caltagirone, "Industrial Control Threat Intelligence," 2018. [Online]. Available: https://www.dragos.com/wp-content/uploads/Industrial-Control-Threat-Intelligence-Whitepaper.pdf.

[522] S. A. Hewlett, M. Marshall and L. Sherbin, "How Diversity Can Drive Innovation," December 2013. [Online]. Available: https://hbr.org/2013/12/how-diversity-can-drive-innovation.

Appendix A Foundational SOC Books

As discussed in the introduction, there are a number of books relevant to general SOC operations that provide a good starting point for additional reading. A non-exhaustive list of these books includes:

- *Intrusion Detection* by R. Bace [219]
- *The Tao of Network Security Monitoring: Beyond Intrusion Detection* by R. Bejtlich [481]
- *Practice of Network Security Monitoring* by R. Bejtlich [482]
- *Organizational Models for Computer Security Incident Response Teams* by G. Killcrece, K.P. Kossakowski, M. Zajicek and R. Ruefle [5]
- *The Modern Security Operations Center* by J. Muniz [483]
- *Blue Team Handbook: Incident Response Edition* by D. Murdoch [484]
- *Blue Team Handbook: SOC, SIEM, and Threat Hunting* by D. Murdoch [485]
- *Network Intrusion Detection (3rd Edition)* by S. Northcutt [486]
- *Designing and Building a Security Operations Center* by D. Nathans [487]
- *Incident Response and Computer Forensics, 2nd ed.* by C. Prosise, K. Mandia, and M. Pepe [488]
- *Incident Response: A Strategic Guide to Handling System and Network Security Breaches* by E. E. Schultz and R. Shumway [489]
- *Intelligence-Driven Incident Response* by S. Roberts and R. Brown [490]
- *Incident Response* by K. R. Van Wyk and R. *Forno* [491]
- *Handbook for Computer Security Incident Response Teams (CSIRTs)* by M. West-Brown, D. Stikvoort, K. Kossakowski, G. Gillcrece, R. Ruefle and M. Zajicekm [6]
- *Blue Team Field Manual (BTFM)* by A. White, B. Clark [492]

Appendix B How to Get Started with Building a SOC

"Strategy 3: Build a SOC Structure to Match Your Organizational Needs" establishes the criteria on which to judge whether the constituency should in-source some or all their security operations, or if an MSSP or other outsourcing arrangement is best. If partial or complete insourcing is selected, then the constituency must make the necessary arrangements to stand up a SOC. This section serves as a high-level guide to building that organization. For more information on standing up a SOC, see: [483], [493], [494], [495].

B.1 Key Design Decisions and Steering Committee

As with any other business project, the SOC will follow the "downward funnel" of narrowing clarity regarding cost, schedule, and performance. In both the initial founding and maturity phases, the SOC and its stakeholders will face several choices that need to be made or refined. Amongst them:

- Staffing model (no, partial, or complete outsourcing)
- Organizational placement
- Technology mix, including whether the SOC will favor on-prem or cloud-based tooling
- Physical location, use of virtual or offsite staffing, and physical build out
- Offered capabilities and functions

In each of the phases discussed below, the SOC and its stakeholders should review, refine, and share design choices. Transparency will help everyone understand the "hows" and "whys" of the SOC as it takes shape.

One of the easiest and best things that SOC founders can do is form a steering committee. This committee will:

- Make critical decisions and recommendations, such as those above
- Represent the business interests of key SOC stakeholders
- Be accountable to constituency leadership and business/mission owners for overall SOC success

Some SOCs form a steering committee that endures past initial SOC formation and may evolve into virtual team consulted on SOC success in a routine basis and may be consulted during major incidents.

B.2 Founding: 0 to 6 Months

The effort to create a SOC may stem from different circumstances. Security operations may have already been carried out in an incomplete or fragmented arrangement, and some external stimulus (perhaps a major incident) compelled its creation. Or perhaps an incumbent SOC capability, such as an MSSP, is ending support and the constituency must resource the new operating model differently. In the first six months, the following outcomes should be achieved:

- Form the SOC steering committee.
- Set high-level expectations.
  - Define the constituency.
  - Ensure upper management support.
  - Gather and set expectations for all major design decisions, discussed in previous section
- Ensure that expectations and mission scope are set and agreed to by major stakeholders both when writing the SOC charter, and at milestones along the SOC's path to maturity
- Survey the constituency for resourcing and existing capability that can be brought into the SOC
  - Leverage existing technologies, resources, and budget to help get started.
  - Tip: do not let the initial influx of resources detract from the importance of a permanent budget line for people, capital improvements, and sustainment.
- Collect security operations best practices from literature and other SOCs.
- Identify existing functions, services, tools, and people that could be assimilated into the SOC.
- Secure funding for people and technology, based on a rough order of magnitude budget.
- If possible, begin the hiring process (See "Strategy 4: Hire AND Grow Quality Staff"), especially for lead analysts and engineers who can support the initial build-out of the SOC.
- Select an approach for SOC physical and virtual presence.
  - The SOC may exist purely as a collection of staff operating in diverse locations with no physical ops floor, either to start or permanently.
  - If the SOC will have a physical ops floor, it may be appropriate to immediately find a location (See "Strategy 3: Build a SOC Structure to Match Your Organizational Needs"), procure construction and contractor resources, and begin build-out.
  - Some SOCs that invest in a physical ops floor may first select a temporary operating location, such as in existing office space occupied by IT operations and/or cybersecurity.

B.3 Build-Out: 6 to 12 Months

Ideally, the SOC will have time to focus on buildout before it is strongly held to certain operations or output requirements. During this time, the steering committee should ensure the vision and mission for the SOC are understood and agreed to in detail by all involved. Likewise, the SOC should begin operations in a nascent form, perhaps a "soft opening" akin to commercial retail where the SOC is open for business, but not widely advertising that fact.

- Write and socialize the SOC CONOPS.
    - Determine the initial team org chart (See "Strategy 3: Build a SOC Structure to Match Your Organizational Needs"), with staff names assigned to each area in the initial operating model.
    - Define, in detail, the SOC's initial offering of services and functions, working with constituents to identify areas where the SOC can provide the most added value.
- Begin hiring staff in large numbers, aiming for 50 percent capacity in the 6- to 12-month window.
- Identify team members delivering above expectations early on and enable them to propel all those around them toward higher maturity, particularly through improving processes and automation.
- Make (or improve) contact with other SOCs, especially those in adjoining constituencies such as federated or tiered models, and in similar areas of government, education, commerce, or geographic region.
- Build requirements for, evaluate, acquire, and pilot essential monitoring capabilities (See "Strategy 7: Select and Collect the Right Data").
    - There may very well be monitoring capability, such as an EDR or firewalls already in place; it is important that the SOC assimilate these technologies early, and as they are ready and able, along with the necessary budget and people to sustain and grow them.
    - If it does not already exist, deploy a pilot monitoring capability in select areas already familiar to the SOC and the steering committee, thereby giving the first hires an initial monitoring capability to focus on.
- Build requirements for, evaluate, and deploy initial data aggregation and analytic capabilities such as a SIEM, SOAR or log management (See "Strategy 8: Leverage Tools to Support Analyst Workflow") if there is no incumbent capability.
- Perform the majority of the build-out of the SOC enclave (See "Strategy 8: Leverage Tools to Support Analyst Workflow").
- As soon as the SOC starts handling alerts or incoming reports from constituents, it is going to need some kind of ticketing system. The SOC may need to make a quick choice to rally around some nascent or prototypical capability while it contemplates its longer-term workflow needs (Section 8.4).
- If the SOC has a temporary ops floor, modify that space to support SOC needs and begin performing initial operations. If creating a virtual SOC, begin testing connections and procedures for a distributed workforce.

- Focus on technologies that match the threat and environment and act as a force multiplier; avoid getting caught up in "technology for its own sake"; extract the maximum amount of value from a modest set of tools.
    - For example, having a flashy, well-organized operations floor can be a leverage point for operations harmonization, employee morale, and attracting attention and resources from various stakeholders. It can also eat up many resources. It is wise to ensure SOC tradecraft and operational maturity is given due attention.
- Ensure strong quality control of what leaves the SOC from day one. Gaining trust and credibility is a big challenge, considering that rookie mistakes can easily undermine progress and stakeholder trust. Consider:
    - Immediately establishing operational quality control over outbound escalations and tickets, in lieu of codified process.
    - Empower many parties in the SOC to draft and iterate SOPs rapidly so that lessons learned are quickly integrated into ops, and staff feel empowered and involved in making the SOC better. Bias toward "good enough" and "better" and bias against perfection.

B.4 Initial Operating Capability: 12–18 Months

If ops floor construction and tool acquisition have proceeded according to plan, the SOC should now have at least a part of the physical space ready for operations (if using a physical space) and SOC capabilities and the connections to those capabilities should be in place. In addition, members of the SOC team should now be showing up for duty. The SOC should now begin to have a stronger sense of identity and recognition as a coherent and growing team. Steps for this phase include:

- Finish hiring staff, aiming for 90 percent capacity in the 12- to 18-month window. Be prepared to hire less experienced staff as well more experienced practitioners and create development plans for all (See "Strategy 4: Hire AND Grow Quality Staff").
- Leverage newly acquired tools and ops floor space to begin creating a monitoring and analysis framework, ensuring that key information and tools are at the analysts' fingertips (See Appendix D).
- Ensure development of various SOPs and other knowledgebase are well underway and have expanded to most areas of routine operations (See Appendix C).
- Begin development and socialization of lower-level authorities (See "Strategy 2: Give the SOC the Authority to do Its Job").
- Begin regular analyst consumption and fusion of cyber intel into monitoring systems (See "Strategy 6: Illuminate Adversaries with Cyber Threat Intelligence"), supporting an initial SA capability.
- If not already started, build, and maintain a SOC partner and trusted agent "rolodex" that covers major operational touchpoints, starting with some members of the SOC steering committee (See "Strategy 3: Build a SOC Structure to Match Your Organizational Needs").
- Spin up operations with greater scale and scope:

- Deploy production sensor capabilities to the initial or expanded set hosts, network monitoring, cloud resource types (See "Strategy 7: Select and Collect the Right Data").
  - Assess the priority for instrumentation of non-traditional IT, OT, and mobile, depending on the SOC's readiness to support and the prevalence and criticality of this technology use. Some effort may be appropriate this early on for some SOCs; others may be able to defer this investment to a later date.
  - If capability does not already exist, begin gathering log data at high volume; if it does, ensure feeds critical to security operations are part of the mix (See "Strategy 8: Leverage Tools to Support Analyst Workflow").
  - Begin advertising the SOC more broadly to constituents; if the SOC has not already established a constituency-facing Web presence, now is a good time to do so.
  - Ensure analysts are consistently leveraging the incident tracking/case management capability established earlier and make considerations for how well this platform is tracking against requirements and expectations (See "Strategy 8: Leverage Tools to Support Analyst Workflow").
  - Assume sustained detection, analysis, and response operations (See "Strategy 5: Prioritize Incident Response").
- If not already done so, initiate a SOC metrics program for both internal quality measures as well as measurement of overall delivery (See "Strategy 10: Measure Performance to Improve Performance").

B.5 Full Operating Capability: 18 Months and More

Each SOC has its own definition of full operating capacity (FOC), but at this stage, the SOC should generally be able to perform the full scope of the mission defined by the steering committee in the first six months. Whereas, in the beginning of operations, many tasks were performed in an ad hoc manner, the SOC should have routine processes in place for their core functions, alerts, and incident types, consistent with a growing set of SOPs.

- If necessary and not already established, expand working hours (possibly to 24x7 operations) (See "Strategy 3: Build a SOC Structure to Match Your Organizational Needs").
- Establish practices to maximize quality staff retention and growth (See "Strategy 4: Hire AND Grow Quality Staff").
- Demonstrate and communicate the value added to constituency mission by SOC's handling of cyber incidents and solicit feedback from constituents (See "Strategy 9: Communicate Clearly, Collaborate Often, Share Generously").
- Adjust operations procedures and capabilities as necessary, given the deltas between the initial vision of the SOC and the operational, resourcing, and policy realities.
- Deploy monitoring and log collection capabilities to an expanded set of monitoring points as appropriate. Measure depth and breadth of coverage, and ensure resources are assigned to driving down gaps (See "Strategy 7: Select and Collect the Right Data").

- Build up data filtering, correlation, triage, and analysis automation techniques (See "Strategy 8: Leverage Tools to Support Analyst Workflow").
- Expand SOC influence into areas of policy, user awareness, and training, if appropriate.
- Establish regular sharing of cyber intel and tippers with partner SOCs and SA with constituents (See "Strategy 6: Illuminate Adversaries with Cyber Threat Intelligence").
- If not already done so, perform different kind of exercise with the SOC, such as a tabletop major breach event with the extended steering committee, or a Red Team operation to test the SOC's overall capability (See "Strategy 11: Turn Up the Volume by Expanding SOC Functionality").

Even though a SOC has achieved its FOC, it will almost certainly take a bit longer for it to become fully mature, since its mission and ops tempo are always changing. Refer to the SOC Services list in "Fundamentals," the table on selecting SOC Functions and Services and the discussion on SOC maturing in "Strategy 3: Build a SOC Structure to Match Your Organizational Needs," and the further descriptions of some the expanded SOC services in "Strategy 11: Turn up the Volume by Expanding SOC Functionality" for ideas on where to take your SOC next. At the same time, make sure to continue to focus on the core SOC mission as discussed in "Strategy 5: Prioritize Incident Response."

Appendix C What Documentation Should the SOC Maintain?

This appendix lists some major written artifacts that most mature SOCs will want to consider maintaining. First are the high-level documents that define core SOC authorities, mission scope, and the "hows" and "whats" of its mission in detail. The next subsection describes the artifacts that the SOC will use to support routine operations. Finally, the last subsection articulates the artifacts that will help the SOC support sound operations, maintenance, and evolution of its technologies and tools. The documents listed in this appendix are representative of documents the SOC itself is responsible for maintaining.

These documents should be considered living artifacts. SOC leadership should evaluate its own mission needs against this potential document library and consider how often it needs to revise each—some every two to three years, others, maybe quarterly. Also, although the term documents are used, many of these artifacts may live in different digital forms. SOPs, for example, may be developed or stored in online shared repositories. Only some of these artifacts, such as the SOC authorities and mission and vision statements, are likely to be very formal documents signed by an executive.

C.1 SOC Programmatics

Most SOCs should have the following documents to help them scope their mission, defend their role when needed, and crisply articulate its services and capabilities to its customers.

Table C-1. SOC Programmatics

Name	What it says	Why the SOC needs it
Charter authority	The scope of the SOC's mission and responsibilities and the responsibilities of other groups with respect to the SOC, signed by the chief executive of the constituency	This is the number one tool the SOC will need to execute its mission in a politically contested environment. While most groups cooperate willingly, sometimes the SOC will need written authorities to be successful.
Additional authorities	Detailed authorities and clarification about SOC mission and touch points that fall outside the charter. It fills in certain details that the charter leaves out (e.g., what the SOC can do in response to or prevention of an incident) or describes additional capabilities taken on after the charter was signed. Can be signed by someone in the SOC's management chain. This document often contains an order of magnitude more detail on the "whos" and "whats" of the SOC and its relationships to others, in contrast to the charter.	These documents will be used in a similar manner to the SOC charter; they clarify what the SOC can and should do and what other orgs are obligated to do in helping the SOC. Good examples include incident escalation and roles and responsibilities, such as RACI matrices [496].

Name	What it says	Why the SOC needs it
Mission and vision	Two crisp statements/slogans saying what the SOC does and what it is aiming for in the future	Helps orient members of the SOC toward a common set of objectives and, in just a few sentences, helps external parties understand what the SOC does.
CONOPS	Covers not only the "what" and "why" of the SOC mission, but also the "how" and "who." This includes the roles and responsibilities of each of the SOC's sections, the technologies it uses, and its ops tempo, inputs, and outputs. While it may articulate escalation flowcharts for major incidents, it does not get down to minute details of specific checklists.	This is the single best place for anyone, including constituents, to understand how the SOC functions, without necessarily covering incident or job specifics. Some SOCs choose to split this document into two pieces: one part for internal consumption and another for reference by other parties.
Short mission presentation	15- to 30-minute slide presentation about the SOC: its mission, structure, how it executes the mission, and key successes. This can also be delivered as a pre-recorded video.	Used to describe the SOC to non-technical audiences in conferences or for quick demos for visiting VIPs. Helps gain trust among stakeholders and partners.
Long mission presentation	Longer (usually one hour or more) technical presentation highlighting SOC successes and TTPs, monitoring architecture, and future initiatives.	Used for technical audiences such as other SOCs. Key for making connections, increasing credibility, and providing trust and transparency with constituents and partners.
Strategic plans	Details regarding future plans for SOC TTPs, technologies and technology investments, sensoring/coverage plans, metrics and KPIs. These can range from informal PowerPoints to OKRs [497], to long documents. Some SOCs will plan only 6 months out while others feel it appropriate to define plans in writing for three plus years.	This will help drive the SOC toward heightened maturity and capability, both for itself and its customers.

C.2 Operations

These documents primarily support routine, efficient, effective, and consistent incident handling.

Table C-2. Operations Documentation

Name	What it says	Why the SOC needs it
Shift schedule and on-call roster	The shift schedule for the SOC, at least two weeks into the future, including who will be on each shift position and who from each section is the designated "on call" person for times of the day or week that that section is not staffed.	So that staff knows who their relief will be and whom to call if they have questions about what happened on the previous shift.

Name	What it says	Why the SOC needs it
Incoming incident reporting form	Constituents fill this out when reporting an incident to the SOC. It captures all incident details the submitter can capture who/what/when/where, what systems were involved, symptoms were observed, time/date, and whom to call for follow-up. This form should be available to constituents in an easily accessible electronic form, usually on the SOC website, and linked from many other sites and training materials. See Appendix E of [4] for examples.	Provides a consistent means for the constituency (users, help desk, sysadmins, ISSOs, etc.) to report potential incidents to the SOC. The SOC will likely update this form based on incident feedback annually.
Incoming incident/ ticket/tip handling SOP	Instructions for handling incoming incident tips from constituents: what data to capture, what to do next, whom to call, thresholds for further escalation, and the like.	Ensures the right information is captured and correctly escalated. Triage analysts should closely follow this SOP every day; it ensures incident coordinators can respond effectively.
Detection handling form	For each detection, such as specific EDR or SIEM alert type or ID, there should be a handling guide that describes the intent of the detection, the expectations for how it should be investigated, and what to do if there is a true positive. Some of this may be simply linked to detection platform documentation such as a detection metadata repository, SOAR automations, SIEM detections, or EDR detection documentation.	This ensures quality and repeatability in alert triage, analysis, and escalation.
Escalation SOP	Sets thresholds and escalation paths for whom incident responders pass incidents to (security, CISO, IT operations, customer relations, etc.). May be released to the constituency so everyone can understand who the SOC calls and under what circumstances.	Members of the constituency may be very sensitive to who gets to know about which incidents at what stage of the incident lifecycle. This ensures consistency in those agreements.
Shift Pass-down Form	Defines what information must be captured by the incoming and outgoing shift. In non-24x7 shops, this may still be used, even though there is no real-time "handoff" from one day to the next.	Ensures nothing gets dropped and major events are recorded; enforces accountability.
Artifact-Handling Process	Defines the process and steps SOC members must follow in accepting, collecting, storing, handling, and releasing digital and physical artifacts. May reference other legal guidelines for evidence handling. This document should be formulated in cooperation with legal counsel.	Ensures that the SOC's hard work stands up in court, in the event an incident leads to legal action.

Name	What it says	Why the SOC needs it
Confidentiality Agreement/Code of Conduct	Concise statement of the rules that define the expected behavior and prohibited activities of SOC staff, above and beyond other agreements they signed as part of the constituency. It will usually articulate the need for SOC staff to maintain strict confidentiality about case and privacy data and to avoid snooping outside the scope of legitimate monitoring duties. This document should be reviewed by legal counsel before approval.	It has been said that with great power comes great responsibility. Should a SOC team member do something seriously wrong, this document supports corrective actions against an employee. It also demonstrates to external groups that the SOC takes its job very seriously and holds its people to a high standard.
Training Materials, Technical Qualification Tests, and Process	Articulates staff in-processing, necessary training, periodic recertification, and qualification tests. Leverages many of the documents in this table.	Serves two key functions: (1) orients new staff on the SOC mission, structure, CONOPS, and SOPs and (2) ensures that each team member is proficient with SOC tools.
Constituency Knowledgebase	A structured knowledge store wherein analysts gather curated information about constituency assets, systems, networks, users, and other entities of interest. If an existing asset/entity store, or service/system inventory exists, the SOC may consider leveraging that. See "Strategy 11: Turn Up the Volume by Expanding SOC Functionality."	Analysts and other members of the SOC will accumulate "tribal knowledge" regarding their constituency. It is important that this knowledge be recorded such analysts can pool their knowledge and limit the impact from personnel changeover.
Continuity of Operations (COOP)	Documents that specify how the SOC will respond and operate in the circumstance of expected outages or system unavailability, ranging from routine weather events and IT outages to more catastrophic events. Will usually include items about how to rotate operations, transfer people and command, alternate site locations, timelines, and escalation procedures. These will cover not only incident handling and staffing, but should incorporate survivability and planning for SOC sensors, analytics, data persistence, and enclave systems as well.	Ensures the SOC operates in an orderly, coordinated, and predictable manner in the case of COOP events. It is important to think through these scenarios in advance, rather than in the heat of the moment.
Exercise plans	Schedule for routine and upcoming tests and exercises relevant to the SOC, such as: phishing training, red/purple teaming, COOP exercises, tabletop exercise, and the like. Some of these items, particularly red team operations, should be kept confidential and unavailable to analysts. Some SOCs may therefore choose to keep two separate lists of exercise: those that the analysts should know about, and those they do not.	These will help the SOC perform capacity planning for operations, deconflict operational events, and ensure SOC TTPs are being evaluated in a rigorous manner.

C.3 Engineering and System Administration

This set of documents help the SOC structure its activities around how it operates, maintains, updates, and evolves its toolset and technologies.

Table C-3. Engineering and System Administration Documentation

Name	What it says	Why the SOC needs it
Monitoring Architecture	Articulates the details on where the SOC's monitoring capabilities are located, and how that data is collected and stored. Should depict a detailed path from the end network all the way to the analyst. Some SOCs break this into two pieces: (1) a generic depiction of how the constituency is instrumented, and (2) a detailed diagram showing exact sensor locations; the former can be shared, the later should not.	Helps SOC members understand how networks, hosts, and cloud resources are instrumented. Being clear on exactly where a sensor is tapped is critical because there are always subtle blind spots due to DMZ and routing complexities. It also helps SOC sensor and sysadmins troubleshoot downed feeds when they occur.
Internal CM Process	Defines how changes are made to SOC systems and documents (e.g., hardware and software installs/upgrades, sensor and SIEM signature changes, and SOP updates). For some SOCs, this may be tied to agile scrum or CI/CD processes and automation.	Ensures that rigor and consistency are enforced, with notification and visibility across the SOC for changes, while balancing agility in ops. For instance, a SOC should be able to turn a piece of cyber intel into a new analytic or signature push in a matter of hours, but not without proper analyst notification.
Systems and Sensors Maintenance and Build Instructions	A series of documents that discuss how to maintain all key SOC systems and how to rebuild them in the event of corruption, hardware failure, or virtualized cluster scale out/rebuild.	While vendor manuals always help, a SOC will have many customizations, especially for homegrown solutions. For example, joining a system to a SAN requires work with at least three different products. It is easiest to distill this into a few pages of instructions rather than pointing sysadmins at 1,000 or more pages of product manuals.
Operational, Functional, and System Requirements	Detailed listing of SOC tool requirements. Contains everything from sensor fleet management specs to capabilities of malware analysis tools. Can articulate needs at three levels: (1) operational (what business needs to be done), (2) functional (what features are needed), and (3) system (what are the specifics of the implementation).	Helps support rigor and formalism in acquisition for all SOC capabilities, if necessary. Gives the engineers a concrete set of needs that must be satisfied. Helps the SOC ensure it is getting what it needs, especially if the engineers are not part of the SOC, or do not themselves have previous experience in security operations.

Name	What it says	Why the SOC needs it
Budget and current spending (capital and operational expenditures)	Allocates money for SOC staffing, soft- ware/ hardware licensing, cloud expenditures, refresh and maintenance, expansion, and capital improvements for the SOC, during the current fiscal year and for one or more years into the future. Recognizes different categories of money and considers both inflation and expected changes in SOC capabilities.	A SOC must plan and budget for its capabilities just like any other organization. Having this at hand (along with a crisp list of successes) will help the SOC defend its budget against constant scrutiny and potential cuts, perform capacity planning, and be prepared for future needs.
Unfunded Requirements	Succinct one- or two-page description of each capability not currently built into the budget. Will include what the SOC wants, what benefit the capability will provide, how much it will cost, and what will happen if the SOC does not get it.	Having these at hand will help the SOC claim money when it becomes available, sometimes on very short notice. Notifications of available funds often come out at random times; being the first one to respond with well-written unfunded requirements frequently means winning the funds offered.
Sensor and SIEM Detections/ Analytics/Content List(s)	List of all the analytics, detections, queries, signatures, and content deployed to each SOC sensor or analytic system. This is usually contained within the tools themselves, but where possible, be extracted and stored in another location for reference and CM tracking. In cases where detections, analytics, signatures, etc. are trackable as code, the SOC may consider a code repository (such as git) to track this. Custom signatures and analytics are especially important to document: what they look for and what analysts should do when their alerts pop up.	Helps analysts know what they are looking at and what to do with each fired event. This list should be scrubbed by sensor managers and other key SOC stakeholders on a regular basis, perhaps quarterly. The SOC should strongly consider mapping its detections and analytics to the ATT&CK framework.
SOC System Inventory	System host name, IP, MAC address, hardware type, location, and serial/barcode of all SOC high-value assets. Inventory lists should also incorporate any cloud based SOC assets and tools, and their appropriate cloud tenancy, identity plane, region, and resource identifiers. If the SOC is part of a larger organization that has an asset tracking database, it may consider using that, keeping in mind it should be careful not to widely advertise the location or function of sensitive systems like sensors and SIEMs.	The SOC must be able to keep track of what it owns so nothing gets lost; inventory must be refreshed on schedule.

Name	What it says	Why the SOC needs it
Network Diagrams	Depicts the detailed network architecture of the constituency, usually showing user networks and server farms as clouds connected by firewalls and routers. Typically broken down into a series of large Visio diagrams or PDFs that can be printed on a large-format printer. Regardless of whether a SOC maintains these diagrams, it should consider overlaying its sensor placement for internal tracking purposes.	Constituencies maintain network diagrams to a varying degree of currency and accuracy. If the SOC does not get what it needs from the NOC or IT operations, it may be necessary for the SOC itself to maintain constituency network diagrams. These help members of the SOC understand the size and shape of the constituency, how data gets from point A to point B, where external connections are, and the connection between subnets and mission/business functions.

Appendix D What Should an Analyst Have Within Reach?

Speed and efficiency are important tenants of SOC operations. One way to enhance both aspects is to have easily available the needed communication methods, data, tools, and documentation for operations. The table below lists what resources are needed, how important (generally) they are during steady state operations, and why the operator needs them. Importance may vary depending on the mission and SOC size, it is provided as a starting point for consideration only.

D.1 Communications

Table D-1. Communications

What	Importance	Why
General Internet access	High	Access to security websites, news, public-facing email, general troubleshooting, and external collaboration.
Access to constituency main business network with email, office automation software	High	An analyst requires regular communication with constituents and the ability to conduct general business.
Unattributed/unfiltered Internet access[13]	Low	Was that user really surfing porn? Where did this piece of malware come from? These questions need to be answered daily, without placing the constituency at risk or tipping off the adversary.
Analyst collaboration forum/ SharePoint/wiki and unstructured file share limited to SOC only	High	Members of the SOC will capture lots of unstructured or semi structured information not directly related to a given incident. Having both organized (wiki, SharePoint) and unorganized file share means of pooling these resources is important. In addition, a real-time chat room may be helpful. All these resources should be stored in the SOC enclave.
Access to user-submitted incident reports; read and (possibly) write access to the SOC externally facing website/portal	High	A central point of communication between the SOC and the constituency, this is often where constituency users go to submit potential incident information.

---

[13] A number of details must be considered when deploying and supporting a truly unattributed Internet connection that are beyond the scope of this book. An unattributed Internet connection assumes that it cannot be traced back to the constituency, and there are none or less-restrictive content filtering technologies on it, unlike the constituency's main Internet gateway that one expects has a robust content filtering solution in place.

What	Importance	Why
Incident tracking/ticketing database limited to SOC only	High	Analysts must be able to record pertinent details about incidents, attach events or other digital artifacts, and escalate that information daily to other members of the SOC.
Write access to current shift pass-down log and read access to past pass-down logs	Medium	Analysts should record their actions and events of note during their shift and summarize them at the end. That way, analysts on later shifts can review what happened and understand what issues require follow-up. Also helps support accountability.
Access to collaboration forums shared with sister SOCs	Medium	Messaging boards and collaboration forums where multiple SOCs share cyber intel, tippers, incident reports, and general cyber news are immensely helpful, especially during major incidents.
Standard real-time voice communications	High	This could be a public/commercial telephone (PSTN) or VoIP. Analysts will spend significant time on the phone every day.
Encrypted out of band voice communication	Medium	The SOC typically requires a secure communications channel to people in the constituency or to other external organizations that sits out of band of ordinary, unencrypted, "in-band" communication paths. For some SOCs, this could be achieved via various VoIP and mobile apps; for others, such as those in the government, this may be via Secure Telephone Equipment (STE or equivalent).
Real-time news feeds	Low	This help provide SA about events that may impact the constituency.

D.2 Data and Tools for All Analysts

Table D-2. Data and Tools for All Analysts

What	Importance	Why
Access to SOC network from a robust workstation that is not used to connect to the main constituency network or the Internet	High	As discussed in "Strategy 8: Leverage Tools to Support Analyst Workflow," the SOC's monitoring infrastructure should be placed in a well-protected enclave. Therefore, the analyst should access the bulk of SOC monitoring tools from high-performance workstations on the SOC network.
Multilevel desktop consolidation system or KVM (keyboard, video, mouse) switch, if more than two or three different desktop systems are needed	Low	Most SOCs can get the job done with two workstations for each analyst: one for the SOC network and the other for Internet and constituency network access. For SOCs that require more than two workstations, a KVM switch may be appropriate.
General user privileges to the SIEM, SOAR, big data, and/or log aggregation systems	High	A SIEM usually serves as the hub for alert triage and event analysis; an analyst should be able to spend more time with the SIEM console than any other tool or system.
General user privileges to network sensor consoles or other out-of-band monitoring systems	Low	Some sensor consoles (such as an EDR) may have details beyond what SIEM can (or should) capture; in this case, analysts will also need access to these.

What	Importance	Why
General user privileges to all inband monitoring device consoles, (e.g., EDR and SIEM)	High	Same reason as out-of-band other consoles: they may offer more detail or options than what is collected by the SIEM.
Complete and current detection/ signature list or repository for all sensors, with detection descriptions and detection syntax	Medium	The analysts should know the policy that is currently deployed to all monitoring equipment, including the description of each alert and the exact syntax of the signature, if available.
Complete content list with descriptions for all production SIEM and SOAR content, automation, and workflows	Medium	Same rationale as sensor signatures: the analyst should be able to understand the provenance and pedigree of the alerts they triage.
For every network sensor alert triaged, the raw event details, the signature (or signature description) that triggered it, and other contextual data as appropriate	High	Without this data, alerts are meaningless. The SIEM or sensor console should contain or link directly to all these items. (See "Strategy 7: Select and Collect the Right Data" and "Strategy 8: Leverage Tools to Support Analyst Workflow")
Vulnerability scan results and/or on-demand vulnerability scanning tool	Low	Did that SIEM alert concern a system that was actually vulnerable? What OS and services is this system running? Analysts will ask these questions regularly, and vulnerability scanners have the answers. Having access to both historical regular scan results and an on-demand scanning capability is best, but either one will help.
Network maps depicting major subnets and interface/peering points for the constituency (e.g., firewalls and network sensor monitoring locations)	Low	Analysts must understand the network they are monitoring. It helps to have a few key network maps (such as Internet gateways) posted on the wall of the ops floor. If the SOC is not responsible for network mapping, it is best to have read-only access to where these maps are stored on the constituency network.
Read-only access to asset-tracking database	Medium	Complements vulnerability scanning data, especially when scan results are stale or unavailable. May also capture information about system owner, contact info, or supported mission.
Real-time network availability status dashboard	Low	It helps to get a feed of planned and unplanned network and system outage events across the constituency, provided by the NOC or IT ops.
Current firewall rule sets for all production firewalls in constituency	Low	Did that attack make it through the firewall? Is it of concern that a given network is wide open? Current firewall rule sets answer these questions. Having read-only access to firewall rule sets helps, rather than having to ask for a download from the firewall managers.

D.3 Documents, Records and Miscellaneous

Table D-3. Documents, Records, and Miscellaneous

What	Importance	Why
All supporting authority documents listed in "Strategy 2: Give the SOC the Authority to do Its Job"	Low	Knowing what the SOC has written authority to do or not do is important. SOC leads will refer to these on a semiregular basis. It is best to have them consolidated in one place.
Employee directory(ies) for entire constituency	High	An analyst requires regular communication with others in the constituency.
Contact information for all parties that are part of the SOC's incident escalation chain	High	Analyst will need to call TAs, sysadmins, security officers and champions, and other parties that are both inputs and outputs to the SOC's incident escalation flowchart. These calls are often time sensitive, so, having up-to-date contact information readily available is key.
Personal (home and cell) contact information for all members of the SOC	High	Members of the SOC will get called after hours on a regular basis. Systems break, and analysts call in sick. Some SOCs put this information on a laminated card clipped to analysts' security badge.
Documents listed in "Appendix C" except budget and requirements	SOC Programmatics: Medium Operations: High Engineering and System Administration: High	Analysts will need to refer to SOPs, CONOPS, and other materials from time to time. While some of these documents are already listed in this table, having all the operationally relevant ones at hand is a good idea.
Password management	High	The SOC will have or manage access to dozens if not hundreds of systems. The SOC should pick a secure password store for these credentials, particularly for "break glass" administrative accounts, with backup and redundancy as appropriate.
Secure physical document and media storage (where applicable)	Varies	The SOC will have to store sensitive data, forensic images, case data, and other materials in an appropriate manager. Chances are, at least one safe will be needed.
Vendor technical documentation on SOC technologies	Medium	The SOC should have references (either in hard copy or soft copy) for all SOC monitoring and analytics systems that analysts use.
Emergency "go bag"	Medium	Includes everything the SOC will need during a building evacuation or similar event if the SOC has a physical ops floor. This includes contact information for all SOC members, rally location, shift schedule, flashlight, satellite phone, COOP activation playbook, and so forth. The ops lead on duty should grab it on the way out the door during an emergency or fire drill.

Appendix E Data Sources and Sensoring Technology

This appendix provides a landscape of typical SOC data sources, other than those related to vulnerability scanning, cyber threat intelligence, and asset data. The purpose of presenting this landscape is to help the SOC approach a very wide and diverse field in a way that enables it to:

- Understand the opportunities and possibilities across the data landscape
- Begin to consider on cost vs value tradeoffs when contemplating things like volume and velocity of data
- Recognize data feed value in terms of both forensic "ground truth" as well as detective capability, both on its own and in the presence of a platform like SIEM or some other correlation or analytic engine

Every SOC will make their own choices on which data to collect, and no two SOC's data strategies are exactly the same. It is worth reiterating that the SOC is increasingly subject to regulatory compliance and other laws that will mandate the collection and retention of certain log types, such as due to local, state, provincial, or national laws. In the US, those might include SOX, PCI, HIPAA; in the US Government, that might also include mandates such as those due to Executive Order. These will take precedence over voluntary, mission-driven choices by the SOC, such as those articulated in this book. For one such highly detailed example, see [498]. The SOC is encouraged to seek legal advice on its compliance and legal obligations.

Assumptions and caveats for the items in the table are:

- The number of devices listed assumes an enterprise of 5,000–50,000 users and IPs, with a centralized SOC organizational model in mind.
    - The number of actual hosts or devices in a given enterprise will vary. The numbers listed in this appendix assist in forming an estimate of data volume—given the number of devices and the typical number of events per device.
    - The number of devices refers to the number of end systems generating this data, not systems that serve as the collection point. For example, the table list tens of thousands of devices for AV, even though their data is likely rolled up to one or a few AV management servers.
- In addition to the items listed in "Volume depends on," the volume of every feed depends on the following four items:
    - The volume of activity seen by the device generating the events
    - How that data feed is tuned at the generating device and upstream
    - The detail and verbosity available from the given source devices

- - How the end device(s) are configured.
- "Subjective value" gives a sense of the likely quality of each data type, assuming it has been tuned properly.
  - The actual value of a given data feed will vary within the context of the constituency and enterprise, implementation, mission, and product implementation.
  - Value uses the following rating scale:
    - 4 = Excellent
    - 3 = Good
    - 2 = Fair
    - 1 = Poor
    - 0 = None/Not applicable
  - The table shows the value of a given data feed under the following three circumstances:
    - "Tip-off" means the data will help direct the analyst's attention, without additional enrichment from correlation by a downstream analytic platform such as a SIEM.
    - "Tip-off with correlation" means a given log entry will provide a good incident tip-off, assuming it is enriched or correlated with other data.
    - "Supporting context" means it helps an analyst establish the ground truth of an incident.
- "Fields of interest" describes the specialized fields commonly found in that data source, which are of interest for correlation or forensics purposes. It is assumed that the following standard fields are included in most or all data sources:
  - Date and time, at least down to the second, possibly with time zone
  - Source IP, port, protocol, and hostname, if applicable
  - Destination IP, port, and hostname, if applicable
  - IP and/or hostname of the device that originated the event
  - Event name, possibly with a detailed description.
- Many rows in this table apply similarly across on-prem, IaaS, SaaS, and PaaS equivalent capabilities, such as:
  - Firewalls, inclusive of general-purpose all-in-one network protection appliances
  - Databases, data warehouses, noSQL and large-scaled structured data storage/ retrieval systems
  - Web servers and services
  - Mail proxies, servers, filtering, and hosting
  - Host (OS) monitoring and logging
  - Key/secret stores (e.g., Azure KeyVault and Amazon Secrets Manger)
  - Identity management and related systems including LDAP, RAS, VPN, Active Directory and Windows domain controllers
  - Storage (e.g., SAN/NAS and cloud-based storage topologies such as Azure blob and Amazon S3)
  - Custom applications

What	What is revealed and use cases	Typical # of events per day per device	Volume dependencies	Typical # of devices	Subjective Value As...			Fields of interest
					Tip-off (raw)	Tip-off w/ correlation	Context	
Host data								
EDR alerts	Attacks, other activity seen at the host; see Section 7.3.2	10s–1,000s	Tuning of signature set	10,000s	4	4	4	Process name/ ID, action taken (allow/block), file name
EDR host bulk telemetry	See "OS logs"[15], below	100s– 10,000s	Sensor vendor	10,000s	0	4	4	
System and file integrity checker	Detailed changes to key host configuration files, settings; see Section 7.3.3	10s– 10,000s	Volume of automated admin tasks	1,000s	3	4	2	File name, hash
Anti-Virus/Anti-Malware	Hits on known, easily identifiable malware; see Section 7.3.4	10s	Exposure to downloaded files	10,000s	3	4	1	File name, virus name, username, action taken
Intellectual property and DLP	Attempts to infiltrate or exfiltrate documents and other information from the enterprise; see Section 7.3.5	1s–100s	Policies on removable storage, file transfer use	10,000s	2	4	4	Username,[16] removable storage unique ID, proto- col or device used
User activity monitoring	Everything a user does while logged in; see Section 7.3.6	100s– 10,000s	# Of users under scrutiny	10,000s	1	4	4	Username, file name, process name, action taken

---

[14] Based on enterprise serving between 5000 and 50,000 users

[15] Most of the best-of-breed EDR products will generate OS-level telemetry roughly equivalent to that of the OS itself, with that caveat an EDR may add extra fields and enrichments, and that it may not "backhaul" every event as ordinary OS log collection would

[16] Many DLP events will not be labeled with the username of the person plugging in or removing USB mass storage devices. This is a good example of how user session tracking in the SIEM might provide a compelling use case, albeit a potentially complicated one.

What	What is revealed and use cases	Typical # of events per day per device	Volume dependencies	Typical # of devices	Subjective Value As…			Fields of interest
					Tip-off (raw)	Tip-off w/ correlation	Context	
Windows Doman Controller[17]	Authentication, access control events for all systems on domain;[18] variety of use cases with careful interpretation	10,000s–1,000,000s	Number of systems in domain & DCs, Audit policy	1s–10s	2	4	4	Username, user privileges, success/fail, other Windows specifics
OS Logs: Windows security event provider (WEC/WEF), Linux auditd, macOS logging, etc.	Privileged system actions, Login/logoff, process creation, OS-specific actions, can be very high EPS if includes process creation events	100s–10,000s	Audit policy	10,000s	2	4	4	Username, user privileges, success/fail, process name, filename, other Windows specifics
Network connection events generated by the host, e.g., "HostFlow"	Can be some of the most voluminous events generated by a host, but also very useful if used with care	1,000s–10,000s	Amount of system activity	10,000s	0	4	4	Initiating/ receiving process, IP, ports
Network device and network service data								
NIDS/NIPS[19]	Attacks seen traversing the network; see Section 7.4.1	100s–10,000s	Signature set, features	10s–100s	2	3	1	Allow/block, vulnerability ID

---

[17] If the SOC does not have the budget or sophistication to instrument hosts with an EDR or ordinary OS-level logging, logging at the domain controller (or its equivalent, inclusive of cloud-managed identities) is a very good way to get at least some host insight

[18] Activity on systems in the domain that occurs under a nondomain account often will not get rolled up to the domain controller. This is very important if an attacker uses a local system account to do something bad and you were expecting to see that information show up in the domain controller logs. However, domain controllers usually serve a logical consolidation point for logs, whereas instrumenting each end host, even just servers, can be costly. If used properly, domain controller logs can be a smoking gun for spotting insider activity.

[19] This includes alerts generated from purpose-built NIDS/NIPS devices and all-in-one network protection appliances and "next generation" firewalls

| What | What is revealed and use cases | Typical # of events per day per device | Volume dependencies | Typical # of devices | Subjective Value As... | | | Fields of interest |
					Tip-off (raw)	Tip-off w/ correlation	Context	
NetFlow[20]	High-level statistics on all network traffic seen; see Section 7.4.2	1000s–1,000,000s	Location of tap	10s–100s	0	4	4	Bytes in/out
Traffic Metadata Collection	Detailed information on network traffic seen, see Section 7.4.3	1,000s–1,000,000s	Location of tap Activated log types	10s–100s	1	4	4	Application-specific details like DNS request, URL, cipher suite, etc.
Full network session capture (PCAP)	Full details of entire network conversation; see Section 7.4.4	N/A[21]	PCAP collection filters	1s–100s	0	0	3	Everything about each layer of the protocol stack
Content detonation devices and services	Malware found in streams of Web or email data; see Section 7.4.5	10s–100s	Amount of malware ingress to the network	1s	4	4	4	Malware name, file name, source content, hash, vulnerability ID
WIPS	Man in the middle attacks, rogue access points, rogue endpoints, denial of service attempts, etc.; see Section 7.4.6	10s–1,000s	Tuning of signature set	10s–100s	3	4	2	AP name
Network firewall	Activity and bandwidth seen across firewall, NAT records, and possibly FTP and HTTP traffic details	100,000s	Firewall rule set, volume/diversity of traffic	10s	1	3	3	Session ID for NATing, bytes in/out

---

[20] Data from these tools can often support many of the same uses cases as firewalls, and vice-versa. Some NetFlow collection systems also produce metadata on popular protocols such as HTTP and SMTP, making Web content filter and email gateway logs somewhat redundant.

[21] PCAP is not delivered in events; however, the volume generally dwarfs all other data sources listed in the table.

What	What is revealed and use cases	Typical # of events per day per device	Volume dependencies	Typical # of devices	Subjective Value As...			Fields of interest
					Tip-off (raw)	Tip-off w/ correlation	Context	
Web content filter/proxy	Details of all proxied traffic, usually HTTP; tracking Web usage, malware sites	10,000s–1,000,000s	Filtering rules in use	1s–10s	3	4	3	Allow/block, URLs, referrer, user agent, Web site category, website reputation score
Router/ switch	Link, port up/down, router changes, location of MAC address attached to network	100s–10,000s	Verbosity of logging level enabled	100s–1,000s	1	2	2	Bytes in/out, MAC
DNS[22]	Major events on the DNS server such as zone transfers; DNS requests from internal servers can reveal malware beaconing.	10,000s–1,000,000s	Verbosity, DNS caching in place	1s–10s	0	4	3	Contents of DNS query and response
DHCP	Records of DHCP lease requests/ renewals; what systems were on the network, when, and where	100s–10,000s	DHCP lease timeout	10s–1,000s	0	1	2	DHCP lease info, lease MAC
Identity and access control								
Network Access Control (NAC)	Results of any system attempting to gain logical access to the network	100s–10,000s	Complexity of policy, openness of network	10s–100s	2	3	3	System details (OS, patch level), MAC, allow/quarantine
Remote Access System and VPN	Attempts to gain remote access to the enterprise	10s–1,000s	Size of remote worker pool, # of partner orgs	1s–10s	0	3	3	Username, remote IP, client version

---

[22] Turning on per-query logging on many DNS servers and services can be very taxing to on-prem DNS servers. Consider achieving the same objective through Traffic Metadata collection of DNS traffic.

What	What is revealed and use cases	Typical # of events per day per device	Volume dependencies	Typical # of devices	Subjective Value As...			Fields of interest
					Tip-off (raw)	Tip-off w/ correlation	Context	
Single sign-on (SSO) and identity access management	Consolidated tracking of logical user access to enterprise resources, common usernames spanning disparate systems	1,000s–100,000s	Number of systems SSO-enabled	1s–10s	0	3	4	Username, translated (real) identity, user attributes
Physical access control (badge reader)	Physical access to enterprise facilities (badge in, possibly badge out); insider threat	1s–100s	Penetration of deployment, requirement for each per- son to swipe in and out	1s–100s	0	3	3	User ID, room #
PaaS, SaaS, applications, storage, and other systems								
OT operational alerts	Attacks to IoT	10s–1,000s	Tuning of signature set	1,000s–10,000s	3	4	2	Process name/ ID, action taken (allow/block), file name
Cloud control plane logs	Actions of users on/against cloud control plane such as authentication, resource changes/add/remove	1000s–1,000,000s+	Size of cloud tenant instance	<10	2	4	4	Cloud resource instance identifier/GUID, username, claims, target resource, before/after state
Web server logs	Malicious use/abuse, IP theft, insider threat	1000s–1,000,000s+	Audit policy	10s–100s	0	3	4	URL/URI, source IP, HTTP verb, (other portions of) HTTP header
Email gateway, email server, and cloud-based email host services	Details of email that goes in and out of enterprise; insider threat, data leakage	1,000s–100,000s	Quantity of spam, email traffic	1s–10s	1	4	4	To/from address, subject, attachment name, allow/block/quarantine

| What | What is revealed and use cases | Typical # of events per day per device | Volume dependencies | Typical # of devices | Subjective Value As... | | | Fields of interest |
					Tip-off (raw)	Tip-off w/ correlation	Context	
Databases, data warehouses, noSQL, big data systems, database firewalls[23]	Malicious use/abuse, IP theft, insider threat	1000s–1,000,000s+	Audit policy	10s–1000s	1	3	4	Username, command/query executed
Secret stores, including hardware-backed secret vaults	Malicious use/abuse, secret/token theft, insider threat	100s–100,000s	Application & secret design, use	10s–1000s	1	3	4	Secret name, IP
Storage-NAS/SAN, cloud	IP theft, insider threat	10,000s–1,000,000s+	Audit policy	10s–1000s	0	4	4	File/storage name & path, username, IP, CRUD operation
COTS application, cloud SaaS applications, application APIs, and custom-built apps	Application-specific actions, logical user access and changes to objects and data; insider threat monitoring; account compromise and data leakage	10s–100,000s	User population application type and complexity	1s–100s	?[24]	4	4	Username, action, object name

---

[23] Recognizing all logging is volume-driven, database, web server, and storage logging can be exquisitely problematic on busy workloads. A good way to bring a busy SQL server, web server, or SAN to a crawl is to enable logging of every query, file access, or HTTP GET. As always, care should be used here.

[24] Custom and COTS applications offer interesting opportunities for monitoring because they often closely support core missions of the organization. However, they often write logs in a variety of formats, requiring custom log parsers and human interpretation. A SOC that has resources to leverage even a few of these can really hit a home run when they uncover malicious activity that manual human review couldn't.

Appendix F Abbreviations and Acronyms

Acronym	Definition
AAR	After Action Review
ACFS	ASM Cluster File Service
ACL	Access Control List
AD	Active Directory
ADFS	Active Directory Federation Services
ADB	Android Debug Bridge
ADVM	ASM Dynamic Volume Manager
API	Application Program Interface
APT	Advanced Persistent Threat
ASM	Automatic Storage Management
AV	Anti-Virus
AWS	Amazon Web Services
B2B	Business-to-Business
B2G	Business-to-Government
BaaS	Breach as a Service
BACnet	Building Automation Control Network
BAS	Breach and Attack Simulation
BIOS	Basic Input/Output System
BYOD	Bring Your Own Device
C2	Command and Control
CASB	Cloud Access Security Broker
CCTV	Closed Circuit Television
CDR	Call Detail Record
CEE	Common Event Expression
CEF	Common Event Format
CEI/CBE	Common Event Infrastructure/Common Base Event
CEO	Chief Executive Officer
CERT	Computer Emergency Response Team
CI/CD	Continuous Integration/Continuous Delivery
CIDF	Common Intrusion Detection Framework
CIFS	Common Internet File System

Acronym	Definition
CIM	Common Information Model
CIO	Chief Information Officer
CIR	Cloud Incident Response
CIRC	Computer Incident Response Center or Capability
CIRT	Computer Incident Response Team
CISA	Cybersecurity and Infrastructure Security Agency
CISO	Chief Information Security Officer
CMM	Capability Maturity Model
CMU SEI	Carnegie Mellon University Software Engineering Institute
CND	Computer Network Defense
CNSS	Committee on National Security Systems
CONOPS	Concept of Operations
COO	Chief Operating Officer
COOP	Continuity of Operation
COTS	Commercial Off the Shelf
CPU	Central Processing Unit
CS	Computer Science
CSA	Cloud Security Alliance
CSIRC	Computer Security Incident Response Center or Capability
CSIRT	Computer Security Incident Response Team
CSO	Chief Security Officer
CSOC	Cybersecurity Operations Center
CSRF	Cross-Site Request Forgery
CTI	Cyber Threat Intelligence
CTO	Chief Technology Officer
CVSS	Common Vulnerability Scoring System
DBIR	Data Breach Investigations Report
DDoS	Distributed Denial of Service
DHCP	Dynamic Host Configuration Protocol
DHS	Department of Homeland Security
DISA	Defense Information Systems Agency
DLL	Dynamic Link Library
DLP	Data Loss Prevention
DMZ	Demilitarized Zone
DNP3	Distributed Network Protocol 3
DNS	Domain Name System

Acronym	Definition
DoD	Department of Defense
DoJ	Department of Justice
DoS	Denial of Service
E-ISAC	Electric Sector Information Sharing and Analysis Center
EDR	Endpoint Detection and Response
EFI	Extensible Firmware Interface
EFS	Amazon Elastic File System
ELF	Executable and Linking Format
ENISA	European Union Agency for Cybersecurity
ETL	Extract, Transform, and Load
FaaS	Function as a Service
FBI	Federal Bureau of Investigation
FFRDC	Federally Funded Research and Development Center
FIRST	Forum of Incident Response and Security Teams
FISMA	Federal Information Security Management Act
FOC	Full Operating Capacity
FOSS	Free and Open-Source Software
FTE	Full Time Equivalent
G2G	Government-to-Government
GDPR	European Union General Data Protection Regulation
GIAC	Global Information Assurance Certification
GPO	Group Policy Object
GPS	Global Positioning System
GRE	Generic Routing Encapsulation
GUI	Graphical User Interface
HIPAA	Health Insurance Portability and Accountability Act
HMI	Human Machine Interfaces
HOPE	Hackers On Planet Earth
HTTP	Hypertext Transfer Protocol
I/O	Input/Output
IaaS	Infrastructure as a Service
ICAM	Identity, Credential, and Access Management
ICMP	Internet Control Message Protocol
ICS	Industrial Control Systems
IDS	Intrusion Detection System
iDRAC	Integrated Dell Remote Access Controller

Acronym	Definition
IFTTT	If-This-Then-That
iLO	Integrated Lights Out
IOC	Indicator of Compromise
IODEF	Incident Object Description and Exchange Format
IoT	Internet of Things
IP	Internet Protocol [address]
IP	Intellectual Property
IR	Incident Response
ISCM	Information Security Continuous Monitoring
ISO	International Organization for Standardization
ISP	Internet Service Provider
ISSM	Information Systems Security Manager
ISSO	Information Systems Security Officer
IT	Information Technology
JWT	JSON Web Token JWT
KPI	Key Performance Indicators
KVM	Keyboard, Video, Mouse
LDAP	Lightweight Directory Access Protocol
MDM	Mobile Device Management
ML	Machine Learning
MOA	Memorandum of Agreement
MOU	Memorandum of Understanding
MPLS	Multiprotocol Label Switching
MSSP	Managed Security Services Provider
MTD	Mobile Threat Defense
MTTD	Mean Time to Detect
MTTR	Mean Time to Recovery
NAC	Network Access Control
NAS	Network Area Storage
NAT	Network Address Translation
NDA	Non-Disclosure Agreement
NIDS	Network-Based Intrusion Detection System
NIPS	Network Intrusion Prevention System
NIST	National Institute of Standards and Technology
NOC	Network Operations Center
NOSC	Network Operations and Security Center

Acronym	Definition
NTP	Network Time Protocol
O&M	Operation and Maintenance
OAUTH	Open Authorization
OCIO	Office of the Chief Information Officer
OCISO	Office of the Chief Information Security Officer
OODA Loop	Observe, Orient, Decide, and Act Loop
OS	Operating System
OPC-UA	Open Platform Communications Unified Architecture
OSI	Open Systems Interconnection
OSINT	Open-Source Intelligence
OSSEM	Open-Source Security Events Metadata
OT	Operational Technology
PaaS	Platform as a Service
PACS	Physical Access Control Systems
PCAP	Packet Capture
PCI	Payment Card Industry
PCI DSS	Payment Card Industry Data Security Standards
PDF	Portable Document Format
PE	Portable Executable
PFS	Perfect Forward Secrecy
PID	Process Identification Number
PIR	Post Incident Review
PKI	Public Key Infrastructure
POC	Point of Contact
RAID	Redundant Array of Independent Disks
RAM	Random Access Memory
RAT	Remote Access Tool
RDBMS	Relational Database Management System
RF	Radio Frequency
RFC	Request For Comments
RX	Receive
SA	Situational Awareness
SaaS	Software as a Service
SAML	Security Assertion Markup Language
SAN	Storage Area Network
SATCOM	Satellite Communications

Acronym	Definition
SCADA	Supervisory Control and Data Acquisition
SCCM	System Center Configuration Manager
SIEM	Security Information and Event Management
SLA	Service Level Agreement
SLO	Service Level Objective
SMB	Server Message Block
SMS	Short Message Service
SMTP	Simple Mail Transfer Protocol
SOAR	Security Orchestration, Automation and Response
SOC	Security Operations Center
SOP	Standard Operating Procedure
SPAN	Switched Port Analyzer
SQL	Structured Query Language
SSD	Solid-State Drive
SSL	Secure Socket Layer
SSO	Single Sign On
STE	Secure Telephone Equipment
STIG	Security Technical Implementation Guide
STIX	Structured Threat Information eXpression
Sysadmin	System Administrator
TAXII	Trusted Automated eXchange of Intelligence Information
TCP	Transmission Control Protocol
TLS	Transport Layer Security
TPM	Trusted Platform Module
TTP	Tactics, Techniques, and Procedures
TTX	Tabletop Exercises
TX	Transmit
UBA	User Behavior Analytics
UDP	User Datagram Protocol
UEBA	User Entity Behavior Analytics
UI	User Interface
VA	Vulnerability Assessment
VLAN	Virtual Local Area Network
VM	Virtual Machine
VoIP	Voice over Internet Protocol
VPN	Virtual Private Network

Acronym	Definition
VRF	Virtual Routing and Forwarding
VTC	Video Teleconferencing Capability
WAF	Web Application Firewall
WAN	Wide Area Network
WEC	Windows Event Collection
WEF	Windows Event Forwarding
WIPS	Wireless Intrusion Prevention System
WLAN	Wireless Local Area Network
XML	EXtensible Markup Language
XSS	Cross-Site Scripting
YAF	Yet Another Flowmeter
YARA	Yet Another Recursive/Ridiculous Acronym
ZFS	Z File System

Index